Bayesian Reasoning in Data Analysis

A Critical Introduction

Bayesian Reasoning in Data Analysis

A Critical Introduction

Giulio D'Agostini

Physics Department
University of Rome "La Sapienza", Italy

World Scientific
New Jersey • London • Singapore • Hong Kong

Published by

World Scientific Publishing Co. Pte. Ltd.

5 Toh Tuck Link, Singapore 596224

USA office: 27 Warren Street, Suite 401-402, Hackensack, NJ 07601

UK office: 57 Shelton Street, Covent Garden, London WC2H 9HE

Library of Congress Cataloging-in-Publication Data

D'Agostini, G. (Giulio).
Bayesian reasoning in data analysis : a critical introduction / Giulio D'Agostini.
p. cm.
Includes bibliographical references and index.
ISBN 981-238-356-5 (alk. paper)
1. Bayesian statistical decision theory. I. Title.

QA279.5 .D28 2003 2003045082
519.5'42--dc21

British Library Cataloguing-in-Publication Data

A catalogue record for this book is available from the British Library.

First published 2003
Reprinted 2005

Printed in Singapore by World Scientific Printers (S) Pte Ltd

To my parents, who always told me
"crediti quello che ti puoi credere"
(*"believe what you find it reasonable to believe"*).

Preface

This book is primarily addressed to physicists and other scientists and engineers who need to evaluate uncertainty in measurement. However, a large portion of its contents should be of interest to anyone who has to deal with probability and uncertainty, has an elementary background in 'standard statistics' and is wondering what this 'new statistics reasoning' is all about.

Although, like myself, you might never have heard about Bayes' theorem during your studies, in recent years you will almost certainly have encountered, with increasing frequency, the term 'Bayesian' in articles, books and the media. The so-called Bayesian methods are being employed in the most disparate fields of research and application, from engineering to computer science, economics, medicine and even forensic science. Some people are going so far as to talk of a 'paradigm shift', in the Kuhnian sense, although it is a strange revolution indeed which has its roots in the centuries-old ideas of the founding fathers of probability - the likes of Bernoulli, Bayes, Laplace and Gauss.

The gist of Bayesian statistics is not difficult to grasp. At its base is the intuitive idea that probability quantifies the 'degree of belief' in an event (in this context, an *event* is whatever can be precisely described by a proposition). Now, degrees of belief can be framed in a mathematical structure which allows the probability of an event A to be calculated on the basis of the probability of other events logically connected to that event A. In particular, the probability of event A changes if other events are assumed to be 'true', provided these other events are 'stochastically dependent' on event A. This is the essence of Bayes' theorem. As a consequence, Bayesian statistics allows the probability of a hypothesis-event to be continually updated on the basis of new observation-events that depend on that hypothesis-event.

Most likely this is not the way you were taught (elements of) probability theory. At most you might have been warned about the existence of a 'subjective probability' as an 'interpretation of probability', helpful in many fields, but definitively not applicable in Science where 'you want to be objective'. This is what I was taught in my training and was the approach I adopted in research, until I suddenly realized that there was something wrong with those ideas and with the methods which resulted from them. The breakthrough came when I myself had to teach probability and data analysis. Here is how several years later I reported my personal experience in the bulletin of the International Society for Bayesian Analysis (ISBA Newsletter, March 2000).

> *It is well known that the best way to learn something is to teach it. When I had to give the Laboratory of Physics course to Chemistry students and introduce elements of probability and statistics applied to data analysis, I did as most new teachers do: I started repeating what I had learned years before, more or less using the same lecture notes. This worked well for explaining the experiments, but when I moved to probability, the situation was quite embarrassing. In the very first lecture I realized that I was not convinced of what I was saying. I introduced probability as the ratio between favorable and possible cases, but I had no courage to add 'if the cases are equally probable'. I cheated by saying 'if the cases are equally possible' and moved rapidly to examples. The students had no time to react, the examples were well chosen, and I was able to survive that lesson and the following weeks.*
>
> *The problem returned when we came to the evaluation of measurement uncertainty, a typical application of statistics in scientific disciplines. I had to acknowledge that the reasoning physicists actually use in practice is quite in contradiction with the statistics theory we learn and teach. The result was that, whereas I had started the semester saying that subjective probability was not scientific, I ended it teaching probability inversion applied to physics quantities.*

I cannot speak of a 'conversion' to Bayesianism, because at that time (spring 1993) I had no alternative framework at my disposal. All books and lecture notes I had were strictly 'standard'. Just one book for economics students, which arrived more or less by chance on my desk, contained, as a kind of side remark, some examples of applications of Bayes' theorem. The problems were so trivial that anyone with a bit of imagination could have solved them just by building contingency tables, without needing to resort to probability theory and that 'strange theorem'. Subsequently, I worked out more intriguing examples, I extended Bayes' theorem to continuous variables and applied it to typical measurements in physics. Only months later did I discover that a Bayesian community existed, that my results had been known for two centuries and that there were (and there *are*) heated debates between 'Bayesians' and 'frequentists'. I was a bit disappointed to learn that my wheel had already been invented centuries earlier (and it was turning so fast I could barely keep up with it!). But I was also glad to realize that I was in good company. Only at this stage did I start to read the literature and to clarify my ideas.

I consider this initial self-learning process to have been very important because, instead of being 'indoctrinated' by a teacher, as had happened with my frequentistic training, I was instinctively selecting what was more in tune with my intuitive ideas and my fifteen years of experience in frontier physics.

I am sorry to be bothering you with these autobiographical notes, but I think they will help in understanding the spirit in which this book has been written and its idiosyncratic style. (I hope, at least, you will find it less tedious than the average statistics book.) For instance, you will find continual, sometimes sarcastic, criticisms of 'conventional' statistical methods. I even refuse to call these methods 'classical', because this appellative is misleading too (the term 'classical' usually refers to the approach of the founding fathers of a subject area, but the reasoning of the pioneers of probability theory was closer to what we nowadays call Bayesian). You might wonder why I am so doggedly critical of these conventional methods. The reason is that I feel I have been cheated by names and methods which seem to mean something they do not. I therefore make it a central issue in this book to show, by reasoning and examples, why many standard statistical recipes are basically wrong, even if they can often produce reasonable results. I simply apply scientific methodology to statistical reasoning in the same way as we apply it in Physics and in Science in general. If, for example, experiments show that Parity is violated, we can be disappointed, but we simply give up the principle of Parity Conservation, at least in the kind of interactions in which it has been observed that it does not hold. I do not understand why most of my colleagues do not behave in a similar way with the Maximum Likelihood principle, or with the 'prescriptions' for building Confidence Intervals, both of which are known to produce absurd results. At most, these methods should be used for special well-controlled cases, under well-stated assumptions.

To continue with my story, some months after having taught my first course, in an attempt to invent interesting problems for students, I wrote a little program for spectrum deconvolution ('unfolding' in our jargon) which several particle and astro-particle physics teams subsequently found useful for analyzing their data. This gave me more confidence in Bayesian ideas (but I continued to avoid the adjective 'subjective', which I still considered negative at that time) and I began to give seminars and mini-courses on the subject. In particular, lectures I gave in 1995 to graduate students at University of Rome 'La Sapienza' and to summer students at the Deutsches Elektronen-Synchroton in Hamburg, Germany, encouraged me to write the 'Bayesian Primer' (DESY-95-242, Roma1 N. 1070), which forms the core of this book. I took advantage of the 'academic training' course I gave to researchers of the European Organization for Nuclear Physics in 1998 to add some material and turn the Primer into CERN Report 99-03. The final step towards producing this book was taken in 2002, thanks to the interest of World Scientific in publishing an expanded version of the previous reports.

Instead of completely rewriting the Primer, producing a thicker report which would have been harder to read sequentially, I have divided the text into three Parts.

- Part 1 is devoted to a critical review of standard statistical methods and to a general overview of the proposed alternative. It contains references to the other two Parts for details.
- Part 2 is an extension of the original Primer, subdivided into chapters for easier reading.
- Part 3 contains further comments concerning the general aspects of probability, as well as other applications.

The advantage of this structure is that the reader should be able to get an overall view of problems and proposed solutions and then decide if he or she wants to enter into details. I hope this organization of the contents will suit the typical reader, whom I find it hard to imagine wishing to read sequentially a tome of over three hundred pages! This structure also allows the book to be read at several levels. For example, most of chapters 1, 2, 3, 5 and 10, which are the most important as far as the basic ideas are concerned, do not require advanced mathematical skills and can be understood by the general reader. However, organizing things in this manner has inevitably led to some repetition. I have tried to keep repetitions to a minimum, but *repetita juvant*, especially in this subject where the real difficulty lies not in understanding the formalism, but in shaking off deep-rooted prejudices.

A comment about the title of this book is in order. A title closer to the spirit of the approach proposed here would have been "Probabilistic reasoning ... ". In fact, the term 'Bayesian' might seem somewhat narrow, as if I am implying that the methods illustrated here always require explicit use of Bayes' theorem. However, in common usage, 'Bayesian' has come to mean 'based on the intuitive idea of probability'. Thus, what is known as the Bayesian approach is effectively a theory of uncertainty which is applicable universally. Within it, 'probability' has the same meaning for everybody: precisely that meaning which the human mind has developed naturally and which frequentists have tried to kill. Therefore, I have kept the term 'Bayesian' in the title, with the hope of attracting the attention of those who are curious about what 'Bayesian' might mean.

This book is based on the work of several years, during which I have had the opportunity to interact, directly or indirectly, with a large variety of persons, most of them physicists and physics students of many nationalities, but also mathematicians, statisticians, metrologists and science historians and philosophers. In particular, the interest shown by those who attended the lectures, and also the criticisms of those who had strong prejudices towards the approach I was presenting, has been highly stimulating. I take this opportunity to thank them all. Special acknowledgements go to Romano Scozzafava for many discussions about the fundamental aspects of probability theory. The many clarifications about DIN and ISO recommendations received by Klaus Weise of the PTB Braunschweig (Germany) have been particularly useful. I would like to thank Paolo Agnoli, Pia Astone, Peppe Degrassi, Mike Disney, Volker Dose, Dino Esposito, Fritz Fröhner, Ken Hanson, Frank Lad, Daniela Monaldi, Gianni Penso, Mirko Raso, Stefan Schlenstedt, Myron Tribus and Günter Zech for discussions and comments on the text, as well on the old version of the Primer [1] and the CERN Report [2] on which this book is based. Finally I would like to thank Jim Mc Manus for his help in finding ways to better express my ideas in English (apart from this sentence) and Bruno Pellizzoni for technical support with many of the drawings.

Rome, April 2005

Email: `giulio.dagostini@roma1.infn.it`
URL: `http://www.roma1.infn.it/~dagos/`

Contents

Part 1

Critical review and outline of the Bayesian alternative

Chapter 1

Uncertainty in physics and the usual methods of handling it

"In almost all circumstances, and at all times,
we find ourselves in a state of uncertainty.
Uncertainty in every sense.
Uncertainty about actual situations, past and present...
Uncertainty in foresight: this would not be eliminated
or diminished even if we accepted, in its most absolute form,
the principle of determinism; in any case, this is no longer in fashion.
Uncertainty in the face of decisions: more than ever in this case...
Even in the field of tautology (i.e of what is true or false by mere
definition, independently of any contingent circumstances) we always
find ourselves in a state of uncertainty ... (for instance,
of what is the seventh, or billionth, decimal place of π ...)..."
(Bruno de Finetti)

1.1 Uncertainty in physics

It is fairly well accepted among physicists that any conclusion which results from a measurement is affected by a degree of uncertainty. Let us remember briefly the reasons which prevent us from reaching certain statements. Figure 1.1 sketches the activity of physicists (or of any other scientist). From experimental data one wishes to determine the value of a given quantity, or to establish which theory describes the observed phenomena better. Although they are often seen as separate, both tasks may be viewed as two sides of the same process: going from observations to hypotheses. In fact, they can be stated in the following terms.

A: Which values are (more) compatible with the definition of the measurand, under the condition that certain numbers have been ob-

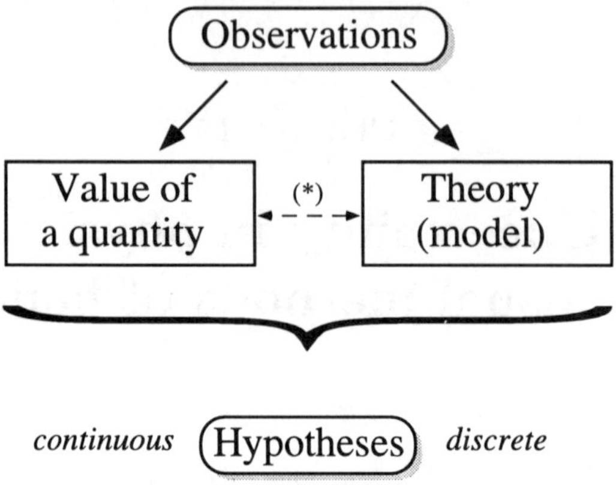

Fig. 1.1 From observations to hypotheses. The link between value of a quantity and theory is a reminder that sometimes a physics quantity has meaning only within a given theory or model. The arrows *observations* → *hypotheses* should not give the impression that the observation alone produces Knowledge (see Secs. 2.4 and 2.8).

served on instruments (and subordinated to all the available knowledge about the instrument and the measurand)?

B: Which theory is (more) compatible with the observed phenomena (and subordinated to the credibility of the theory, based also on aesthetics and simplicity arguments)?

The only difference between the two processes is that in the first the number of hypotheses is virtually infinite (the quantities are usually supposed to assume continuous values), while in the second it is discrete and usually small.

The reasons why it is impossible to reach the ideal condition of certain knowledge, i.e. only one of the many hypotheses is considered to be true and all the others false, may be summarized in the following, well-understood, scheme.

A: As far as the determination of the value of a quantity is concerned, one says that *"uncertainty is due to measurement errors"*.

B: In the case of a theory, we can distinguish two subcases:

(B_1) The law is probabilistic, i.e. the observations are not just a logical consequence of the theory. For example, tossing a regular coin, the three sequences of heads (h) and tails (t)

```
hhhhhhhhhhhhhhhhhhhhhhhhh
hhttttthhhhthhtthhhththht
ttttttttttttttttttttttttt
```

have the same probability of being observed (as any other sequence). Hence, there is no way of reaching a firm conclusion about the regularity of a coin after an observed sequence of any particular length.[1]

(B_2) The law is deterministic. But this property is only valid in principle, as can easily be understood. In fact, in all cases the actual observations also depend on many other factors external to the theory, such as initial and boundary conditions, influence factors, experimental errors, etc. All unavoidable uncertainties on these factors mean that the link between theory and observables is of a probabilistic nature in this case too.

1.2 True value, error and uncertainty

Let us start with case **A**. A first objection would be *"What does it mean that uncertainties are due to errors? Isn't this just tautology?"*. Well, the nouns 'error' and 'uncertainty', although currently used almost as synonyms, are related to different concepts. This is a first hint that in this subject there is neither uniformity of language, nor of methods. For this reason the metrological organizations have made great efforts to bring some order into the field [3, 4, 5, 6, 7]. In particular, the International Organization for Standardization (ISO) has published a *"Guide to the expression of uncertainty in measurement"* [5], containing definitions, recommendations and practical examples. Consulting the 'ISO Guide' we find the following definitions.

- Uncertainty: *"a parameter, associated with the result of a measurement, that characterizes the dispersion of the values that could reasonably be attributed to the measurement."*
- Error: *"the result of a measurement minus a true value of the measurand."*

[1] But after observation of the first sequence one would strongly suspect that the coin had two heads, if one had no means of directly checking the coin. The concept of probability will be used, in fact, to quantify the degree of such suspicion.

One has to note the following.

- The ISO definition of uncertainty defines the concept; as far as the operative definition is concerned, they recommend the 'standard uncertainty', i.e. the standard deviation (σ) of the possible values that the measurand may assume (each value is weighted with its 'degree of belief' in a way that will become clear later).
- It is clear that the error is usually unknown, as follows from the definition.
- The use of the article 'a' (instead of 'the') when referring to 'true value' is intentional, and rather subtle (see point *1* of next section).

Also the ISO definition of true value differs from that of standard textbooks. One finds, in fact:

- true value: *"a value compatible with the definition of a given particular quantity."*

This definition may seem vague, but it is more practical and pragmatic, and of more general use, than *"the value obtained after an infinite series of measurements performed under the same conditions with an instrument not affected by systematic errors."* For instance, it holds also for quantities for which it is not easy to repeat the measurements, and even for those cases in which it makes no sense to speak about repeated measurements under the same conditions.

1.3 Sources of measurement uncertainty

It is worth reporting the sources of uncertainty in measurement as listed by the ISO Guide:

"
1 *incomplete definition of the measurand;*
2 *imperfect realization of the definition of the measurand;*
3 *non-representative sampling — the sample measured may not represent the measurand;*
4 *inadequate knowledge of the effects of environmental conditions on the measurement, or imperfect measurement of environmental conditions;*
5 *personal bias in reading analogue instruments;*
6 *finite instrument resolution or discrimination threshold;*

7 inexact values of measurement standards and reference materials;
8 inexact values of constants and other parameters obtained from external sources and used in the data-reduction algorithm;
9 approximations and assumptions incorporated in the measurement method and procedure;
10 variations in repeated observations of the measurand under apparently identical conditions."

These do not need to be commented upon. Let us just give examples of the first two sources.

(1) If one has to measure the gravitational acceleration g at sea level, without specifying the precise location on the earth's surface, there will be a source of uncertainty because many different — even though 'intrinsically very precise' — results are consistent with the definition of the measurand.[2] What is then 'the' true value?
(2) The magnetic moment of a neutron is, in contrast, an unambiguous definition, but there is the experimental problem of performing experiments on isolated neutrons.

In terms of the usual jargon, one may say that sources 1–9 are related to *systematic effects* and 10 to *statistical effects*. Some caution is necessary regarding the sharp separation of the sources, which is clearly somehow artificial. In particular, all sources 1–9 may contribute to 10, because each of them depends upon the precise meaning of the clause *"under apparently identical conditions"* (one should talk, more precisely, about 'repeatability conditions' [5]). In other words, if the various effects change during the time of measurement, without any possibility of monitoring them, they contribute to the random error.

1.4 Usual handling of measurement uncertainties

The present situation concerning the treatment of measurement uncertainties can be summarized as follows.

[2]It is then clear that the definition of true value implying an indefinite series of measurements with ideal instrumentation gives the illusion that the true value is unique. The ISO definition, instead, takes into account the fact that measurements are performed under real conditions and can be accompanied by all the sources of uncertainty in the above list.

- Uncertainties due to statistical errors are currently treated using the frequentistic concept of 'confidence interval', although
  - there are well-known cases — of great relevance in frontier physics — in which the approach is not applicable (e.g. small number of observed events, or measurement close to the edge of the physical region);
  - the procedure is rather unnatural, and in fact the interpretation of the results is unconsciously subjective (as will be discussed later).
- There is no satisfactory theory or model to treat uncertainties due to systematic errors[3] consistently. Only *ad hoc* prescriptions can be found in the literature and in practice (*"my supervisor says ..."*): *"add them linearly"*; *"add them linearly if ..., else add them quadratically"*; *"don't add them at all"*.[4] The fashion at the moment is to add them quadratically if they are considered to be independent, or to build a covariance matrix of statistical and systematic contribution to treat the general case. In my opinion, besides all the 'theoretically' motivated excuses for justifying this praxis, there is simply the reluctance of experimentalists to combine linearly 10, 20 or more contributions to a global uncertainty, as the (out of fashion) 'theory' of maximum bounds would require.[5]

The problem of interpretation will be treated in the next section. For the moment, let us see why the use of standard propagation of uncertainty, namely

$$\sigma^2(Y) = \sum_i \left(\frac{\partial Y}{\partial X_i}\right)^2 \sigma^2(X_i) + \textit{correlation terms}\,, \qquad (1.1)$$

is not justified (especially if contributions due to systematic effects are included). This formula is derived from the rules of probability distributions, making use of linearization (a usually reasonable approximation for routine applications). This leads to theoretical and practical problems.

---

[3]To be more precise one should specify 'of unknown size', since an accurately assessed systematic error does not yield uncertainty, but only a correction to the raw result.

[4]By the way, it is a good and recommended practice to provide the complete list of contributions to the overall uncertainty [5]; but it is also clear that, at some stage, the producer or the user of the result has to combine the uncertainty to form his idea about the interval in which the quantity of interest is believed to lie.

[5]And in fact, one can see that when there are only two or three contributions to the 'systematic error', there are still people who prefer to add them linearly.

- X_i and Y should have the meaning of random variables.
- In the case of systematic effects, how do we evaluate the input quantities $\sigma(X_i)$ entering in the formula in a way which is consistent with their meaning as standard deviations?
- How do we properly take into account correlations (assuming we have solved the previous questions)?

It is very interesting to go to your favorite textbook and see how 'error propagation' is introduced. You will realize that some formulae are developed for random quantities, making use of linear approximations, and then suddenly they are used for physics quantities without any justification.[6] A typical example is measuring a velocity $v \pm \sigma(v)$ from a distance $s \pm \sigma(s)$ and a time interval $t \pm \sigma(t)$. It is really a challenge to go from the uncertainty on s and t to that of v without considering s, t and v as random variables, and to avoid thinking of the final result as a probabilistic statement on the velocity. Also in this case, an intuitive interpretation conflicts with standard probability theory.

1.5 Probability of observables versus probability of 'true values'

The criticism about the inconsistent interpretation of results may look like a philosophical quibble, but it is, in my opinion, a crucial point which needs to be clarified. Let us consider the example of n independent measurements of the same quantity under identical conditions (with n large enough to simplify the problem, and neglecting systematic effects). We can evaluate the arithmetic average $\overline{x}$ and the standard deviation σ. The well-known result on the true value μ is

$$\mu = \overline{x} \pm \frac{\sigma}{\sqrt{n}} . \tag{1.2}$$

[6]Some others, including some old lecture notes of mine, try to convince the reader that the propagation is applied to the observables, in a very complicated and artificial way. Then, later, as in the 'game of the three cards' proposed by professional cheaters in the street, one uses the same formulae for physics quantities, hoping that the students do not notice the logical gap.

The reader will have no difficulty in admitting that the large majority of people interpret Eq. (1.2) as if it were[7]

$$P(\overline{x} - \frac{\sigma}{\sqrt{n}} \leq \mu \leq \overline{x} + \frac{\sigma}{\sqrt{n}}) = 68\% \,. \tag{1.3}$$

However, conventional statistics says only that[8]

$$P(\mu - \frac{\sigma}{\sqrt{n}} \leq \overline{X} \leq \mu + \frac{\sigma}{\sqrt{n}}) = 68\% \,, \tag{1.4}$$

a probabilistic statement about $\overline{X}$, given μ, σ and n. Probabilistic statements concerning μ are not foreseen by the theory (*"μ is a constant of unknown value"*[9]), although this is what we are, intuitively, looking for: Having observed the *effect* $\overline{x}$ we are interested in stating something about the possible true value responsible for it. In fact, when we do an experiment, we want to increase our knowledge about μ and, consciously or not, we want to know which values are more or less believable. A statement concerning the probability that an observed value falls within a certain interval around μ is meaningless if it cannot be turned into an expression which states the quality of the knowledge about μ itself. Since the usual probability theory does not help, the probability inversion is performed intuitively. In routine cases it usually works, but there are cases in which it fails (see Sec. 1.7).

[7] There are also those who express the result, making the trivial mistake of saying *"this means that, if I repeat the experiment a great number of times, then I will find that in roughly 68% of the cases the observed average will be in the interval* $[\overline{x} - \sigma/\sqrt{n},\ \overline{x} + \sigma/\sqrt{n}]$*."* (Besides the interpretation problem, there is a missing factor of $\sqrt{2}$ in the width of the interval. See Sec. 6.6 for details.)

[8] The capital letter to indicate the average appearing in Eq. (1.4) is used because here this symbol stands for a random variable, while in Eq. (1.3) it indicated numerical value that it can assume. For the Greek symbols this distinction is not made, but the different role should be evident from the context.

[9] It is worth noting the paradoxical inversion of role between μ, about which we are in a state of uncertainty, considered to be a constant, and the observation $\overline{x}$, which has a certain value and which is instead considered a random quantity. This distorted way of thinking produces the statements to which we are used, such as speaking of *"uncertainty (or error) on the observed number"*: If one observes 10 on a scaler, there is no uncertainty on this number, but on the quantity which we try to infer from the observation (e.g. λ of a Poisson distribution, or a rate).

1.6 Probability of the causes

Generally speaking, what is missing in the usual theory of probability is the crucial concept of probability of hypotheses and, in particular, probability of causes: *"the essential problem of the experimental method"* (Poincaré):

> *"I play at écarté with a gentleman whom I know to be perfectly honest. What is the chance that he turns up the king? It is 1/8. This is a problem of the probability of effects. I play with a gentleman whom I do not know. He has dealt ten times, and he has turned the king up six times. What is the chance that he is a sharper? This is a problem in the probability of causes. It may be said that it is the essential problem of the experimental method"* [8].
>
> *"... the laws are known to us by the observed effects. Trying to deduct from the effects the laws which are the causes, it is solving a problem of probability of causes"* [9].

A theory of probability which does not consider probabilities of hypothesis is unnatural and prevents transparent and consistent statements about the causes which may have produced the observed effects from being assessed.

1.7 Unsuitability of frequentistic confidence intervals

According to the standard theory of probability, statement (1.3) is nonsense, and, in fact, good frequentistic books do not include it. They speak instead about 'confidence intervals', which have a completely different interpretation [that of Eq. (1.4)], although several books and many teachers suggest an interpretation of these intervals as if they were probabilistic statements on the true values, like Eq. (1.3). But it seems to me that it is practically impossible, even for those who are fully aware of the frequentistic theory, to avoid misleading conclusions. This opinion is well stated by Howson and Urbach in a paper to Nature [10]:

> *"The statement that such-and-such is a 95% confidence interval for μ seems objective. But what does it say? It may be imagined that a 95% confidence interval corresponds to a 0.95 probability that the unknown parameter lies in the confidence range. But in the classical approach, μ is not a random variable, and so has no probability. Nevertheless, statisticians regularly say that one can be '95% confident' that the parameter lies in the confidence interval. They never say why."*

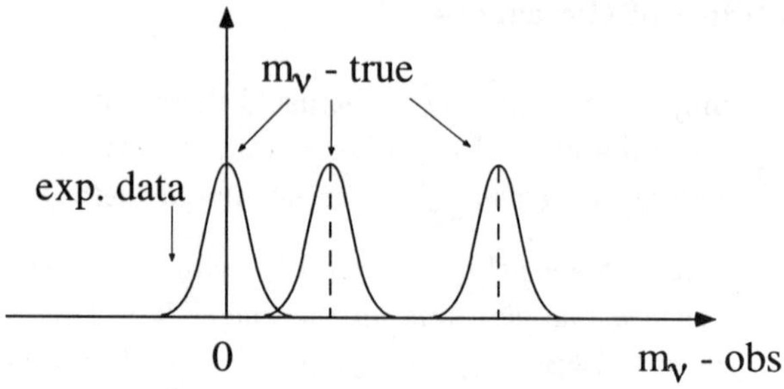

Fig. 1.2 Negative neutrino mass?

The origin of the problem goes directly to the underlying concept of probability. The frequentistic concept of confidence interval is, in fact, a kind of artificial invention to characterize the uncertainty consistently with the frequency-based definition of probability. But, unfortunately — as a matter of fact — this attempt to classify the state of uncertainty (on the true value) trying to avoid the concept of probability of hypotheses produces misinterpretation. People tend to turn arbitrarily Eq. (1.4) into Eq. (1.3) with an intuitive reasoning that I like to paraphrase as 'the dog and the hunter': We know that a dog has a 50% probability of being 100 m from the hunter; if we observe the dog, what can we say about the hunter? The terms of the analogy are clear:

$$\begin{aligned} \text{hunter} &\leftrightarrow \text{true value} \\ \text{dog} &\leftrightarrow \text{observable}\,. \end{aligned}$$

The intuitive and reasonable answer is *"The hunter is, with 50% probability, within 100 m of the position of the dog."* But it is easy to understand that this conclusion is based on the tacit assumption that 1) the hunter can be anywhere around the dog; 2) the dog has no preferred direction of arrival at the point where we observe him. Any deviation from this simple scheme invalidates the picture on which the inversion of probability Eq. (1.4) $\rightarrow$ Eq. (1.3) is based. Let us look at some examples.

Example 1: Measurement at the edge of a physical region.

An experiment, planned to measure the electron-neutrino mass with a resolution of $\sigma = 2\,\mathrm{eV}/c^2$ (independent of the mass, for simplicity, see Fig. 1.2), finds a value of $-4\,\mathrm{eV}/c^2$ (i.e. this value comes out of the analysis of real data treated in exactly the same way as that of

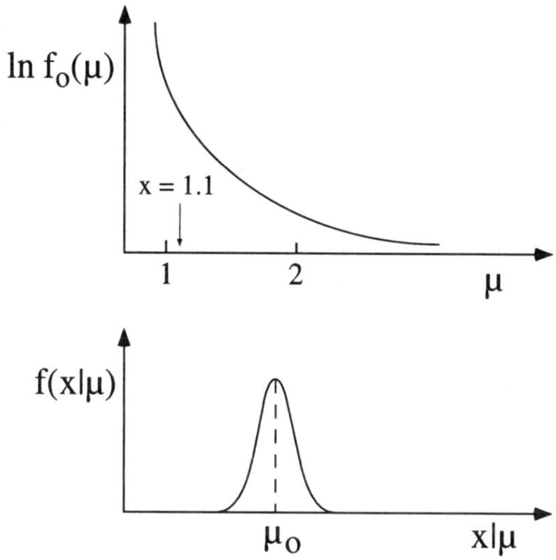

Fig. 1.3 Case of highly asymmetric expectation on the physics quantity.

simulated data, for which a $2\,\mathrm{eV}/c^2$ resolution was found).
What can we say about m_ν?

$$m_\nu = -4 \pm 2\ \mathrm{eV}/c^2\ ?$$

$$P(-6\,\mathrm{eV}/c^2 \leq m_\nu \leq -2\ \mathrm{eV}/c^2) = 68\%\ ?$$

$$P(m_\nu \leq 0\ \mathrm{eV}/c^2) = 98\%\ ?$$

No physicist would sign a statement which sounded like he was 98% sure of having found a negative mass!

Example 2: Non-flat distribution of a physical quantity.
Let us take a quantity μ that 'we know',[10] from previous knowledge, to be distributed as in Fig. 1.3. It may be, for example, the energy of bremsstrahlung photons or of cosmic rays. We know that an observable value X will be normally distributed around the true value μ, independently of the value of μ. We have performed a measurement and obtained $x = 1.1$, in arbitrary units. What can we say about the

[10]Those who make an easy use of this engaging expression are recommended to browse Wittgenstein's *"On certainty"*.

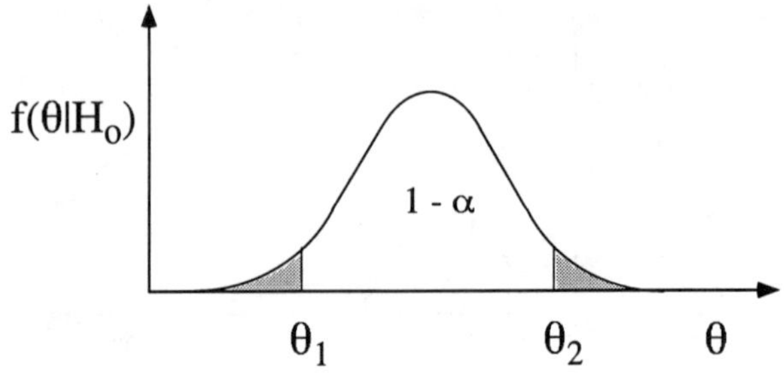

Fig. 1.4 Hypothesis test scheme in the frequentistic approach.

true value μ that has caused this observation? Also in this case the formal definition of the confidence interval does not work. Intuitively, we feel that there is more chance that μ is on the left of (1.1) than on the right. In the jargon of the experimentalists, *"there are more migrations from left to right than from right to left"*.

Example 3: High-momentum track in a magnetic spectrometer.
The previous examples deviate from the simple dog–hunter picture only because of an asymmetric possible position of the 'hunter'. The case of a very-high-momentum track in a central detector of a high-energy physics (HEP) experiment involves asymmetric response of a detector for almost straight tracks and non-uniform momentum distribution of charged particles produced in the collisions. Also in this case the simple inversion scheme does not work.

To sum up the last two sections, we can say that *"intuitive inversion of probability*

$$P(\ldots \leq \overline{X} \leq \ldots) \Longrightarrow P(\ldots \leq \mu \leq \ldots), \tag{1.5}$$

besides being theoretically unjustifiable, yields results which are numerically correct only in the case of symmetric problems." I recommend Ref. [11] to those interested in a more detailed analysis of the many problems with (the many variations of) standard statistical methods to compute 'confidence intervals'.

1.8 Misunderstandings caused by the standard paradigm of hypothesis tests

Similar problems of interpretation appear in the usual methods used to test hypotheses. I will briefly outline the standard procedure and then give some examples to show the kind of paradoxical conclusions that one can reach.

A frequentistic hypothesis test follows the scheme outlined below (see Fig. 1.4). [11]

(1) Formulate a hypothesis $H_\circ$ (the 'null' hypothesis).
(2) Choose a test variable θ of which the probability density function (p.d.f.) $f(\theta \mid H_\circ)$ is known (analytically or numerically) for a given $H_\circ$.
(3) Choose an interval $[\theta_1, \theta_2]$ such that there is high probability that θ falls inside the interval:

$$P(\theta_1 \leq \theta \leq \theta_2) = 1 - \alpha\,, \tag{1.6}$$

with α typically equal to 1% or 5%.
(4) Perform an experiment, obtaining $\theta = \theta_m$.
(5) Draw the following conclusions :

- if $\theta_1 \leq \theta_m \leq \theta_2$ $\Longrightarrow$ $H_\circ$ accepted;
- otherwise $\Longrightarrow$ $H_\circ$ rejected (with a significance level α).

The usual justification for the procedure is that the probability α is so low that it is practically impossible for the test variable to fall outside the interval. Then, if this event happens, we have good reason to reject the hypothesis.

One can recognize behind this reasoning a revised version of the classical 'proof by contradiction' (see, e.g., Ref. [13]). In standard dialectics, one assumes a hypothesis to be true and looks for a logical consequence which is manifestly false in order to reject the hypothesis. The 'slight difference' is that in the hypothesis test scheme, the false consequence is replaced by an improbable one. The argument may look convincing, but it has no grounds. Moreover, since in many cases the probability of observing a particular 'consequence' can be very small (and 'then' the hypothesis

[11] For a short and clear introduction about meaning and historical origin of the standard hypothesis testing paradigma see Ref. [12].

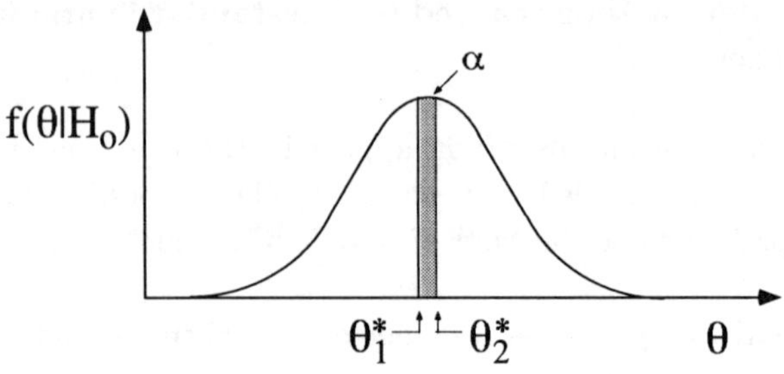

Fig. 1.5 Would you accept this scheme to test hypotheses?

under test should be falsified whatever one observes), statisticians had the brilliant idea of considering the 'probability of the tail(s)',[12] i.e. conclusions do not depend anymore on what has been observed, but also on all *non-observed* events which are considered rarer than the observed one. This procedure seems to have solved the problem but, from the logical point of view, is unacceptable, while, in practical applications, the perception is that of 'something that works' (though 'by chance', as it will be discussed in Sec. 10.8). In order to analyze the problem well, we need to review the logic of uncertainty. For the moment a few examples are enough to indicate that there is something troublesome behind the procedure.

Example 4: Choosing the rejection region in the middle of the distribution.

Imagine choosing an interval $[\theta_1^*, \theta_2^*]$ around the expected value of θ (or around the mode) such that

$$P(\theta_1^* \leq \theta \leq \theta_2^*) = \alpha\,, \tag{1.7}$$

with α small (see Fig. 1.5). We can then reverse the test, and reject the hypothesis if the measured θ_m is inside the interval. This strategy is clearly unacceptable, indicating that the rejection decision cannot be based on the argument of practically impossible observations (smallness

[12] At present, 'p-values' (or 'significance probabilities') are also ***"used in place of hypothesis tests as a means of giving more information about the relationship between the data and the hypothesis than does a simple reject/do not reject decision"*** [14]. They consist in giving the probability of the 'tail(s)', as also usually done in physics, although the name 'p-values' has not yet entered our lexicon (to my knowledge, the first statistic book for physicists using the term 'p-values' is Ref. [15]. Anyhow, they produce the same interpretation problems of the hypothesis test paradigm (see also example 8 of next section).

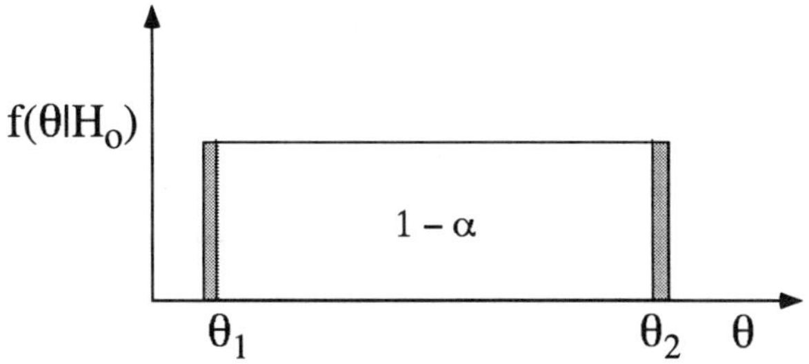

Fig. 1.6 Would you accept this scheme to test hypotheses?

of α).

One may object that the reason is not only the small probability of the rejection region, but also its distance from the expected value. Figure 1.6 is an example against this objection. Although the situation is not as extreme as that depicted in Fig. 1.5, one would need a certain amount of courage to say that the $H_\circ$ is rejected if the test variable falls by chance in 'the bad region'.

Example 5: Has the student made a mistake?

A teacher gives to each student an individual sample of 300 random numbers, uniformly distributed between 0 and 1. The students are asked to calculate the arithmetic average. The *prevision*[13] of the teacher can be quantified with

$$\mathrm{E}\left[\overline{X}_{300}\right] = \frac{1}{2} \tag{1.8}$$

$$\sigma\left[\overline{X}_{300}\right] = \frac{1}{\sqrt{12}} \cdot \frac{1}{\sqrt{300}} = 0.017\,, \tag{1.9}$$

with the random variable $\overline{X}_{300}$ normally distributed because of the central limit theorem. This means that there is 99% probability that an average will come out in the interval $0.5 \pm (2.6 \times 0.017)$:

$$P(0.456 \leq \overline{X}_{300} \leq 0.544) = 99\%\,. \tag{1.10}$$

Imagine that a student obtains an average outside the above interval (e.g. $\overline{x} = 0.550$). The teacher may be interested in the probability that the student has made a mistake (for example, he has to decide if it is

[13]By *prevision* [16], I mean a probabilistic 'prediction', which corresponds to what is usually known as expectation value (see Sec. 5.2).

worthwhile checking the calculation in detail). Applying the standard methods one draws the conclusion that

> *"the hypothesis $H_\circ$ = 'no mistakes' is rejected at the 1% level of significance"*,

i.e. one receives a precise answer to a different question. In fact, the meaning of the previous statement is simply

> *"there is only a 1% probability that the average falls outside the selected interval, if the calculations were done correctly"*.

But this does not answer our natural question,[14] i.e. that concerning the probability of mistake, and not that of results far from the average if there were no mistakes. Moreover, the statement sounds as if one would be 99% sure that the student has made a mistake! This conclusion is highly misleading.

If you ask the students (before they take a standard course in hypothesis tests) you will realize of a crucial ingredient extraneous to the logic of hypothesis tests:

> *"It all depends on who has made the calculation!"*

In fact, if the calculation was done by a well-tested program, the probability of mistake would be zero. And students know rather well their probability of making mistakes.

Example 6: A bad joke to a journal.[15]

A scientific journal changes its publication policy. The editors announce that results with a significance level of 5% will no longer be accepted. Only those with a level of $\leq$ 1% will be published. The rationale for the change, explained in an editorial, looks reasonable and it can be shared without hesitation: *"We want to publish only good results."*

1000 experimental physicists, not convinced by this severe rule, conspire against the journal. Each of them formulates a wrong physics hypothesis and performs an experiment to test it according to the accepted/rejected scheme.

Roughly 10 physicists get 1% significant results. Their papers are accepted and published. It follows that, contrary to the wishes of the

[14] Personally, I find it is somehow impolite to give an answer to a question which is different from that asked (*"What time is it?"* – *"My cat is sick"*). At least one should apologize for being unable to answer the original question. However, textbooks usually do not do this, and people get confused.

[15] Example taken from Ref. [17].

editors, the first issue of the journal under the new policy contains only wrong results!
The solution to the kind of paradox raised by this example seems clear: The physicists knew with certainty that the hypotheses were wrong. So the example looks like an odd case with no practical importance. But in real life who knows in advance with certainty if a hypothesis is true or false?

1.9 Statistical significance versus probability of hypotheses

The examples in the previous section have shown the typical ways in which significance tests are misinterpreted. This kind of mistake is commonly made not only by students, but also by professional users of statistical methods. There are two different probabilities playing a role:

$P(H\,|\,\mathbf{data})$: the probability of the hypothesis H, conditioned by the observed data. This is the probabilistic statement in which we are interested. It summarizes the status of knowledge on H, achieved in conditions of uncertainty: it might be the probability that the W mass is between 80.00 and 80.50 GeV, that the Higgs mass is below 200 GeV, or that a charged track is a π^- rather than a K^-.

$P(\mathbf{data}\,|\,H)$: the probability of the observables under the condition that the hypothesis H is true.[16] For example, the probability of getting two consecutive heads when tossing a regular coin, the probability that a W mass is reconstructed within 1 GeV of the true mass, or that a 2.5 GeV pion produces a $\geq 100\,$pC signal in an electromagnetic calorimeter.

Unfortunately, conventional statistics considers only the second case. As a consequence, since the very question of interest remains unanswered, very often (practically always, according to my experience) significance levels are incorrectly treated as if they were probabilities of the hypothesis. For example, *"H refused at 5% significance"* may be understood to mean the same as *"H has only 5% probability of being true."*

It is important to note the different consequences of the misunderstanding caused by the arbitrary probabilistic interpretation of confidence intervals and of significance levels. Measurement uncertainties on directly

---

[16] This should not be confused with the probability of the actual data, which is clearly 1, since they have been observed.

measured quantities obtained by confidence intervals are at least numerically correct in most routine cases, although arbitrarily interpreted. In hypothesis tests, however, the conclusions may become seriously wrong. This can be shown with the following examples.

Example 7: AIDS test.

An Italian citizen is chosen at random to undergo an AIDS test. Let us assume that the analysis used to test for HIV infection has the following performances:

$$P(Positive\,|\,HIV) \approx 1, \tag{1.11}$$

$$P(Positive\,|\,\overline{HIV}) = 0.2\%\,, \tag{1.12}$$

where HIV stands for "infected" and $\overline{HIV}$ for healthy. The analysis may declare healthy people 'Positive', even if only with a very small probability.

Let us assume that the analysis states 'Positive'. Can we say that, since the probability of an analysis error Healthy → Positive is only 0.2%, then the probability that the person is infected is 99.8%? Certainly not. If one calculates on the basis of an estimated 100 000 infected persons out of a population of 60 million, there is a 55% probability that the person is healthy![17] Some readers may be surprised to read that, in order to reach a conclusion, one needs to have an idea of how 'reasonable' the hypothesis is, independently of the data used: a mass cannot be negative; the spectrum of the true value is of a certain type; students often make mistakes; physical hypotheses happen to be incorrect; the proportion of Italians carrying the HIV virus is fortunately low. The notion of prior reasonableness of the hypothesis is fundamental to the approach we are going to present, but it is something to which physicists put up strong resistance (although in practice they often instinctively use this intuitive way of reasoning continuously and correctly). In the following I will try to show that 'priors' are rational and unavoidable, although their influence may become negligible when there is strong experimental evidence in favor of a given hypothesis.

[17]The result will be a simple application of Bayes' theorem, which will be introduced later. A crude way to check this result is to imagine performing the test on the entire population. Then the number of persons declared Positive will be all the HIV infected plus 0.2% of the remaining population. In total 100 000 infected and 120 000 healthy persons. The general, Bayesian solution is given in Sec. 3.12.1.

Example 8: Probabilistic statements about the 1997 HERA high-Q^2 events.

A very instructive example of the misinterpretation of probability can be found in the statements which commented on the 'excess' of events observed by the e–p experiments at the HERA collider (DESY Laboratory, Hamburg, Germany) in the high-Q^2 region. For example, the official DESY statement [18] was:[18]

> *"The two HERA experiments, H1 and ZEUS, observe an excess of events above expectations at high x (or $M = \sqrt{x\,s}$), y, and Q^2. For* $Q^2 > 15\,000$ GeV^2 the joint distribution has a probability of less than one per cent to come from Standard Model NC DIS processes."

(Standard Model refers to the most believed and successful model of particle physics; NC and DIS stand for Neutral Current and Deep inelastic Scattering, respectively; Q^2 is inversely proportional to the region of space inside the proton probed by the electron beam.) Similar statements were spread out in the scientific community, and finally to the press. For example, a message circulated by INFN stated (it can be understood even in Italian)

> *"La probabilità che gli eventi osservati siano una fluttuazione statistica è inferiore all' 1%."*

Obviously these two statements led the press (e.g. Corriere della Sera, 23 Feb. 1998) to announce that scientists were highly confident that a great discovery was just around the corner.[19]

[18]One might think that the misleading meaning of that sentence was due to unfortunate wording, but this possibility is ruled out by other statements which show clearly a quite odd point of view of probabilistic matter. In fact the 1998 activity report [19] insists that *"the likelihood that the data produced are the result of a statistical fluctuation ... is equivalent to that of tossing a coin and throwing seven heads or tails in a row"* (replacing 'probability' by 'likelihood' does not change the sense of the message). Then, trying to explain the meaning of a statistical fluctuation, the following example is given: *"This process can be simulated with a die. If the number of times a die is thrown is sufficiently large, the die falls equally often on all faces, i.e. all six numbers occur equally often. The probability for each face is exactly a sixth or 16.66%, assuming the die is not loaded. If the die is thrown less often, then the probability curve for the distribution of the six die values is no longer a straight line but has peaks and troughs. The probability distribution obtained by throwing the die varies about the theoretical value of 16.66% depending on how many times it is thrown."*

[19]One of the odd claims related to these events was on a poster of an INFN exhibition at Palazzo delle Esposizioni in Rome: *"These events are absolutely impossible within the current theory ... If they will be confirmed, it will imply that...."* Some friends of mine who visited the exhibition asked me what it meant that "something impossible needs to be confirmed".

The experiments, on the other hand, did not mention this probability. Their published results [20] can be summarized, more or less, as *"there is a $\lessapprox 1\%$ probability of observing such events or rarer ones within the Standard Model"*.
To sketch the flow of consecutive statements, let us indicate by SM *"the Standard Model is the only cause which can produce these events"* and by tail the *"possible observations which are rarer than the configuration of data actually observed"*.

(1) Experimental result: $P(data + tail \,|\, SM) \lesssim 1\%$.
(2) Official statements: $P(SM \,|\, data) \lesssim 1\%$.
(3) Press: $P(\overline{SM} \,|\, data) \gtrsim 99\%$, simply applying standard logic to the outcome of step 2. They deduce, correctly, that the hypothesis $\overline{SM}$ (= hint of new physics) is almost certain.

One can recognize an arbitrary inversion of probability. But now there is also something else, which is more subtle, and suspicious: *"why should we also take into account data which have not been observed?"*[20] Stated in a schematic way, it seems natural to draw conclusions on the basis of the observed data:

$$\mathbf{data} \longrightarrow P(H \,|\, data)\,,$$

although $P(H \,|\, data)$ differs from $P(data \,|\, H)$. But it appears strange that unobserved data should also play a role. Nevertheless, because of our educational background, we are so used to the tacit inferential scheme of the kind

$$\mathbf{data} \longrightarrow P(H \,|\, data + tail)\,,$$

that we even have difficulty in understanding the meaning of this objection (see Ref. [13] for an extensive discussion).

I have considered this case in detail because I was personally involved in one of the HERA experiments. There are countless examples of this kind of claim in the scientific community, and I am very worried when I think that this kind of logical mistake might be applied in other fields of research on which our health and the future of the Planet depends. Recent frontier

[20]This is as if the conclusion from the AIDS test depended not only on $P(Positive \,|\, \overline{HIV})$ and on the prior probability of being infected, but also on the probability that this poor guy experienced events rarer than a mistaken analysis, like sitting next to Claudia Schiffer on an international flight, or winning the lottery, or being hit by a meteorite.

physics examples of misleading probabilistic claims of discovery concern the Higgs boson (*"It is a 2.6 sigma effect. So there's still a 6 in 1000 chance that what we are seeing are background events, rather than the Higgs"* [21]), the muon magnetic moment (*"We are now 99 percent sure that the present Standard Model cannot describe our data"* [22]) and the neutrino properties (*"The experimenters reported a three-sigma discrepancy in sin2qW, which translates to a 99.75 percent probability that the neutrinos are not behaving like other particles"* [23]).

Since I am aware that many physicists, used the usual hypothesis test scheme, have difficulty to realize that this kind of reasoning is wrong, let us make finally another example, conceptually very similar to the previous ones, but easier to understand intuitively.

Example 9: Probability that a particular random number comes from a generator.

The value $x = 3.01$ is extracted from a Gaussian random-number generator having $\mu = 0$ and $\sigma = 1$. It is well known that

$$P(|X| > 3) = 0.27\%\,,$$

but we cannot state that the value X has 0.27% probability of coming from that generator, or that the probability that the observation is a statistical fluctuation is 0.27%. In this case, the value comes with 100% probability from that generator, and it is at 100% a statistical fluctuation. This example helps to illustrate the logical mistake one can make in the previous examples. One may speak about the probability of the generator (let us call it A) only if another generator B is taken into account. If this is the case, the probability depends on the parameters of the generators, the observed value x and on the probability that the two generators enter the game. For example, if B has $\mu = 6.02$ and $\sigma = 1$, it is reasonable to think that

$$P(A \,|\, x = 3.01) = P(B \,|\, x = 3.01) = 0.5\,. \qquad (1.13)$$

Let us imagine a variation of the example: The generation is performed according to an algorithm that chooses A or B, with a ratio of probability 10 to 1 in favor of A. The conclusions change: Given the same observed value $x = 3.01$, one would tend to infer that x is most probably due to A. It is not difficult to be convinced that, even if the value is a bit closer to the center of generator B (for example $x = 3.3$), there will still be a tendency to attribute it to A. This natural way of reasoning is exactly what is meant by 'Bayesian', and will be illustrated starting from next chapter. It should be noted that we are only considering the observed data ($x = 3.01$ or $x = 3.3$), and not other values which could be observed ($x \geq 3.01$, for instance). This example shows also that we cannot simply extend the proof of contradiction from impossible to improbable events, as discussed in Sec. 1.8.

I hope these examples might at least persuade the reader to take the question of principles in probability statements seriously. Anyhow, even if we ignore philosophical aspects, there are other kinds of more technical inconsistencies in the way the standard paradigm is used to test hypotheses. These problems, which deserve extensive discussion, are effectively described in an interesting American Scientist article [13].

At this point I imagine that the reader will have a very spontaneous and legitimate objection: *"but why does this scheme of hypothesis tests usually work?"*. I will comment on this question in Sec. 10.8, but first we must introduce the alternative scheme for quantifying uncertainty.

Chapter 2

A probabilistic theory of measurement uncertainty

> *"If we were not ignorant there would be no probability, there could only be certainty. But our ignorance cannot be absolute, for then there would be no longer any probability at all. Thus the problems of probability may be classed according to the greater or less depth of our ignorance."*
> *(Henri Poincaré)*

2.1 Where to restart from?

In the light of the criticisms made in the previous chapter, it seems clear that we would be advised to completely revise the process which allows us to learn from experimental data. Paraphrasing Kant [24], one could say that (substituting the words in *italics* with those in parentheses):

> *"All metaphysicians (physicists) are therefore solemnly and legally suspended from their occupations till they shall have answered in a satisfactory manner the question, how are synthetic cognitions a priori possible (is it possible to learn from observations)?"*

Clearly this quotation must be taken in a playful way (at least as far as the invitation to suspended activities is concerned...). But, joking apart, the quotation is indeed more pertinent than one might initially think. In fact, Hume's criticism of the problem of induction, which interrupted the 'dogmatic slumber' of the great German philosopher, has survived the subsequent centuries.[1] We shall come back to this matter in a while.

[1] For example, it is interesting to report Einstein's opinion [25] about Hume's criticism: *"Hume saw clearly that certain concepts, as for example that of causality, cannot be deduced from the material of experience by logical methods. Kant, thoroughly convinced*

In order to build a theory of measurement uncertainty which does not suffer from the problems illustrated above, we need to ground it on some kind of first principles, and derive the rest by logic. Otherwise we replace a collection of formulae and procedures handed down by tradition with another collection of cooking recipes.

We can start from two considerations.

(1) In a way which is analogous to Descartes' *cogito*, the only statement with which it is difficult not to agree — in some sense the only certainty — is that

> *"the process of induction from experimental observations to statements about physics quantities (and, in general, physical hypotheses) is affected, unavoidably, by some degree of uncertainty".*

(2) The natural concept developed by the human mind to quantify the plausibility of the statements in situations of uncertainty is that of probability.[2]

---

of the indispensability of certain concepts, took them — just as they are selected — to be necessary premises of every kind of thinking and differentiated them from concepts of empirical origin. I am convinced, however, that this differentiation is erroneous." In the same Autobiographical Notes [25] Einstein, explaining how he came to the idea of the arbitrary character of absolute time, acknowledges that *"The type of critical reasoning which was required for the discovery of this central point was decisively furthered, in my case, especially by the reading of David Hume's and Ernst Mach's philosophical writings."* This tribute to Mach and Hume is repeated in the 'gemeinverständlich' of special relativity [26]: *"Why is it necessary to drag down from the Olympian fields of Plato the fundamental ideas of thought in natural science, and to attempt to reveal their earthly lineage? Answer: In order to free these ideas from the taboo attached to them, and thus to achieve greater freedom in the formation of ideas or concepts. It is to the immortal credit of D. Hume and E. Mach that they, above all others, introduced this critical conception."* I would like to end this parenthesis dedicated to Hume with a last citation, this time by de Finetti [16], closer to the argument of this chapter: *"In the philosophical arena, the problem of induction, its meaning, use and justification, has given rise to endless controversy, which, in the absence of an appropriate probabilistic framework, has inevitably been fruitless, leaving the major issues unresolved. It seems to me that the question was correctly formulated by Hume ... and the pragmatists ... However, the forces of reaction are always poised, armed with religious zeal, to defend holy obtuseness against the possibility of intelligent clarification. No sooner had Hume begun to prise apart the traditional edifice, then came poor Kant in a desperate attempt to paper over the cracks and contain the inductive argument — like its deductive counterpart — firmly within the narrow confines of the logic of certainty."*

[2]Perhaps one may try to use instead fuzzy logic or something similar. I will only try to show that this way is productive and leads to a consistent theory of uncertainty which does not need continuous injections of extraneous matter. I am not interested in demonstrating the uniqueness of this solution, and all contributions on the subject are welcome.

In other words we need to build a probabilistic (probabilistic and not, generically, statistic) theory of measurement uncertainty.

These two starting points seem perfectly reasonable, although the second appears to contradict the criticisms of the probabilistic interpretation of the result, raised in Sections 1.4 and 1.5. However this is not really a problem, it is only a product of a distorted (i.e. different from the natural) view of the concept of probability. So, first we have to review the concept of probability. Once we have clarified this point, all the applications in measurement uncertainty will follow and there will be no need to inject *ad hoc* methods or use magic formulae, supported by authority but not by logic.

2.2 Concepts of probability

We have arrived at the point where it is necessary to define better what probability is. This is done in Chapter 3. As a general comment on the different approaches to probability, I would like, following Ref. [27], to cite de Finetti [16]:

> *"The only relevant thing is uncertainty - the extent of our knowledge and ignorance. The actual fact of whether or not the events considered are in some sense determined, or known by other people, and so on, is of no consequence.*
> *The numerous, different opposed attempts to put forward particular points of view which, in the opinion of their supporters, would endow Probability Theory with a 'nobler status', or a 'more scientific' character, or 'firmer' philosophical or logical foundations, have only served to generate confusion and obscurity, and to provoke well-known polemics and disagreements - even between supporters of essentially the same framework.*
> *The main points of view that have been put forward are as follows.*
> *The* classical *view is based on physical considerations of symmetry, in which one should be* obliged *to give the same probability to such 'symmetric' cases. But which 'symmetry'? And, in any case, why? The original sentence becomes meaningful if reversed: the symmetry is probabilistically significant, in someone's opinion, if it leads him to assign the same probabilities to such events.*
> *The* logical *view is similar, but much more superficial and irresponsible inasmuch as it is based on similarities or symmetries which no longer derive from the facts and their actual properties, but merely from sentences which describe them, and their formal structure or language.*
> *The* frequentistic *(or* statistical*) view presupposes that one accepts the*

classical view, in that it considers an event as a class of individual events, the latter being 'trials' of the former. The individual events not only have to be 'equally probable', but also 'stochastically independent' ... (these notions when applied to individual events are virtually impossible to define or explain in terms of the frequentistic interpretation). In this case, also, it is straightforward, by means of the subjective approach, to obtain, under the appropriate conditions, in perfectly valid manner, the result aimed at (but unattainable) in the statistical formulation. It suffices to make use of the notion of exchangeability. The result, which acts as a bridge connecting the new approach to the old, has often been referred to by the objectivists as "de Finetti's representation theorem."

It follows that all the three proposed definitions of 'objective' probability, although useless per se, turn out to be useful and good as valid auxiliary devices when included as such in the subjectivist theory."

Also interesting is Hume's point of view on probability, where concept and evaluations are neatly separated. Note that these words were written in the middle of the 18th century [28].

"Though there be no such thing as Chance in the world; our ignorance of the real cause of any event has the same influence on the understanding, and begets a like species of belief or opinion.
There is certainly a probability, which arises from a superiority of chances on any side; and according as this superiority increases, and surpasses the opposite chances, the probability receives a proportionable increase, and begets still a higher degree of belief or assent to that side, in which we discover the superiority. If a dye were marked with one figure or number of spots on four sides, and with another figure or number of spots on the two remaining sides, it would be more probable, that the former would turn up than the latter; though, if it had a thousand sides marked in the same manner, and only one side different, the probability would be much higher, and our belief or expectation of the event more steady and secure. This process of the thought or reasoning may seem trivial and obvious; but to those who consider it more narrowly, it may, perhaps, afford matter for curious speculation.
...
Being determined by custom to transfer the past to the future, in all our inferences; where the past has been entirely regular and uniform, we expect the event with the greatest assurance, and leave no room for any contrary supposition. But where different effects have been found to follow from causes, which are to appearance exactly similar, all these various effects must occur to the mind in transferring the past to the future, and enter into our consideration, when we determine the prob-

ability of the event. Though we give the preference to that which has been found most usual, and believe that this effect will exist, we must not overlook the other effects, but must assign to each of them a particular weight and authority, in proportion as we have found it to be more or less frequent."

2.3 Subjective probability

I would like to sketch the essential concepts related to subjective probability,[3] for the convenience of those who wish to have a short overview of the subject, discussed in detail in Part 2. This should also help those who are not familiar with this approach to follow the scheme of probabilistic induction which will be presented in the next section, and the summary of the applications which will be developed in the rest of this text.

- Essentially, one assumes that the concept of probability is primitive, i.e. close to that of common sense (said with a joke, probability is what everybody knows before going to school and continues to use afterwards, in spite of what one has been taught[4]).
- Stated in other words, probability is a measure of the *degree of belief that an event will occur.*
- Probability is related to the state of uncertainty, and not (only) to the outcome of repeated experiments.
- The value of probability ranges between 0 and 1 from events which go from false to true (see Fig. 3.1 in Sec. 3.2).
- Since the more one believes in an event the more money one is prepared to bet, the 'coherent' bet can be used to define the value of probability in an operational way (see Sec. 3.2).
- From the condition of *coherence* one obtains, as theorems, the basic rules of probability (usually known as axioms) and the 'formula of conditional probability' (see Secs. 3.5.2, and 10.3 for further clarifications).
- There is, in principle, an infinite number of ways to evaluate the probability, with the only condition being that they must satisfy coherence. We can use symmetry arguments, statistical data (past frequencies), Monte Carlo simulations, quantum mechanics[5] and so on. What is im-

[3]For an introductory and concise presentation of the subject see also Ref. [29].

[4]This remark — not completely a joke — is due to the observation that most physicists interviewed are convinced that Eq. (1.3) is legitimate, although they maintain that probability is the limit of the frequency (see more details in Ref. [30]).

[5]Without entering into the open problems of quantum mechanics, let us just say

portant is that if we get a number close to one, we are very confident that the event will happen; if the number is close to zero we are very confident that it will not happen; if $P(A) > P(B)$, then we believe in the realization of A more than in the realization of B.

- It is easy to show that the usual 'definitions' suffer from circularity (Sec. 3.1), and that they can be used only in very simple and stereotypical cases. For example, it is remarkable Poincaré's criticism [8] concerning the combinatorial definition:

  > *"The definition, it will be said, is very simple. The probability of an event is the ratio of the number of cases favorable to the event to the total number of possible cases. A simple example will show how incomplete this definition is: ...*
  > *... We are therefore bound to complete the definition by saying '... to the total number of possible cases, provided the cases are equally probable.' So we are compelled to define the probable by the probable. How can we know that two possible cases are equally probable? Will it be by convention? If we insert at the beginning of every problem an explicit convention, well and good! We then have nothing to do but to apply the rules of arithmetic and algebra, and we complete our calculation, when our result cannot be called in question. But if we wish to make the slightest application of this result, we must prove that our convention is legitimate, and we shall find ourselves in the presence of the very difficulty we thought we had avoided."*

  In the subjective approach these 'definitions' can be easily recovered as 'evaluation rules' under appropriate conditions. As far as the combinatorial evaluation is concerned, the reason is quite intuitive and it is already contained in the original Laplace's 'Laplace definition' (see Sec. 3.1). The frequency based evaluation will be reobtained in Sec. 7.1.
- Subjective probability becomes the most general framework, which is valid in all practical situations and, particularly, in treating uncertainty in measurements.
- Subjective probability does not mean arbitrary [6]; on the contrary, since

that it does not matter, from the cognitive point of view, whether one believes that the fundamental laws are intrinsically probabilistic, or whether this is just due to a limitation of our knowledge, as hidden variables *à la Einstein* would imply. [31, 32] If we calculate that process A has a probability of 0.9, and process B 0.4, we will believe A much more than B.

[6] Perhaps this is the reason why Poincaré [8], despite his many brilliant intuitions,

the normative role of coherence morally obliges a person who assesses a probability to take personal responsibility, he will try to act in the 'most objective way' (as perceived by common sense).

- The word 'belief' can hurt those who think, naïvely, that in science there is no place for beliefs. This point will be discussed in more detail in Sec. 10.4 (see also Ref. [33], while a more extensive and historical account can be found in Ref. [34]).
- Objectivity is recovered if rational individuals share the same culture and the same knowledge about experimental data, as happens for most textbook physics; but one should speak, more appropriately, of inter-subjectivity.
- The utility of subjective probability in measurement uncertainty has already been recognized[7] by the aforementioned ISO Guide [5], after many internal discussions [35] (see Ref. [36] and references therein):

  > *"In contrast to this frequency-based point of view of probability an equally valid viewpoint is that probability is a measure of the degree of belief that an event will occur... Recommendation INC-1... implicitly adopts such a viewpoint of probability."*

- In the subjective approach random variables (or, better, *uncertain numbers*) assume a more general meaning than that they have in the frequentistic approach: a random number is just any number in respect of which one is in a condition of uncertainty. For example:

  (1) if I put a reference weight (1 kg) on a balance with digital indication to the centigramme, then the random variable is the value (in grammes) that I am expected to read (X): 1000.00, 999.95... 1000.03 ...?

  (2) if I put a weight of unknown value and I read 576.23 g, then the random value (in grammes) becomes the mass of the body (μ): 576.10, 576.12... 576.23 ... 576.50...?

---

above all about the necessity of the priors (*"there are certain points which seem to be well established. To undertake the calculation of any probability, and even for that calculation to have any meaning at all, we must admit, as a point of departure, a hypothesis or convention which has always something arbitrary on it ...*), concludes to *"... have set several problems, and have given no solution ..."*. The coherence makes the distinction between arbitrariness and 'subjectivity' and gives a real sense to subjective probability.

[7] One should not feel obliged to follow this recommendation as a metrology rule. It is however remarkable to hear that, in spite of the diffused cultural prejudices against subjective probability, the scientists of the ISO working groups have arrived at such a conclusion.

In the first case the random number is linked to observations, in the second to true values.

- The different values of the random variable are classified by a function $f(x)$ which quantifies the degree of belief of all the possible values of the quantity.
- All the formal properties of $f(x)$ are the same as in conventional statistics (average, variance, etc.).
- All probability distributions are conditioned to a given state of information: in the examples of the balance one should write, more correctly,

$$f(x) \longrightarrow f(x \,|\, \mu = 1000.00)$$
$$f(\mu) \longrightarrow f(\mu \,|\, x = 576.23)\,.$$

- Of particular interest is the special meaning of conditional probability within the framework of subjective probability. Also in this case this concept turns out to be very natural, and the subjective point of view solves some paradoxes of the so-called 'definition' of conditional probability (see Sec. 10.3).
- The subjective approach is often called Bayesian, because of the central role of Bayes' theorem, which will be introduced in Sec. 2.6. However, although Bayes' theorem is important, especially in scientific applications, one should not think that this is the only way to evaluate probabilities. Outside the well-specified conditions in which it is valid, the only guidance is that of coherence.
- Considering the result of a measurement, the entire state of uncertainty is held in $f(\mu)$; then one may calculate intervals in which we think there is a given probability to find μ, value(s) of maximum belief (mode), average, standard deviation, etc., which allow the result to be summarized with only a couple of numbers, chosen in a conventional way.

2.4 Learning from observations: the 'problem of induction'

Having briefly shown the language for treating uncertainty in a probabilistic way, it remains now to see how one builds the function $f(\mu)$ which describes the beliefs in the different possible values of the physics quantity. Before presenting the formal framework we still need a short introduction on the link between observations and hypotheses.

Every measurement is made with the purpose of increasing the knowledge of the person who performs it, and of anybody else who may be interested in it. This may be the members of a scientific community, a physician who has prescribed a certain analysis or a merchant who wants to buy a certain product. It is clear that the need to perform a measurement indicates that one is in a state of uncertainty with respect to something, e.g. a fundamental constant of physics or a theory of the Universe; the state of health of a patient; the chemical composition of a product. In all cases, the measurement has the purpose of modifying a given state of knowledge. One would be tempted to say 'acquire', instead of 'modify', the state of knowledge, thus indicating that the knowledge could be created from nothing with the act of the measurement. Instead, it is not difficult to realize that, in all cases, it is just an updating process, in the light of new facts and of some reason.

Let us take the example of the measurement of the temperature in a room, using a digital thermometer — just to avoid uncertainties in the reading — and let us suppose that we get 21.7 °C. Although we may be uncertain on the tenths of a degree, there is no doubt that the measurement will have squeezed the interval of temperatures considered to be possible before the measurement: those compatible with the physiological feeling of 'comfortable environment'. According to our knowledge of the thermometer used, or of thermometers in general, there will be values of temperature in a given interval around 21.7 °C which we believe more and values outside which we believe less.[8] It is, however, also clear that if the thermometer had indicated, for the same physiological feeling, 17.3 °C, we might think that it was not well calibrated. There would be, however, no doubt that the instrument was not working properly if it had indicated 2.5 °C! The three cases correspond to three different degrees of modification of the knowledge. In particular, in the last case the modification is null.[9]

The process of learning from empirical observations is called induction by philosophers. Most readers will be aware that in philosophy there exists the unsolved 'problem of induction', raised by Hume. His criticism can be summarized by simply saying that induction is not justified, in the sense that observations do not lead necessarily (with the logical strength of a mathematical theorem) to certain conclusions. The probabilistic approach

[8]To understand the role of implicit prior knowledge, imagine someone having no scientific or technical education at all, entering a physics laboratory and reading a number on an instrument. His scientific knowledge will not improve at all, apart from the triviality that a given instrument displayed a number (not much knowledge).

[9]But also in this case we have learned something: the thermometer does not work.

adopted here seems to be the only reasonable way out of such a criticism.

2.5 Beyond Popper's falsification scheme

People very often think that the only scientific method valid in physics is that of Popper's falsification scheme. There is no doubt that, if a theory is not capable of explaining experimental results, it should be rejected or modified. But, since it is impossible to demonstrate with certainty that a theory is true, it becomes impossible to decide among the infinite number of hypotheses which have not been falsified. Adopting the falsification method literally would produce stagnation in research. A probabilistic method allows, instead, for a scale of credibility to be provided for classifying all hypotheses taken into account (or credibility ratios between any pair of hypotheses). This is close to the natural development of science, where new investigations are made in the direction which seems the most credible, according to the state of knowledge at the moment at which the decision on how to proceed was made.

As far as the results of measurements are concerned, the falsification scheme is absolutely unsuitable. Taking it literally, one should be authorized only to check whether or not the value read on an instrument is compatible with a true value, nothing more. It is understandable then that, with this premise, one cannot go very far.

We will show in Sec. 3.8 that falsification is just a subcase of the Bayesian inference.

2.6 From the probability of the effects to the probability of the causes

The scheme of updating knowledge that we will use is that of Bayesian statistical inference, widely discussed in Part 2 (in particular Secs. 3.5 and 5.1.1). I wish to make a less formal presentation of it here, to show that there is nothing mysterious behind Bayes' theorem, and I will try to justify it in a simple way.

It is very convenient to consider true values and observed values as causes and effects (see Fig. 2.1, imagining also a continuous set of causes and many possible effects). The process of going from causes to effects is called 'deduction'.[10] The possible values x which may be observed are

[10]To be correct, the deduction we are talking about is different from the classical one.

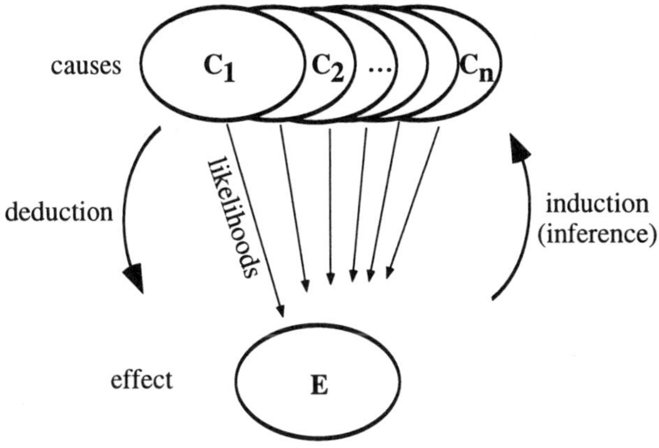

Fig. 2.1 Deduction and induction.

classified in belief by

$$f(x \mid \mu)\,.$$

This function is traditionally called 'likelihood' and summarizes all previous knowledge on that kind of measurement (behavior of the instruments, of influence factors, etc. – see list in Sec. 1.3). Often, if one deals only with random error, the $f(x \mid \mu)$ is a Gaussian distribution around μ, but in principle it may have any form.

Once the likelihood is determined (in other words, we have modelled the performance of the detector) we can build $f(\mu \mid x)$, under the hypothesis that x will be observed.[11] In order to arrive at the general formula in an heuristic way, let us consider only two values of μ. If they seem to us equally possible, it will seem natural to be in favor of the value which gives the highest likelihood that x will be observed. For example, assuming $\mu_1 = -1$, $\mu_2 = 10$, considering a normal likelihood with $\sigma = 3$, and having observed $x = 2$, one tends to believe that the observation is most likely caused by μ_1. If, on the other hand, the quantity of interest is positively defined, then μ_1 switches from most probable to impossible cause; μ_2 becomes certain.

We are dealing, in fact, with probabilistic deduction, in the sense that, given a certain cause, the effect is not univocally determined.

[11]It is important to understand that $f(\mu \mid x)$ can be evaluated before one knows the observed value x. In fact, to be correct, $f(\mu \mid x)$ should be interpreted as beliefs of μ under the hypothesis that x is observed, and not only as beliefs of μ after x is observed. Similarly, $f(x \mid \mu)$ can also be built after the data have been observed, although for teaching purposes the opposite has been suggested.

There are, in general, intermediate cases in which, because of previous knowledge (see, e.g., Fig. 1.3 and related text), one tends to believe *a priori* more in one or other of the causes. It follows that, in the light of a new observation, the degree of belief of a given value of μ will depend on

- the likelihood that μ will produce the observed effect;
- the degree of belief attributed to μ before the observation, quantified by $f_\circ(\mu)$.

Assuming[12] linear dependence on each contribution, we have finally[13]:

$$f(\mu \,|\, x) \propto f(x \,|\, \mu)\, f_\circ(\mu)\,.$$

This is one of the ways to write Bayes' theorem.

2.7 Bayes' theorem for uncertain quantities: derivation from a physicist's point of view

Let us show a little more formally the concepts illustrated in the previous section. This is a proof of the Bayes' theorem alternative to the proof applied to events, given in Part 2. It is now applied directly to uncertain quantities, and it should be closer to the physicist's reasoning than the standard proof. For teaching purposes I explain it using time ordering, but this is unnecessary, as it will be explained in Part 2.

- Before doing the experiment we are uncertain of the values of μ and x: we know neither the true value, nor the observed value. Generally speaking, this uncertainty is quantified by $f(x, \mu)$.
- Under the hypothesis that we observe x, we can calculate the conditional probability

$$f(\mu \,|\, x) = \frac{f(x,\mu)}{f(x)} = \frac{f(x,\mu)}{\int f(x,\mu)\,\mathrm{d}\mu}\,,$$

  just using probability rules (see Chapter 4 for a reminder – note the convention that the limit of integrals are omitted if they extend to all possible values of the variable of interest).

[12] Bayes' theorem will show that this assumption is indeed correct.

[13] Note the use of the same symbol $f()$ for all p.d.f.'s, though they refer to different quantities, with different status of information, and have different mathematical expressions.

- Usually we don't have $f(x, \mu)$, but this can be calculated by $f(x \,|\, \mu)$ and $f(\mu)$:

$$f(x, \mu) = f(x \,|\, \mu)\, f(\mu)\,.$$

- If we do an experiment we need to have a good idea of the behavior of the apparatus; therefore $f(x \,|\, \mu)$ must be a narrow distribution, and the most imprecise factor remains the knowledge about μ, quantified by $f(\mu)$, usually very broad. But it is all right that this should be so, because we want to learn about μ.
- Putting all the pieces together we get the standard formula of Bayes' theorem for uncertain quantities:

$$f(\mu \,|\, x) = \frac{f(x \,|\, \mu)\, f(\mu)}{\int f(x \,|\, \mu)\, f(\mu)\, \mathrm{d}\mu}\,.$$

The steps followed in this proof of the theorem should convince the reader that $f(\mu \,|\, x)$ calculated in this way is the best we can say about μ with the given status of information.

2.8 Afraid of 'prejudices'? Logical necessity versus frequent practical irrelevance of the priors

Doubtless, many readers could be at a loss at having to accept that scientific conclusions may depend on prejudices about the value of a physical quantity ('prejudice' currently has a negative meaning, but in reality it simply means 'scientific judgement based on previous experience'). We shall have many opportunities to enter again into discussion about this problem, but it is important to give a general overview now and to make some firm statements on the role of priors.

- First, from a theoretical point of view, it is impossible to get rid of priors; that is if we want to calculate the probability of events of practical interest, and not just solve mathematical games.
- At a more intuitive level, it is absolutely reasonable to draw conclusions in the light of some reason, rather than in a purely automatic way.
- In routine measurements the interval of prior acceptance of the possible values is so large, compared to the width of the likelihood (seen as a function of μ), that, in practice, it is as if all values were equally

possible. The prior is then absorbed into the normalization constant:

$$f(x \mid \mu)\, f_\circ(\mu) \xrightarrow[\text{prior very vague}]{} f(x \mid \mu). \tag{2.1}$$

- If, instead, this is not the case, it is legitimate that the priors influence our conclusions. In the most extreme case, if the experimental information is scarce or doubtful it is absolutely right to believe more in personal prejudices than in empirical data. This could be when one uses an instrument of which one is not very confident, or when one does for the first time measurements in a new field, or in a new kinematical domain, and so on. For example, it is easier to believe that a student has made a trivial mistake than to conceive that he has discovered a new physical effect. An interesting case is mentioned by Poincaré [8]:

  > *"The impossibility of squaring the circle was shown in 1885, but before that date all geometers considered this impossibility as so 'probable' that the Académie des Sciences rejected without examination the, alas! too numerous memoirs on this subject that a few unhappy madmen sent in every year. Was the Académie wrong? Evidently not, and it knew perfectly well that by acting in this manner it did not run the least risk of stifling a discovery of moment. The Académie could not have proved that it was right, but it knew quite well that its instinct did not deceive it. If you had asked the Academicians, they would have answered: 'We have compared the probability that an unknown scientist should have found out what has been vainly sought for so long, with the probability that there is one madman the more on the earth, and the latter has appeared to us the greater.'"*

In conclusion, contrary to those who try to find 'objective priors' which would give the Bayesian theory a nobler status of objectivity, I prefer to state explicitly the naturalness and necessity of subjective priors [33]. If rational people (e.g. physicists), under the guidance of coherence (i.e. they are honest, first of all with themselves), but each with unavoidable personal experience, have priors which are so different that they reach divergent conclusions, it just means that the data are still not sufficiently solid to allow a high degree of intersubjectivity (i.e. the subject is still in the area of active research rather than in that of consolidated scientific culture). On the other hand, the step from abstract objective rules to dogmatism is very short [33].

Turning now to the more practical aspect of presenting a result, I will give some recommendations about unbiased ways of doing this, in cases when priors are really critical (Chapter 13). Nevertheless, it should be clear that:

- since the natural conclusions should be probabilistic statements on physical quantities, someone has to turn the likelihoods into probabilities, and those who have done the experiment are usually the best candidates for doing this;
- taking the spirit of publishing unbiased results — which is in principle respectable — to extremes, one should not publish any result, but just raw data tapes.

2.9 Recovering standard methods and short-cuts to Bayesian reasoning

Before moving on to applications, it is necessary to answer an important question: *"Should one proceed by applying Bayes' theorem in every situation?"* The answer is no, and the alternative is essentially implicit in Eq. (2.1), and can be paraphrased with the example of the dog and the hunter of Sec. 1.7, when we discussed the arbitrariness of probability inversion performed unconsciously by (most of)[14] those who use the scheme of confidence intervals. The same example will also be used in Sec. 5.3, when discussing the reason why Bayesian estimators appear to be distorted (a topic discussed in more detail in Sec. 10.6). This analogy is very important, and, in many practical applications, it allows us to bypass the explicit use of Bayes' theorem when priors do not influence significantly the result (in the case of a normal model the demonstration can be seen in Sec. 6.3).

Figure 2.2 shows how it is possible to recover standard methods from a Bayesian perspective. One sees that the crucial link is with the Maximum Likelihood Principle, which, in this approach is just a subcase (see Secs. 5.2 and 8.1). Then, when extra simplifying restrictions are verified, the different forms of the Least Squares are reobtained. In conclusion:

- One is allowed to use these methods if one thinks that the approximations are valid; the same happens with the usual propagation of

[14] Although I don't believe it, I leave open the possibility that there really is someone who has developed some special reasoning to avoid, deep in his mind, the category of the probable when figuring out the uncertainty on a true value.

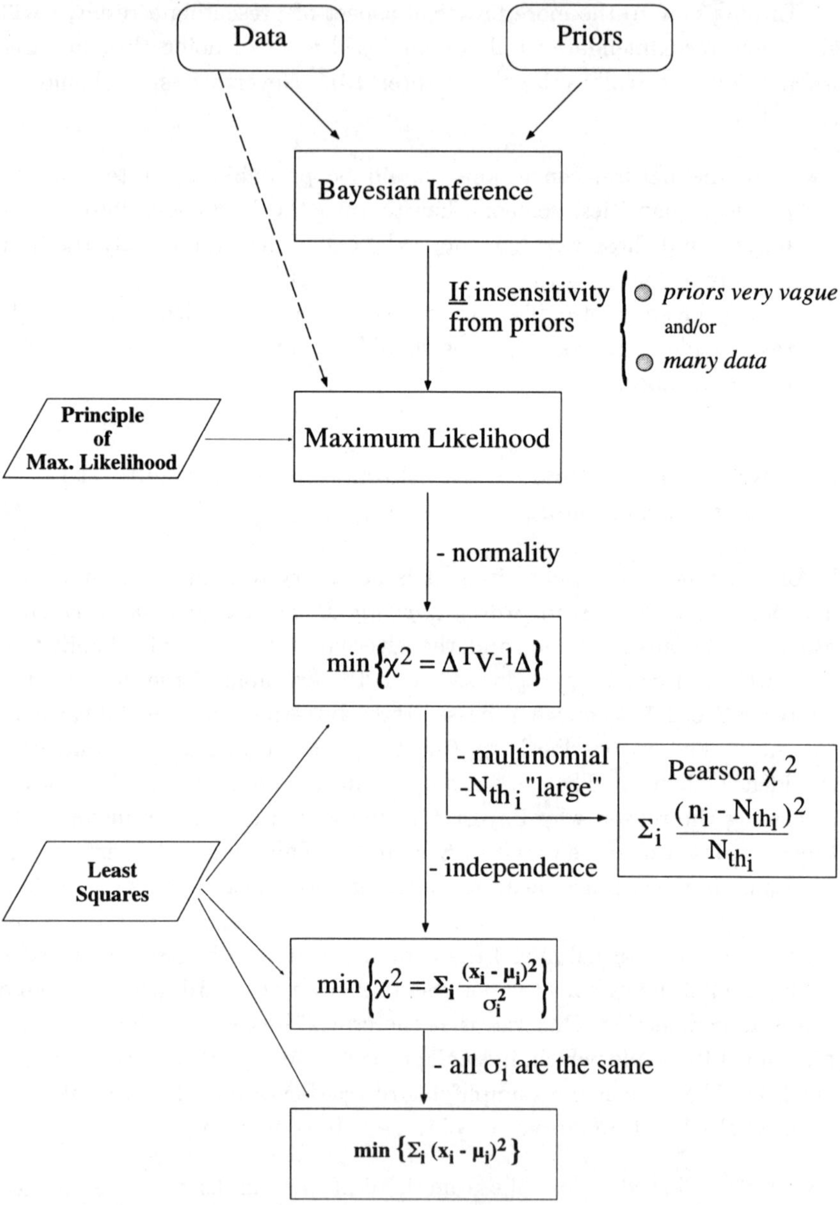

Fig. 2.2 Relation between Bayesian inference and standard data analysis methods. The top-down flow shows subsequent limiting conditions. For an understanding of the relation between the 'normal' χ^2 and the Pearson χ^2 Ref. [37] is recommended. Details are given in Chapter 8.

uncertainties and of their correlations, outlined in the next section.

- One keeps the Bayesian interpretation of the results; in particular, one is allowed to talk about the probability distributions of the true values, with philosophical and practical advantages.
- Even if the priors are not negligible, but the final distribution is roughly normal (in case of doubt it is recommended to plot it). One can evaluate the expected value and standard deviation from the shape of the distribution, as is well known:

$$\frac{\partial \ln f(\mu \,|\, x)}{\partial \mu} = 0 \Rightarrow E(\mu) \approx \mu_m, \tag{2.2}$$

$$-\left.\frac{\partial^2 \ln f(\mu \,|\, x)}{\partial \mu^2}\right|_{\mu_m} \Rightarrow \approx \frac{1}{Var(\mu)}, \tag{2.3}$$

where μ_m stands for the mode of the distribution. When the prior is very vague the conditions on the derivatives apply to $f(x \,|\, \mu)$, thus recovering, once more, standard methods. Examples of application are shown in Secs. 7.1, 7.7.2 and 8.2.

2.10 Evaluation of measurement uncertainty: general scheme

Now that we have set up the framework, we can draw the general scheme to evaluate uncertainty in measurement in the most general cases. For the basic applications we will refer to Parts II (the "primer") and III. For more sophisticated applications the reader is recommended to search in specialized literature.

2.10.1 *Direct measurement in the absence of systematic errors*

The first step consists in evaluating the uncertainty on a quantity measured directly. The most common likelihoods which describe the observed values are the Gaussian, the binomial and the Poisson distributions.

Gaussian: This is the well-known case of 'normally' distributed errors. For simplicity, we will only consider σ independent of μ (constant r.m.s. error within the range of measurability), but there is no difficulty of principle in treating the general case. The following cases will be analyzed:

- inference on μ starting from a prior much more vague than the width of the likelihood (Sec. 6.2);
- prior width comparable with that of the likelihood (Sec. 6.3): this case also describes the combination of independent measurements;
- observed values very close to, or beyond the edge of the physical region (Sec. 6.7);

Binomial: This distribution is important for efficiencies and, in the general case, for making inferences on unknown proportions. The cases considered include (see Sec. 7.1):

- general case with flat prior leading to the recursive Laplace formula (the problem solved originally by Bayes);
- limit to normality;
- combinations of different datasets coming from the same proportion;
- upper and lower limits when the efficiency is 0 or 1;
- comparison with Poisson approximation.

Poisson: The cases of counting experiments here considered:

- inference on λ starting from a flat distribution;
- upper limit in the case of null observation;
- counting measurements in the presence of a background, when the background rate is well known (Sec. 7.7.5);
- more complicated case of background with an uncertain rate (Sec. 7.7.5);
- dependence of the conclusions on the choice of experience-motivated priors (Sec. 7.7.1);
- combination of upper limits, also considering experiments of different sensitivity (Sec. 7.7.3).
- effect of possible systematic errors (Sec. 7.7.4);

2.10.2 *Indirect measurements*

The case of quantities measured indirectly is conceptually very easy, as there is 'nothing to think'. Since all values of the quantities are associated with random numbers, the uncertainty on the input quantities is propagated to that of output quantities, making use of the rules of probability. Calling μ_1, μ_2 and μ_3 the generic quantities, the inferential scheme is:

$$\begin{matrix} f(\mu_1\,|\,data_1) \\ f(\mu_2\,|\,data_2) \end{matrix} \xrightarrow[\mu_3=g(\mu_1,\mu_2)]{} f(\mu_3\,|\,data_1, data_2)\,. \tag{2.4}$$

The problem of going from the p.d.f.'s of μ_1 and μ_2 to that of μ_3 makes use of probability calculus, which can become difficult, or impossible to do analytically, if p.d.f.'s or $g(\mu_1, \mu_2)$ are complicated mathematical functions. Anyhow, it is interesting to note that the solution to the problem is, indeed, simple, at least in principle. In fact, $f(\mu_3)$ is given, in the most general case, by

$$f(\mu_3) = \int f(\mu_1)\, f(\mu_2)\, \delta(y_3 - g(\mu_1, \mu_2))\, \mathrm{d}\mu_1 \mathrm{d}\mu_2\,, \tag{2.5}$$

where $\delta()$ is the Dirac delta and the integration is over all possible values of μ_1 and μ_2. The formula can be easily extended to many variables, or even correlations can be taken into account (one needs only to replace the product of individual p.d.f.'s by a joint p.d.f.). Equation (2.5) has a simple intuitive interpretation: the infinitesimal probability element $f(\mu_3)\, d\mu_3$ depends on 'how many' (we are dealing with infinities!) elements $d\mu_1 d\mu_2$ contribute to it, each weighed with the p.d.f. calculated in the point $\{\mu_1, \mu_2\}$. An alternative interpretation of Eq. (2.5), very useful in applications, is to think of a Monte Carlo simulation, where all possible values of μ_1 and μ_2 enter with their distributions, and correlations are properly taken into account. The histogram of μ_3 calculated from $\mu_3 = g(\mu_1, \mu_2)$ will 'tend' to $f(\mu_3)$ for a large number of generated events.[15]

In routine cases the propagation is done in an approximate way, assuming linearization of $g(\mu_1, \mu_2)$ and normal distribution of μ_3. Therefore only variances and covariances need to be calculated. The well-known error propagation formulae are recovered (Chapter 8), but now with a well-defined probabilistic meaning.

2.10.3 *Systematic errors*

Uncertainty due to systematic effects is also included in a natural way in this approach. Let us first define the notation (i is the generic index):

- $\boldsymbol{x} = \{x_1, x_2, \ldots x_{n_x}\}$ is the 'n-tuple' (vector) of observables X_i;
- $\boldsymbol{\mu} = \{\mu_1, \mu_2, \ldots \mu_{n_\mu}\}$ is the n-tuple of true values μ_i;
- $\boldsymbol{h} = \{h_1, h_2, \ldots h_{n_h}\}$ is the n-tuple of *influence quantities* H_i.

[15] As we shall see, the use of frequencies is absolutely legitimate in subjective probability, once the distinction between probability and frequency is properly made. In this case it works because of the Bernoulli theorem, which states that for a very large Monte Carlo sample *"it is very improbable that the frequency distribution will differ much from the p.d.f."* (This is the probabilistic meaning to be attributed to 'tend'.)

By influence quantities we mean:

→ all kinds of external factors which may influence the result (temperature, atmospheric pressure, etc.);
→ all calibration constants;
→ all possible hypotheses upon which the results may depend (e.g. Monte Carlo parameters).

From a probabilistic point of view, there is no distinction between $\boldsymbol{\mu}$ and $\boldsymbol{h}$: they are all conditional hypotheses for the $\boldsymbol{x}$, i.e. causes which produce the observed effects. The difference is simply that we are interested in $\boldsymbol{\mu}$ rather than in $\boldsymbol{h}$.[16]

There are alternative ways to take into account the systematic effects in the final distribution of $\boldsymbol{\mu}$:

(1) Global inference on $f(\boldsymbol{\mu}, \boldsymbol{h})$. We can use Bayes' theorem to make an inference on $\boldsymbol{\mu}$ and $\boldsymbol{h}$, as described in Sec. 5.1.1. A subsequent marginalization over $\boldsymbol{h}$ yields the p.d.f. of interest:

$$\boldsymbol{x} \Rightarrow f(\boldsymbol{\mu}, \boldsymbol{h} \,|\, \boldsymbol{x}) \Rightarrow f(\boldsymbol{\mu} \,|\, \boldsymbol{x})\,.$$

This method, depending on the joint prior distribution $f_\circ(\boldsymbol{\mu}, \boldsymbol{h})$, can even model possible correlations between $\boldsymbol{\mu}$ and $\boldsymbol{h}$.

(2) Conditional inference (see Fig. 2.3). Given the observed data, one has a joint distribution of $\boldsymbol{\mu}$ for all possible configurations of $\boldsymbol{h}$:

$$\boldsymbol{x} \Rightarrow f(\boldsymbol{\mu} \,|\, \boldsymbol{x}, \boldsymbol{h})\,.$$

Each conditional result is reweighed with the distribution of beliefs of $\boldsymbol{h}$, using the well-known law of probability:

$$f(\boldsymbol{\mu} \,|\, \boldsymbol{x}) = \int f(\boldsymbol{\mu} \,|\, \boldsymbol{x}, \boldsymbol{h})\, f(\boldsymbol{h})\, \mathrm{d}\boldsymbol{h}\,. \tag{2.6}$$

(3) Propagation of uncertainties. Essentially, one applies the propagation of uncertainty, whose most general case has been illustrated in the

[16] For example, in the absence of random error the reading (X) of a voltmeter depends on the probed voltage (V) and on the scale offset (Z): $X = V - Z$. Therefore, the result from the observation of $X = x$ gives only a constraint between V and Z:

$$V - Z = x\,.$$

If we know Z well (within unavoidable uncertainty), then we can learn something about V. If instead the prior knowledge on V is better than that on Z we can use the measurement to calibrate the instrument.

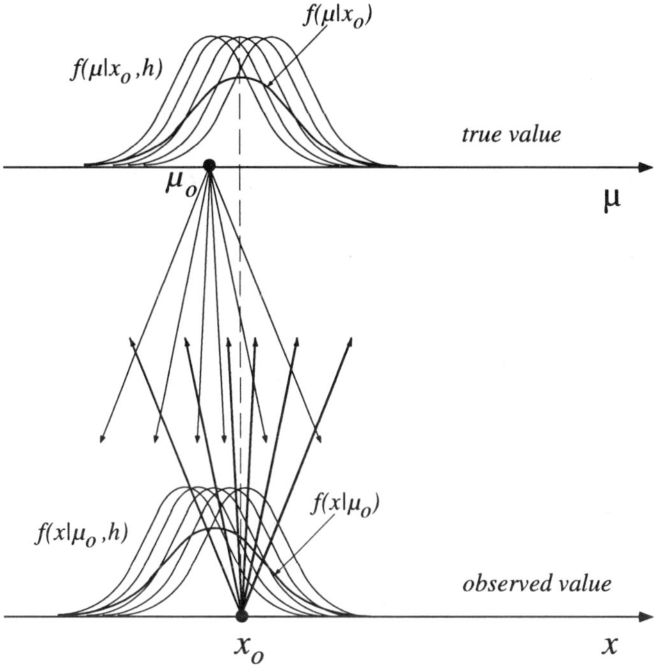

Fig. 2.3 Model to handle the uncertainty due to systematic errors by the use of conditional probability.

previous section, making use of the following model: One considers a 'raw result' on raw values $\boldsymbol{\mu}_R$ for some nominal values of the influence quantities, i.e.

$$f(\boldsymbol{\mu}_R \,|\, \boldsymbol{x}, \boldsymbol{h}_\circ)\,;$$

then (corrected) true values are obtained as a function of the raw ones and of the possible values of the influence quantities, i.e.

$$\mu_i = \mu_i(\mu_{i_R}, \boldsymbol{h})\,,$$

and $f(\boldsymbol{\mu})$ is evaluated by probability rules.

The three ways lead to the same result and each of them can be more or less intuitive to different people, and more or less suitable for different applications. For example, the last two, which are formally equivalent, are the most intuitive for experimentalists, and it is conceptually equivalent to what they do when they vary — within reasonable intervals — all Monte

Carlo parameters in order to estimate the systematic errors.[17] The third form is particularly convenient to make linear expansions which lead to approximate solutions (see Sec. 8.6).

There is an important remark to be made. In some cases it is preferable not to 'integrate' over all h's. Instead, it is better to report the result as $f(\boldsymbol{\mu}\,|\,\{h\})$, where $\{h\}$ stands for a subset of $\boldsymbol{h}$, taken at their nominal values, if:

- $\{h\}$ could be controlled better by the users of the result (for example $h_i \in \{h\}$ is a theoretical quantity on which there is work in progress);
- there is some chance of achieving a better knowledge of $\{h\}$ within the same experiment (for example h_i could be the overall calibration constant of a calorimeter);
- a discrete and small number of very different hypotheses could affect the result. For example, considering the coupling constant α_s between quarks and gluons, we could have

$$f(\alpha_s\,|\,M_1, \mathcal{O}(\alpha_s^2), \ldots) = \ldots \tag{2.7}$$
$$f(\alpha_s\,|\,M_2, \mathcal{O}(\alpha_s^2), \ldots) = \ldots\,, \tag{2.8}$$

  where M_1 and M_2 are two theoretical models, and $\mathcal{O}(\alpha_s^2)$ stands for second order approximation.[18]

If results are presented under the condition of $\{h\}$, one should also report the derivatives of $\boldsymbol{\mu}$ with respect to the result, so that one does not have to redo the complete analysis when the influence factors are better known. A typical example from particle physics in which this is usually done is the possible variation of the result due to the uncertainty on the charm-quark mass. An example in which this idea has been applied thoroughly is given in Ref. [38].

2.10.4 *Approximate methods*

Of extreme practical importance are the approximate methods, which enable us not only to avoid having to use Bayes' theorem explicitly, but also to avoid working with probability distributions. In particular, propagation

[17] But, in order to give a well-defined probabilistic meaning to the result, the variations must be performed according to $f(\boldsymbol{h})$, and not arbitrary.

[18] This is, in fact, the standard way in which this kind of result has often been presented in the past (apart from the inessential fact that only best values and standard deviations are given, assuming normality).

of uncertainty, including that due to statistical effects of unknown size, is done in this way in all routine applications, as has been remarked in the previous section. These methods are discussed in Chapters 8 and 12, together with some words of caution about their uncritical use (see Secs. 8.11, 8.12 and 8.14).

Part 2

A Bayesian primer

Chapter 3

Subjective probability and Bayes' theorem

"The only relevant thing is uncertainty – the extent of our knowledge and ignorance. The actual fact of whether or not the events considered are in some sense determined, or known by other people, and so on, is of no consequence"
(Bruno de Finetti)

"The best way to explain it is, I'll bet you fifty to one that you don't find anything"
(Richard Feynman)

"I do not believe that the Lord is a weak left-hander, and I am ready to bet a large sum that the experiments will give symmetric results"
(Wolfgang Pauli)

"It is a bet of 11,000 to 1 that the error on this result [the mass of Saturn] is not 1/100th of its value"
(Pierre-Simone Laplace)

3.1 What is probability?

The standard answers to this question are

(1) "the ratio of the number of favorable cases to the number of all cases";

(2) "the ratio of the number of times the event occurs in a test series to the total number of trials in the series".

It is very easy to show that neither of these statements can define the concept of probability:

- Definition (1) lacks the clause "if all the cases are equally probable". This has been done here intentionally, because people often forget it. The fact that the definition of probability makes use of the term "probability" is clearly embarrassing. Often in textbooks the clause is replaced by "if all the cases are equally possible", ignoring that in this context "possible" is just a synonym of "probable". There is no way out. This statement does not define probability but gives, at most, a useful rule for evaluating it – assuming we know what probability is, i.e. of what we are talking about. The fact that this definition is labelled "classical" or "Laplace" simply shows that some authors are not aware of what the "classicals" (Bayes, Gauss, Laplace, Bernoulli, etc.) thought about this matter.[1] We shall call this "definition" *combinatorial.*
- Definition (2) is also incomplete, since it lacks the condition that the number of trials must be very large ("it goes to infinity"). But this is a minor point. The crucial point is that the statement merely defines the relative *frequency* with which an event (a "phenomenon") occurred in the past. To use frequency as a measurement of probability we have to assume that the phenomenon occurred in the past, and will occur in the future, with the same probability. But who can tell if this hypothesis is correct? Nobody: we have to guess in every single case. Note that, while in the first "definition" the assumption of equal probability was explicitly stated, the analogous clause is often missing from the second one. We shall call this "definition" *frequentistic.*

We have to conclude that if we want to make use of these statements to assign a numerical value to probability, in those cases in which we judge that the clauses are satisfied, we need a better definition of probability.

3.2 Subjective definition of probability

So, *"what is probability?"* Consulting a good dictionary helps. Webster's states, for example, that "*probability* is the quality, state, or degree of being probable", and then that *probable* means "supported by evidence strong enough to make it likely though not certain to be true". The concept of

[1]For example, even the famous 'Laplace definition' contains, explicitly, the word believe (*croire* in French): *"La probabilité d'un évenement est le rapport du nombre des cas qui lui sont favorables au nombre de tous les cas possibles, lorsque rien ne porte a croire que l'un de ces cas doit arriver plutot que les autres"* [39]. It is instructive to remember that Laplace considered probability theory *"good sense turned into calculation."*

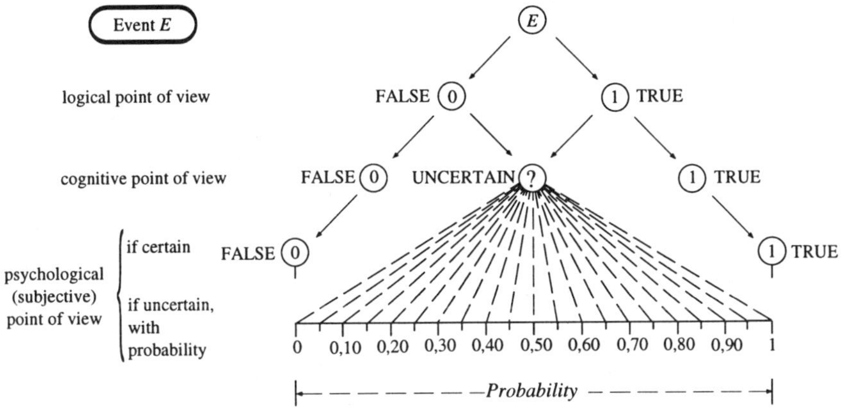

Fig. 3.1 Certain and uncertain events (de Finetti [40]).

probable arises in reasoning when the concept of *certain* is not applicable. If we cannot state firmly whether an *event* is 'true' or 'false', we just say that this is possible or *probable*, where by 'event' we mean the physical or conceptual fact described in words by a 'proposition' (indeed, we shall use the term event as a synonym for *any possible statement*, or *proposition*, relative to past, present or future).

Different events may have different *degree* of probability, depending whether we think that they are more likely to be true or false (see Fig. 3.1). Different expressions could be used to mean exactly the same concept. For example, given two events E_1 and E_2, we could say: we consider E_2 more likely than E_1; you are more confident in E_2; having to choose between E_1 and E_2 to win a price, you would promptly choose E_2. The concept of probability is then simply that of *degree of belief* [2] in an event, or

a measure of the degree of belief that an event will occur,

where the latter is the formulation that one finds often in Bayesian literature (the formulation cited here is that given in the ISO *Guide* [5]).

The use of the future tense does not imply that this definition can only be applied for future events. "Will occur" simply means that the statement

[2]It is worth mentioning the definition used by Schrödinger in Ref. [41]: *"Given the state of our knowledge about everything that could possibly have any bearing on the coming true of a certain event (thus in dubio: of the sum total of our knowledge), the numerical probability p of this event is to be a real number by the indication of which we try in some cases to set up a quantitative measure of the strength of our conjecture or anticipation, founded on the said knowledge, that the event comes true".*

"will be proven to be true", even if it refers to the past. Think for example of "the probability that it was raining in Rome on the day of the battle of Waterloo".

At first sight this definition does not seem to be superior to the combinatorial or the frequentistic ones. At least they give some practical rules to calculate "something". Defining probability as *"degree of belief"* seems too vague to be of any use. We need, then, some explanation of its meaning; a tool to evaluate it - and we will look at this tool (Bayes' theorem) later. We will end this section with some explanatory remarks on the definition, but first let us discuss the advantages of this definition. (See also Ref. [32] for comments of some common misconceptions about subjective probability.)

- It is natural, very general and can be applied to any thinkable event, independently of the feasibility of making an inventory of all (equally) possible and favorable cases, or of repeating the experiment under conditions of equal probability.
- It avoids the linguistic schizophrenia of having to distinguish "scientific" probability from "non scientific" probability used in everyday reasoning (though a meteorologist might feel offended to hear that evaluating the probability of rain tomorrow is "not scientific").
- As far as measurements are concerned, it allows us to talk about the probability of the *true value* of a physical quantity, or of any scientific hypothesis. In the frequentistic frame it is only possible to talk about the probability of the *outcome* of an experiment, as the true value is considered to be a constant. This approach is so unnatural that most physicists speak of "95 % probability that the mass of the top quark is between ...", although they believe that the correct definition of probability is the limit of the frequency (see details in Ref. [30]).
- It is possible to make a very general theory of uncertainty which can take into account any source of statistical or systematic error, independently of their distribution.

To get a better understanding of the subjective definition of probability let us take a look at *odds in betting*. The higher the degree of belief that an event will occur, the higher the amount of money A that someone ("a rational better") is ready to pay in order to receive a sum of money B if the event occurs. Clearly the bet must be acceptable in both directions (*"coherent"* is the correct adjective), i.e. the amount of money A must be smaller or equal to B and not negative (who would accept such a bet?). The cases of $A = 0$ and $A = B$ mean that the events are considered to be false or true, respectively, and obviously it is not worth betting on certainty. They are just limit cases, and in fact they can be treated with standard logic. It seems reasonable[3] that the amount of money A that one is willing to pay grows linearly with the degree of belief. It follows that if someone thinks that the probability of the event E is p, then he will bet $A = p\,B$ to get B if the event occurs, and to lose $p\,B$ if it does not. It is easy to demonstrate that the condition of "coherence" implies that $0 \le p \le 1$.

What has gambling to do with physics? The definition of probability through betting odds has to be considered *operational*, although there is no need to make a bet (with whom?) each time one presents a result. It has the important role of forcing one to make an *honest* assessment of the value of probability that one believes. One could replace money with other forms of gratification or penalization, like the increase or the loss of scientific reputation. Moreover, the fact that this operational procedure is not to be taken literally should not be surprising. Many physical quantities are defined in a similar way. Think, for example, of the textbook definition of the electric field, and try to use it to measure $\boldsymbol{E}$ in the proximity of an electron. A nice example [42] comes from the definition of a poisonous chemical compound: it *would be lethal if ingested.* Clearly it is preferable to keep this operational definition at a hypothetical level, even though it is the best definition of the concept.

3.3 Rules of probability

The subjective definition of probability, together with the condition of *coherence*, requires that $0 \le p \le 1$. This is one of the rules which probability has to obey. It is possible, in fact, to demonstrate that coherence yields to the standard rules of probability, generally known as *axioms*. In other words, *"beliefs follow the same grammar of abstract axiomatic probability."*

There is no single way to derive this important result. de Finetti's coherence [16] is considered the best guidance by many leading Bayesians (see e.g. Refs. [27, 43]). Others, in particular practitioners close to the Jaynes'

[3]This is not always true in real life. There are also other practical problems related to betting which have been treated in the literature. Other variations of the definition have also been proposed, like the one based on the *penalization rule*. A discussion of the problem goes beyond the purpose of this text. Some hints about decision problems will be given in Sec. 3.9. Many authors talk explicitly of 'small amount of money B', such that the perception of the 'value of money' does not differ substantially from one individual to another.

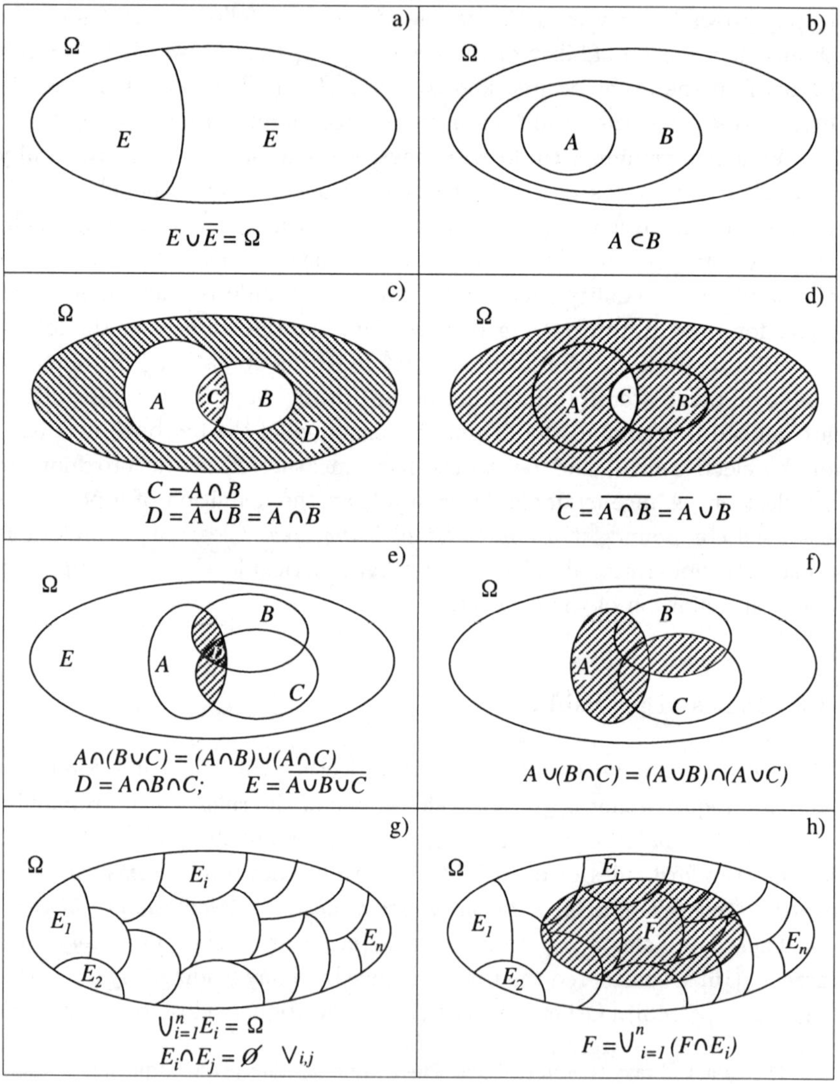

Fig. 3.2 Venn diagrams and set properties.

Maximum Entropy [44] school, feel easier with Cox' logical consistency reasoning [45] (see also Refs. [45, 46, 47], and in particular Tribus' book [48] for accurate derivations and a clear account of meaning and role of information entropy in data analysis). An approach similar to Cox's is followed by Jeffreys [49], another leading figure who has contributed to give new vitality to the methods based on this "new" point of view on probability. Note that Cox and Jeffreys were physicists. Remarkably, also Schrödinger arrived at similar conclusions [41, 50], though his definition of event is very similar to de Finetti's one, i.e. referring only to a verifiable *"state of affairs (or fact or occurrence or happening)"* [41].

Before reviewing the basic rules of probability it is worth clarifying the relationship between the axiomatic approach and the others.

- Combinatorial and frequentistic "definitions" give useful rules for evaluating probability, although they do not, as it is often claimed, define the concept.
- In the axiomatic approach one refrains from defining what the probability is and how to evaluate it: probability is just any real number which satisfies the axioms. It is easy to demonstrate that the probabilities evaluated using the combinatorial and the frequentistic prescriptions do in fact satisfy the axioms.
- The subjective approach to probability, together with the coherence requirement, defines what probability is and provides the rules which its evaluation must obey; these rules turn out to be the same as the axioms.

Since everybody is familiar with the axioms and with the analogy *events* $\Leftrightarrow$ *sets* (see Tab. 3.1 and Fig. 3.2) let us remind ourselves of the *rules of probability* in this form:

Axiom 1 $0 \le P(E) \le 1$;
Axiom 2 $P(\Omega) = 1$ (a certain event has probability 1);
Axiom 3 $P(E_1 \cup E_2) = P(E_1) + P(E_2)$, if $E_1 \cap E_2 = \emptyset$.

From the basic rules the following properties can be derived:

1: $P(E) = 1 - P(\overline{E})$;
2: $P(\emptyset) = 0$;
3: if $A \subseteq B$ then $P(A) \le P(B)$;
4: $P(A \cup B) = P(A) + P(B) - P(A \cap B)$.

We also anticipate here another rule which will be discussed in Sec. 3.5.1:

Table 3.1 Events versus sets.

Events	Sets	
		Symbol
event	set	E
certain event	sample space	Ω
impossible event	empty set	$\emptyset$
implication	inclusion (subset)	$E_1 \subseteq E_2$
opposite event (complementary)	complementary set	$\overline{E} \quad (E \cup \overline{E} = \Omega)$
logical product ("AND")	intersection	$E_1 \cap E_2$
logical sum ("OR")	union	$E_1 \cup E_2$
incompatible events	disjoint sets	$E_1 \cap E_2 = \emptyset$
complete class	finite partition	$\begin{cases} E_i \cap E_j = \emptyset \;\; \forall i \neq j \\ \cup_i E_i = \Omega \end{cases}$

5: $P(A \cap B) = P(A \,|\, B)\, P(B) = P(A)\, P(B \,|\, A)\,.$

3.4 Subjective probability and 'objective' description of the physical world

The subjective definition of probability seems to contradict the aim of physicists to describe the laws of physics in the most objective way (whatever this means ...). This is one of the reasons why many regard the subjective definition of probability with suspicion (but probably the main reason is because we have been taught at university that "probability is frequency"). The main philosophical difference between this concept of probability and an objective definition that *"we would have liked"* (but which does not exist in reality) is that $P(E)$ is not an intrinsic characteristic of the event E, but depends on the *state of information* available to whoever evaluates $P(E)$. The ideal concept of "objective" probability is recovered when everybody has the "same" state of information. But even in this case it would be better to speak of *intersubjective* probability. The best way to convince ourselves about this aspect of probability is to try to ask practical questions and to evaluate the probability in specific cases, instead of seeking refuge in abstract questions. I find, in fact, that — to paraphrase a famous statement about Time — *"Probability is objective as long as I am not asked to evaluate it."* Here are some examples.

Example 1: "What is the probability that a molecule of nitrogen at room temperature has a velocity between 400 and 500 m/s?". The answer

Table 3.2 Results of measurements of the gravitational constant G_N at the end of 1995.

Institute	G_N	$\frac{\sigma(G_N)}{G_N}$	$\frac{G_N - G_N^C}{G_N^C}$
	$\left(10^{-11}\,\frac{\mathrm{m}^3}{\mathrm{kg\cdot s^2}}\right)$	(ppm)	(10^{-3})
CODATA 1986 ("G_N^C")	6.6726 ± 0.0009	128	–
PTB (Germany) 1994	6.7154 ± 0.0006	83	$+6.41 \pm 0.16$
MSL (New Zealand) 1994	6.6656 ± 0.0006	95	-1.05 ± 0.16
Wuppertal (Germany) 1995	6.6685 ± 0.0007	105	-0.61 ± 0.17

appears easy: "take the Maxwell distribution formula from a textbook, calculate an integral and get a number". Now let us change the question: *"I give you a vessel containing nitrogen and a detector capable of measuring the speed of a single molecule and you set up the apparatus (or you let a person you trust do it). Now, what is the probability that the first molecule that hits the detector has a velocity between 400 and 500 m/s?"*. Anybody who has minimal experience (direct or indirect) of experiments would hesitate before answering. He would study the problem carefully and perform preliminary measurements and checks. Finally he would *probably* give not just a single number, but a range of possible numbers compatible with the formulation of the problem. Then he starts the experiment and eventually, after 10 measurements, he may form a different opinion about the outcome of the eleventh measurement.

Example 2: "What is the probability that the gravitational constant G_N has a value between $6.6709 \cdot 10^{-11}$ and $6.6743 \cdot 10^{-11}$ $\mathrm{m^3 kg^{-1} s^{-2}}$?". Before 1994 you could have looked at the latest issue of the Particle Data Book (PDG) [51] and answered that the probability was 95 %. At the end of 1995 three new measurements were available [52] and the four numbers do not agree with each other (see Tab. 3.2). The probability of the true value of G_N being in that range was suddenly dramatically decreased.

Example 3: "What is the probability that the mass of the top quark, or that of any of the supersymmetric particles, is below 20 or 50 GeV/c^2?". Currently it looks as if it must be zero. In the 80's many experiments were intensively looking for these particles in those energy ranges. Because so many people where searching for them, with enormous human and capital investment, it meant that, at that time, the probability was considered rather high: high enough for fake signals to be reported as strong evidence for them.

The above examples show how the evaluation of probability is conditioned by some *a priori* ("theoretical") prejudices and by some facts ("experimental data"). "Absolute" probability makes no sense. Even the classical example of probability 1/2 for each of the results in tossing a coin is only acceptable if: the coin is regular, it does not remain vertical (not impossible when playing on the beach), it does not fall into a manhole, etc.

The subjective point of view is expressed in a provocative way by de Finetti's [16]

"PROBABILITY DOES NOT EXIST".

3.5 Conditional probability and Bayes' theorem

3.5.1 *Dependence of the probability on the state of information*

If the state of information changes, the evaluation of the probability also has to be modified. For example most people would agree that the probability of a car being stolen depends on the model, age and parking site. To take an example from physics, the probability that in a detector a charged particle gives a certain number of Analog to Digital Converter (ADC) counts due to the energy loss in a gas detector can be evaluated in a very general way by making a (huge) Monte Carlo simulation which takes into account all possible reactions (weighted with their cross-sections), all possible backgrounds, changing all physical and detector parameters within *reasonable* ranges, and also taking into account the trigger efficiency. The probability changes if one knows that the particle is a 1 GeV K^+: instead of very complicated Monte Carlo simulation one can just run a single particle generator at fixed energy. The probability changes further if one also knows the exact gas mixture, pressure, etc., up to the latest determination of the pedestal and the temperature of the ADC module. More in general, using Schrödinger words [41],

> *"Since the knowledge may be different with different persons or with the same person at different times, they may anticipate the same event with more or less confidence, and thus different numerical probabilities may be attached to the same event. ... Thus whenever we speak loosely of the 'probability of an event,' it is always to be understood: probability with regard to a certain given state of knowledge."*

3.5.2 *Conditional probability*

Although everybody knows the formula of conditional probability, it is useful to derive it here in a kind of "standard way". A derivation closer to subjectivist spirit will be given in Sec. 10.3, where the meaning of the resulting formula will be described in more detail.

The notation is $P(E \mid H)$, to be read "probability of E given H", where H stands for *hypothesis*. This means: the probability that E will occur under the hypothesis that H has occurred[4].

The event $E \mid H$ can have three values:

TRUE: if E is TRUE and H is TRUE;
FALSE: if E is FALSE and H is TRUE;
UNDETERMINED: if H is FALSE; in this case we are merely uninterested in what happens to E. In terms of betting, the bet is invalidated and none loses or gains.

Then $P(E)$ can be written $P(E \mid \Omega)$, to state explicitly that it is the probability of E whatever happens to the rest of the world (Ω means all possible events). We realize immediately that this condition is really too vague and nobody would bet a cent on such a statement. The reason for usually writing $P(E)$ is that many conditions are implicitly, and reasonably, assumed in most circumstances. In the classical problems of coins and dice, for example, one assumes that they are regular. In the example of the energy loss of the previous section it was implicit ("obvious") that the high voltage was on (at which voltage?) and that the accelerator was operational (under which condition?). But one has to take care: many riddles are based on the fact that one tries to find a solution which is valid under stricter conditions than those explicitly stated in the question [53], and many people make bad business deals by signing contracts in which what "was obvious" was not explicitly stated (or precisely the contrary was stated explicitly, but in 'small print', as in insurance policies...).

In order to derive the formula of conditional probability let us assume for a moment that it is reasonable to talk about "absolute probability"

---

[4] $P(E \mid H)$ should not be confused with $P(E \cap H)$, "the probability that both events occur". For example $P(E \cap H)$ can be very small, but nevertheless $P(E \mid H)$ very high. Think of the limit case

$$P(H) \equiv P(H \cap H) \leq P(H \mid H) = 1 \ :$$

"H given H" is a certain event no matter how small $P(H)$ is, even if $P(H) = 0$ (in the sense of Sec. 4.2).

$P(E) = P(E\,|\,\Omega)$, and let us rewrite

$$\begin{aligned} P(E) \equiv P(E\,|\,\Omega) &\underset{\mathbf{a}}{=} P(E\cap\Omega) \\ &\underset{\mathbf{b}}{=} P\left(E\cap(H\cup\overline{H})\right) \\ &\underset{\mathbf{c}}{=} P\left((E\cap H)\cup(E\cap\overline{H})\right) \\ &\underset{\mathbf{d}}{=} P(E\cap H)+P(E\cap\overline{H})\,, \end{aligned} \tag{3.1}$$

where the result has been achieved through the following steps:

(a) E implies Ω (i.e. $E\subseteq\Omega$) and hence $E\cap\Omega=E$;
(b) the complementary events H and $\overline{H}$ make a *finite partition* of Ω, i.e. $H\cup\overline{H}=\Omega$;
(c) distributive property;
(d) axiom 3.

The final result of (3.1) is very simple: $P(E)$ is equal to the probability that E occurs and H also occurs, plus the probability that E occurs but H does not occur. To obtain $P(E\,|\,H)$ we just get rid of the subset of E which does not contain H (i.e. $E\cap\overline{H}$) and renormalize the probability dividing by $P(H)$, assumed to be different from zero. This guarantees that if $E=H$ then $P(H\,|\,H)=1$. We get, finally, the well-known formula

$$P(E\,|\,H) = \frac{P(E\cap H)}{P(H)} \qquad [P(H)\neq 0]\,. \tag{3.2}$$

In the most general (and realistic) case, where both E and H are conditioned by the occurrence of a third event $H_\circ$, the formula becomes

$$P(E\,|\,H,H_\circ) = \frac{P(E\cap H\,|\,H_\circ)}{P(H\,|\,H_\circ)} \qquad [P(H\,|\,H_\circ)\neq 0]\,. \tag{3.3}$$

Usually we shall make use of Eq. (3.2) (which means $H_\circ=\Omega$) assuming that Ω has been properly chosen. We should also remember that Eq. (3.2) can be resolved with respect to $P(E\cap H)$, obtaining

$$P(E\cap H) = P(E\,|\,H)\,P(H)\,, \tag{3.4}$$

and by symmetry

$$P(E\cap H) = P(H\,|\,E)\,P(E)\,. \tag{3.5}$$

We remind that two events are called *independent* if

$$P(E \cap H) = P(E)\, P(H)\,. \tag{3.6}$$

This is equivalent to saying that $P(E \,|\, H) = P(E)$ and $P(H \,|\, E) = P(H)$, i.e. the knowledge that one event has occurred does not change the probability of the other. If $P(E \,|\, H) \neq P(E)$, then the events E and H are *correlated*. In particular:

- if $P(E \,|\, H) > P(E)$ then E and H are *positively* correlated;
- if $P(E \,|\, H) < P(E)$ then E and H are *negatively* correlated.

3.5.3 *Bayes' theorem*

Let us think of all the possible, mutually exclusive, hypotheses H_i which could condition the event E. The problem here is the inverse of the previous one: what is the probability of H_i under the hypothesis that E has occurred? For example, "what is the probability that a charged particle which went in a certain direction and has lost between 100 and 120 keV in the detector is a μ, a π, a K, or a p?" Our event E is "energy loss between 100 and 120 keV", and H_i are the four "particle hypotheses". This example sketches the basic problem for any kind of measurement: having observed an *effect*, to assess the probability of each of the *causes* which could have produced it. This intellectual process is called *inference*, and it will be discussed in Sec. 5.1.1.

In order to calculate $P(H_i \,|\, E)$ let us rewrite the joint probability $P(H_i \cap E)$, making use of Eqs. (3.4)–(3.5), in two different ways:

$$P(H_i \,|\, E) P(E) = P(E \,|\, H_i)\, P(H_i)\,, \tag{3.7}$$

obtaining

$$\boxed{P(H_i \,|\, E) = \frac{P(E \,|\, H_i)\, P(H_i)}{P(E)}\,,} \tag{3.8}$$

or

$$\boxed{\frac{P(H_i \,|\, E)}{P(H_i)} = \frac{P(E \,|\, H_i)}{P(E)}\,.} \tag{3.9}$$

Since the hypotheses H_i are mutually exclusive (i.e. $H_i \cap H_j = \emptyset$, $\forall\, i, j$) and exhaustive (i.e. $\bigcup_i H_i = \Omega$), E can be written as $\bigcup_i E \cap H_i$, the union of the intersections of E with each of the hypotheses H_i. It follows that

$$\begin{aligned} P(E)\,[\equiv P(E\cap\Omega)] &= P\left(\bigcup_i (E\cap H_i)\right) \\ &= \sum_i P(E\cap H_i) \\ &= \sum_i P(E\,|\,H_i)P(H_i)\,, \end{aligned} \tag{3.10}$$

where we have made use of Eq. (3.4) again in the last step. It is then possible to rewrite Eq. (3.8) as

$$\boxed{P(H_i\,|\,E) = \frac{P(E\,|\,H_i)\,P(H_i)}{\sum_j P(E\,|\,H_j)\,P(H_j)}\,.} \tag{3.11}$$

This is the standard form by which *Bayes' theorem* is known. Equations (3.8) and (3.9) are also different ways of writing it. As the denominator of Eq. (3.11) is nothing but a normalization factor, such that $\sum_i P(H_i\,|\,E) = 1$, formula (3.11) can be written as

$$\boxed{P(H_i\,|\,E) \propto P(E\,|\,H_i)\,P(H_i)\,.} \tag{3.12}$$

Factorizing $P(H_i)$ in Eq. (3.11), and explicitly writing that all the events were already conditioned by $H_\circ$, we can rewrite the formula as

$$\boxed{P(H_i\,|\,E,H_\circ) = \alpha\,P(H_i\,|\,H_\circ)\,,} \tag{3.13}$$

with

$$\alpha = \frac{P(E\,|\,H_i,H_\circ)}{\sum_i P(E\,|\,H_i,H_\circ)\,P(H_i\,|\,H_\circ)}\,. \tag{3.14}$$

These five ways of rewriting the same formula simply reflect the importance that we shall give to this simple theorem. They stress different aspects of the same concept.

- Equation (3.11) is the standard way of writing it, although some prefer Eq. (3.8).
- Equation (3.9) indicates that $P(H_i)$ is altered by the condition E with the same ratio with which $P(E)$ is altered by the condition H_i.
- Equation (3.12) is the simplest and the most intuitive way to formulate the theorem: "the probability of H_i given E is proportional to the *initial* probability of H_i times the probability of E given H_i".

- Equations (3.13)–(3.14) show explicitly how the probability of a certain hypothesis is updated when the *state of information* changes:

  $\boxed{P(H_i \mid H_\circ)}$ [also indicated as $P_\circ(H_i)$] is the *initial*, or *a priori*, probability (or simply *'prior'*) of H_i, i.e. the probability of this hypothesis with the state of information available 'before' the knowledge that E has occurred;

  $\boxed{P(H_i \mid E, H_\circ)}$ [or simply $P(H_i \mid E)$] is the *final*, or *'a posteriori'*, probability of H_i 'after' the new information.

  $\boxed{P(E \mid H_i, H_\circ)}$ [or simply $P(E \mid H_i)$] is called *likelihood.*

  Note that 'before' and 'after' do not really necessarily imply time ordering, but only the consideration or not of the new piece of information.

To better understand the terms 'initial', 'final' and 'likelihood', let us formulate the problem in a way closer to the physicist's mentality, referring to *causes* and *effects*: 'causes' are all the physical sources capable of produing a given *observable* (the effect). The 'likelihood' indicates — as the word suggests — *"the likelihood that a cause will produce a given effect"* (not to be confused with "the likelihood that an effect is due to a given cause" which has a different meaning: A 'likelihood' may be arbitrarily small, but in spite of this, it is certain that an effect is due to a given cause, if there are no other causes capable of producing that effect!).

Using our example of the energy loss measurement again, the causes are all the possible charged particles which can pass through the detector; the effect is the amount of observed ionization; the likelihoods are the probabilities that each of the particles give that amount of ionization. Note that in this example we have fixed all the other sources of influence: physics process, accelerator operating conditions, gas mixture, high voltage, track direction, etc. This is our $H_\circ$. The problem immediately gets rather complicated (all real cases, apart from tossing coins and dice, are complicated!). The real inference would be of the kind

$$P(H_i \mid E, H_\circ) \propto P(E \mid H_i, H_\circ)\, P(H_i \mid H_\circ)\, P(H_\circ)\,. \tag{3.15}$$

For each state $H_\circ$ (the set of all the possible values of the influence parameters) one gets a different result for the final probability[5]. So, instead of

[5]The symbol $\propto$ could be misunderstood if one forgets that the proportionality factor depends on all likelihoods and priors [see Eq. (3.13)]. This means that, for a given hypothesis H_i, as the state of information E changes, $P(H_i \mid E, H_\circ)$ may change if $P(E \mid H_i, H_\circ)$ and $P(H_i \mid H_\circ)$ remain constant, and if some of the other likelihoods get modified by the new information.

getting a single number for the final probability we have a distribution of values. This spread will result in a large uncertainty of $P(H_i \,|\, E)$. This is what every physicist knows: if the calibration constants of the detector and the physics process are not under control, the "systematic errors" are large and the result is of poor quality.

3.5.4 *'Conventional' use of Bayes' theorem*

Bayes' theorem follows directly from the rules of probability, and it can be used apparently in any kind of approach. Let us take an example:

Problem 1: A particle detector has a μ identification efficiency of 95 %, and a probability of identifying a π as a μ of 2 %. If a particle is identified as a μ, then a trigger is fired. Knowing that the particle beam is a mixture of 90 % π and 10 % μ, what is the probability that a trigger is really fired by a μ? What is the signal-to-noise (S/N) ratio?

Solution: The two hypotheses (causes) which could condition the event (effect) T (="trigger fired") are "μ" and "π". They are incompatible (clearly) and exhaustive (90 % + 10 % = 100 %). Then:

$$P(\mu \,|\, T) = \frac{P(T \,|\, \mu)\, P_\circ(\mu)}{P(T \,|\, \mu)\, P_\circ(\mu) + P(T \,|\, \pi)\, P_\circ(\pi)} \tag{3.16}$$

$$(3.17)$$

$$= \frac{0.95 \times 0.1}{0.95 \times 0.1 + 0.02 \times 0.9} = 0.84\,,$$

and $P(\pi \,|\, T) = 0.16$.

The S/N ratio is $P(\mu \,|\, T)/P(\pi \,|\, T) = 5.3$. It is interesting to rewrite the general expression of the S/N ratio if the effect E is observed as

$$S/N = \frac{P(S \,|\, E)}{P(N \,|\, E)} = \frac{P(E \,|\, S)}{P(E \,|\, N)} \cdot \frac{P_\circ(S)}{P_\circ(N)}\,. \tag{3.18}$$

This formula explicitly shows that when there are *noisy conditions*,

$$P_\circ(S) \ll P_\circ(N)\,,$$

the experiment must be *very selective*,

$$P(E \,|\, S) \gg P(E \,|\, N)\,,$$

in order to have a decent S/N ratio.

(How does S/N change if the particle has to be identified by two independent detectors in order to give the trigger? Try it yourself, the answer is $S/N = 251$.)

Problem 2: Three boxes contain two rings each, but in one of them they are both gold, in the second both silver, and in the third one of each type. You have the choice of randomly extracting a ring from one of the boxes, the content of which is unknown to you. You look at the selected ring, and you then have the possibility of extracting a second ring, again from any of the three boxes. Let us assume the first ring you extract is gold. Is it then preferable to extract the second one from the same or from a different box?

Solution: Choosing the same box you have a 2/3 probability of getting a second gold ring. (Try to apply the theorem, or help yourself with intuition; the solution is given in Sec. 3.12.)

The difference between the two problems, from the conventional statistics point of view, is that the first seems to be only meaningful in the frequentistic approach, the second only in the combinatorial one. However, in a deeper analysis, the situation is a bit more complicated and, sticking strictly to the 'definitions', there is trouble in both cases.

- Problem 1 uses frequency derived probabilities (the beam composition and the detector efficiency), obtaining the relative frequencies for each kind of particle when the trigger fired. This seems consistent with the frequentistic scheme. The first trouble comes when the reasoning is applied to a single event in which the trigger has fired (Why not? You could be interested in that event, e.g. because that particular particle has produced a spectacular effect in the apparatus and you are interested in understanding its cause.) Unavoidably – and often unconsciously – physicists will turn these numbers into probability of hypotheses, in the sense of how much they have to believe in the two possibilities (μ or π?).
  The second trouble is more general and subtle, and also affects the meaning of some of the initial data of the problem. Imagine that after the particles have passed the detector they are lost, or they interact in such a way that they are not identifiable on the event-by-event base. What is the meaning of the calculated frequencies? Certainly, it is not like rolling a coin of 'unknown bias'. It is not something that we can relate to real 'random processes' à la von Mises [54] to which the frequentistic definition applies. In sum, then, also in this case the situation is not dissimilar from the single event.

- Problem 2 seems quite, a classical text book exercise. Let us take a closer look at it. As we said earlier, the solution lies in that the probability of finding a gold or silver ring in the same box is 2/3 and 1/3, respectively. The only trouble is, once more, the meaning of these numbers. There are only two events, gold and silver. What are the equiprobable 'possible' and 'favorable' cases which will result in the events? If probability is that famous ratio, at any moment we could be required to list the equiprobable cases which enter into this evaluation. This is already a hard task even in this simple exercise! Again, everybody (mostly intuitively) interprets 2/3 and 1/3 as how much we can be confident in either hypothesis.

In conclusion, even these simple 'standard' problems have a consistent solution only in the Bayesian approach. Moreover, apart from the question of how to interpret the results, in many and important cases of life and science, neither of the two conventional definitions is applicable from the very beginning.

3.6 Bayesian statistics: learning by experience

The advantage of the Bayesian approach (leaving aside the "little philosophical detail" of trying to define what probability is) is that one may talk about the probability of any kind of event, as already emphasized. Moreover, the procedure of updating the probability with increasing information is very similar to that followed by the mental processes of rational people.[6] Let us consider a few examples of "Bayesian use" of Bayes' theorem.

Example 1: Imagine some persons listening to a common friend having a phone conversation with an unknown person X_i, and who are trying to guess who X_i is. Depending on the knowledge they have about the friend, on the language spoken, on the tone of voice, on the subject of conversation, etc., they will attribute some probability to several possible persons. As the conversation goes on they begin to consider some

[6] How many times have you met neighbors far from home and wondered for a while who they are? Think also how difficult it is to understand a person in a noisy environment, or a person who speaks your language poorly, if you do not know what he wants to say (this process happens every time you hear something which sounds illogical or simply 'wrong' and, trusting the good faith of the person, who hopefully wanted to say something meaningful, you try to interpret the message 'correctly'). Ref. [55] shows an interesting study on the relation between perception and Bayesian inference.

possible candidates for X_i, discarding others, then hesitating perhaps only between a couple of possibilities, until the state of information I is such that they are ***practically sure*** of the identity of X_i. This experience has happened to most of us, and it is not difficult to recognize the Bayesian scheme:

$$P(X_i \,|\, I, I_\circ) \propto P(I \,|\, X_i, I_\circ)\, P(X_i \,|\, I_\circ)\,. \tag{3.19}$$

We have put the initial state of information $I_\circ$ explicitly in Eq. (3.19) to remind us that likelihoods and initial probabilities depend on it. If we know nothing about the person, the final probabilities will be very *vague*, i.e. for many persons X_i the probability will be different from zero, without necessarily favoring any particular person.

Example 2: A person X meets an old friend F in a pub. F proposes that the drinks should be paid for by whichever of the two extracts the card of lower value from a pack (according to some rule which is of no interest to us). X accepts and F wins. This situation happens again in the following days and it is always X who has to pay. What is the probability that F has become a cheat, as the number of consecutive wins n increases?

The two hypotheses are: *cheat* (C) and *honest* (H). $P_\circ(C)$ is low because F is an "old friend", but certainly not zero: let us assume 5 %. To make the problem simpler let us make the approximation that a cheat always wins (not very clever...): $P(W_n \,|\, C) = 1$. The probability of winning if he is honest is, instead, given by the rules of probability *assuming* that the chance of winning at each trial is 1/2 ("why not?", we shall come back to this point later): $P(W_n \,|\, H) = 2^{-n}$. The result

$$P(C \,|\, W_n) = \frac{P(W_n \,|\, C)\, P_\circ(C)}{P(W_n \,|\, C)\, P_\circ(C) + P(W_n \,|\, H)\, P_\circ(H)} \tag{3.20}$$

$$= \frac{1 \times P_\circ(C)}{1 \times P_\circ(C) + 2^{-n} \times P_\circ(H)} \tag{3.21}$$

is shown in the following table.

n	$P(C \mid W_n)$ (%)	$P(H \mid W_n)$ (%)
0	5.0	95.0
1	9.5	90.5
2	17.4	82.6
3	29.4	70.6
4	45.7	54.3
5	62.7	37.3
6	77.1	22.9
...	...	...

Naturally, as F continues to win the suspicion of X increases. It is important to make two remarks.

- The answer is always probabilistic. X can never reach absolute certainty that F is a cheat, unless he catches F cheating, or F confesses to having cheated. This is coherent with the fact that we are dealing with random events and with the fact that any sequence of outcomes has the same probability (although there is only one possibility over 2^n in which F is always luckier). Making use of $P(C \mid W_n)$, X can make a decision about the next action to take:
  - continue the game, with probability $P(C \mid W_n)$ of losing with certainty the next time too;
  - refuse to play further, with probability $P(H \mid W_n)$ of offending the innocent friend.
- If $P_\circ(C) = 0$ the final probability will always remain zero: if X fully trusts F, then he just has to record the occurrence of a rare event when n becomes large.

To better follow the process of updating the probability when new experimental data become available, according to the Bayesian scheme

> *"the final probability of the present inference is the initial probability of the next one".*

Let us call $P(C \mid W_{n-1})$ the probability assigned after the previous win.

The sequential application of the Bayes formula yields

$$P(C \mid W_n) = \frac{P(W \mid C)\, P(C \mid W_{n-1})}{P(W \mid C)\, P(C \mid W_{n-1}) + P(W \mid H)\, P(H \mid W_{n-1})} \tag{3.22}$$

$$= \frac{1 \times P(C \mid W_{n-1})}{1 \times P(C \mid W_{n-1}) + 1/2 \times P(H \mid W_{n-1})}, \tag{3.23}$$

where $P(W \mid C) = 1$ and $P(W \mid H) = 1/2$ are the probabilities of each win. The interesting result is that exactly the same values of $P(C \mid W_n)$ of Eq. (3.21) are obtained (try to believe it!).

It is also instructive to see the dependence of the final probability on the initial probabilities, for a given number of wins n.

$P_\circ(C)$ (%)	$P(C \mid W_n)$ (%)			
	$n = 5$	$n = 10$	$n = 15$	$n = 20$
1 %	24	91.1	99.70	99.99
2.5 %	45	96.3	99.88	99.996
5 %	63	98.2	99.94	99.998
10 %	78	99.1	99.97	99.999
50 %	97	99.90	99.997	99.9999

As the number of experimental observations increases the conclusions no longer depend, practically, on the initial assumptions. This is a crucial point in the Bayesian scheme and it will be discussed in more detail later. Another interesting feature we learn from the table is that the results are stable relative to reasonable variations of the prior (see for example the factor of two variations around $P_\circ(C) = 5\%$) (the study of the dependence of the results on the assumptions is referred to as *sensitivity study*, or *sensitivity analysis*).

3.7 Hypothesis 'test' (discrete case)

Although in conventional statistics books this argument is usually dealt with in one of the later chapters, in the Bayesian approach it is so natural that it is in fact the first application, as we have seen in the above examples.

If one needs to compare two hypotheses, as in the example of the signal to noise calculation, the ratio of the final probabilities can be taken as a quantitative result of the test. Let us rewrite the S/N formula (3.18) in the most general case:

$$\frac{P(H_1\,|\,E,H_\circ)}{P(H_2\,|\,E,H_\circ)} = \frac{P(E\,|\,H_1,H_\circ)}{P(E\,|\,H_2,H_\circ)} \cdot \frac{P(H_1\,|\,H_\circ)}{P(H_2\,|\,H_\circ)}, \tag{3.24}$$

where again we have reminded ourselves of the existence of $H_\circ$. The ratio depends on the product of two terms: the ratio of the priors and the ratio of the likelihoods. When there is absolutely no reason for choosing between the two hypotheses, the prior ratio is 1 and the decision depends only on the other term, called *the Bayes factor*. If one firmly believes in either hypothesis, the Bayes factor is of minor importance, unless it is zero or infinite (i.e. one and only one of the likelihoods is vanishing). Perhaps this is disappointing for those who expected objective certainty from a probability theory, but this is in the nature of things.

3.7.1 *Variations over a problem to Newton*

It seems[7] that Isaac Newton was asked to solve the following problem. A man condemned to death has an opportunity of having his life saved and to be freed, depending on the outcome of an uncertain event. The man can choose between three options: a) roll 6 dice, and be free if he gets '6' with one and only one die (A); b) roll 12 dice, and be freed if he gets '6' with exactly 2 dice; c) roll 18 dice, and be freed if he gets '6' in exactly 3 dice. Clearly, he will choose the event about which he is *more confident* (we could also say the event which he considers *more probable*; the event *most likely to happen*; the event which *he believes mostly*; and so on). Most likely the condemned man is not able to solve the problem, but he certainly will understand Newton's suggestion to choose A, which gives him the *highest chance* to survive. He will also understand the statement that A is about 36% more likely than B and 64% more likely than C. [8] The condemned man would perhaps ask Newton to give him some idea how likely the event A is. A good answer would be to make a comparison with a box containing 100 balls, 40 of which are white. He should be as confident of surviving

[7] My source of information is Ref. [56]. It seems that Newton gave the 'correct answer' - indeed, in this stereotyped problem there is *the* correct answer.

[8] The solution is an easy application of the binomial distribution. Using the notation of Eq. (4.18), we have: $P(A) = f(1\,|\,\mathcal{B}_{6,1/6}) = 0.402$, $P(B) = f(2\,|\,\mathcal{B}_{12,1/6}) = 0.296$ and $P(C) = f(3\,|\,\mathcal{B}_{18,1/6}) = 0.245$.

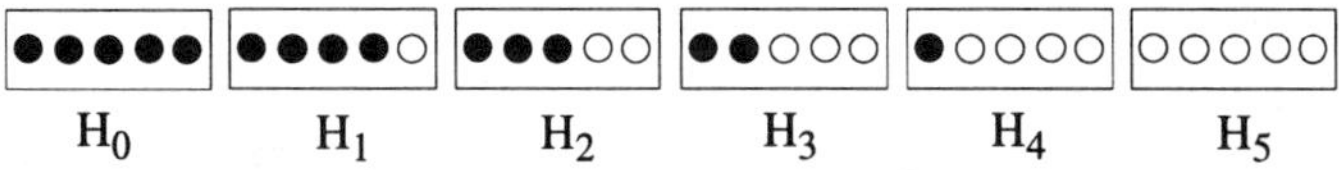

Fig. 3.3 A box has with certainty one of these six black and white ball compositions. The content of the box is inferred by extracting at random a ball from the box then returning it to the box. How confident are you initially of each composition? How does your confidence change after the observation of 1, 5 and 8 consecutive extractions of a black ball? See Ref. [29] for an introduction to the Bayesian reasoning based on this problem.

as of extracting a white ball from the box; i.e. 40% confident of being freed and 60% confident of dying: not really an enviable situation, but better than choosing C, corresponding to only 25 white balls in the box. Note that the fact that any person is able to claim to be more confident of extracting a white ball from the box that contains the largest fraction of white balls, while for the evaluation of the above events one has to 'ask Newton', does not imply a different *perception of the probability* in the two classes of events. It is only because the events A, B and C are complex events, the probability of which is evaluated from the probability of the elementary events (and everybody can figure out what it means that the six faces of a die are equally likely) plus some combinatorics, for which some mathematical education is needed. The condemned man, trusting Newton, will make Newton's beliefs his own beliefs, though he might never understand how Newton arrived at those numbers.

Let us imagine now a more complicated situation, in which you have to make the choice (imagine for a moment you are the prisoner, just to be emotionally more involved in this academic exercise. A box contains with certainty 5 balls, with a white ball content ranging from 0 to 5, the remaining balls being black (see Fig. 3.3, and Ref. [29] for further variations on the problem). One ball is extracted at random, shown to you, and then returned to the box. The ball is **black**. You get freed if you guess correctly the composition of the box. Moreover you are allowed to ask a question, to which the judges will reply correctly if the question is pertinent and such that their answer does not indicate with certainty the exact content of the box.

Having observed a black ball, the only certainty is that H_5 is ruled out. As far as the other five possibilities are concerned, a first idea would be to be more confident about the box composition which has more black balls (H_0), since this composition gives the highest chance of extracting this color. Following this reasoning, the confidence in the various box compositions would be proportional to their black ball content. But it is not difficult to understand that this solution is obtained by assuming that the compositions are considered a priori equally possible. However, this condition was not stated explicitly in the formulation of the problem. How was the box prepared? You might think of an initial situation of six boxes each having a different composition. But you might also think that the balls were picked at random from a large bag containing a roughly equal proportion of white and black balls. Clearly, the initial situation changes. In the second case the composition H_0 is initially so unlikely that, even after having extracted a black ball, it remains not very credible. The observation alone is not enough to state how much one is confident about something.

The use of Bayes' theorem to solve this problem is sketched in Fig. 3.4. The top bar diagram shows the likelihood $P(\text{Black} \mid H_i)$ of observing a black ball assuming each possible composition; The second pair of plots shows the two priors considered in our problem. The final probabilities are shown next. We see that the two solutions are quite different, as a consequence of different priors. So a good question to ask the judges would be how the box was prepared. If they say it was uniform, bet your life on H_0. If they say the five balls were extracted from a large bag, bet on H_2.

Perhaps the judges might be so clement as to repeat the extraction (and subsequent reintroduction) several times. Figure 3.4 shows what happens if five or height consecutive black balls are observed. The evaluation is performed by sequential use of Bayes' theorem

$$P_n(H_i \mid E) \propto P(E_n \mid H_i)\, P_{n-1}(H_i)\,. \tag{3.25}$$

If you are convinced[9] that the preparation procedure is binomial (large bag), you still consider H_1 more likely than H_0, even after five consecutive observations. Only after eight consecutive extractions of a black ball are you mostly confident about H_0 independently of how much you believe in the two preparation procedures (but, obviously, you might imagine – and perhaps even believe in – more fancy preparation procedures which still give different results). After many extractions we are practically sure of the box content, as we shall see in a while, though we can never be certain.

[9]And if you have doubts about the preparation? The probability rules teach us what to do. Calling U (uniform) and B (binomial) the two preparation procedures, with probabilities $P(U)$ and $P(B)$, we have $P(H \mid \text{obs}) = P(H \mid \text{obs}, U)\cdot P(U) + P(H \mid \text{obs}, B)\cdot P(B)$.

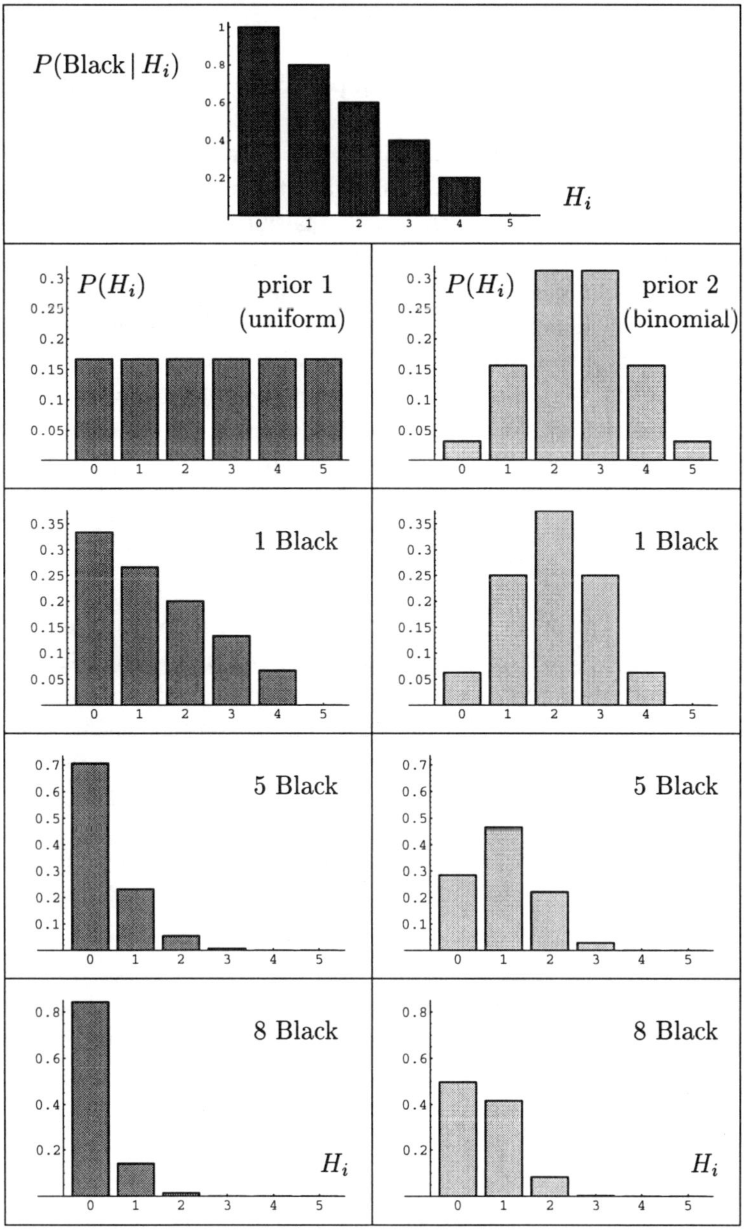

Fig. 3.4 Confidence in the box contents (Fig. 3.3) as a function of prior and observation (see text).

3.8 Falsificationism and Bayesian statistics

If an observation is impossible, given a particular hypothesis, that hypothesis will in turn become impossible ("false") if that observation is made. On the other hand, no hypothesis can be declared 'true' until alternative hypotheses are conceivable. This straightforward application of Bayes' theorem recovers the essence of Popper's falsificationism. However, a few remarks are in order. First, all non-falsified hypotheses, which in falsificationism are in a kind of Limbo, in the Bayesian approach acquire different degrees of beliefs depending on all available information. Second, one has to be very careful to distinguish between what is impossible and what is very improbable. As discussed at length in Sec. 1.8, many erroneous scientific conclusions are the result of adopting statistical methods which are essentially based on a confusion between impossible and improbable.

3.9 Probability versus decision

We have seen in the previous sections that beliefs are used to take decisions. Taking decisions is usually more complicated and more subject-dependent than assessing probability, because the decision depends not only on how much the events are believed to happens, but also on the benefits and costs that the events cause. If someone offers me odds of 10 to 1 on the outcome of tossing an unbiased coin (i.e. I stand to win ten times what I wager), I will consider it an incredible opportunity and will not hesitate to gamble 100 or 1000 dollars. And this would certainly be a good decision, even though I stand a 50% chance of losing (yes, even if I were to lose, I would still consider it to have been a good decision).

The simplest figure of merit to classifying the advantage of a decision is to consider the expected value of gain, i.e. the weighted average of the gains, each weighted with its probability. For example, if I consider many events E_i, to each of which I believe with probability $P(E_i)$, and such that I get a gain $G(E_i)$ from their occurrence (some of the gains could be negative, i.e. losses), my expected gain is

$$\text{Expected gain} = \sum_i P(E_i)\, G(E_i). \tag{3.26}$$

In the case of the above bet, the expected gain is $1/2 \times (-A) + 1/2 \times (+10A) = +4.5A$, where A is the amount of money I wager. Apparently, the bet is very convenient, and the convenience increases with A! But there must be something wrong. This can be easily understood, considering the limiting situation. If the offer were to be valid only on condition that I gambled all my property, I would be crazy to accept it. This subjective and non-linear perception of the value of money is well known. Other complications arise because the effect of the occurrence of some events are not simply money, as in example 2 of Sec. 3.6.

We shall talk no further about decision in the rest of this book and the reader interested to this subject can look into specialized literature, e.g. Refs. [57, 58]. However, I think it is important to separate probability assessing from decision-taking issues. Probability assessment must be based on a cool consideration of beliefs, under the normative rule of the coherent bet, *as if* we really had to accept that bet in either direction and had an infinite budget at our disposal. Subsequently decision issues will arise, and their risks and benefits will then need to be properly taken into account.

3.10 Probability of hypotheses versus probability of observations

In previous sections we have concentrated on the probability of hypotheses in the light of past observations. We could also be interested in the probability of future observations, for example the probability that our 'old friend' from Sec. 3.6 will win the next game. As usual, we apply probability rules:

$$P(W_{n+1}\,|\,W_n) = P(W_{n+1}\,|\,C)\cdot P(C\,|\,W_n) + P(W_{n+1}\,|\,H)\cdot P(H\,|\,W_n)\,. \tag{3.27}$$

Let us see how this probability increases as a function of past wins (Table 3.3). We also include in the same table the relative frequency of past wins, as this could be an alternative way of calculating the probability (frequentistic 'definition'): Little comment needs to be made about the pure frequency based calculation of probability; do we really feel 100% confident that our old friend will win again?

Having made this rough comparison between frequentistic and Bayesian methods for calculating the probability of future observations (for another, more realistic, example of the evolution of probabilities of hypotheses and observables, together with comparisons with the frequentistic approach, see Ref. [29]), it is natural to ask for a comparison of their performance in

Table 3.3 Probability that the 'old friend' will win next time: Bayesian solution compared with relative frequency. In the Bayesian solution the initial probability of 'cheat' is 5%.

n	$P(W_{n+1} \mid W_n)$ (%)	$f(W_n) \stackrel{?}{\to} P(W_{n+1})$ (%)
0	52.5	—
1	54.8	100
2	58.7	100
3	64.7	100
4	72.9	100
5	81.4	100
6	88.6	100
...	...	...

calculating the probability of hypotheses. But this cannot be done, simply because the very concept of probability of hypotheses is prohibited in the frequentistic approach. This would be no great problem if frequentists refrained from assessing levels of confidence in hypotheses. But this is not the case: frequentists deal with hypotheses by means of the popular hypothesis-test scheme and use phrases which sound like degree of confidence in hypotheses, although they say these phrases should not be considered as probabilistic statements. Needless to say, this kind of approach generates confusion, as was pointed out in Sec. 1.8. [10]

3.11 Choice of the initial probabilities (discrete case)

3.11.1 *General criteria*

The dependence of Bayesian inferences on initial probability is considered by opponents as the fatal flaw in the theory. But this criticism is less severe than one might think at first sight. In fact:

---

[10] Some say that Bayesian ideas are just philosophical irrelevances, because in practice frequentistic "CL's" often coincide with Bayesian results. I think, however, that things should be looked at the other way round. Frequentistic "CL's" are usually meaningless unless they coincide with Bayesian results obtained under well-defined conditions. As an outstanding example of this, I point to the case of two CL results, obtained by different reactions, on the same physics quantity (the Higgs boson particle) given by the same experimental teams using the same wording (Ref. [59], page 8). The two results have completely different meanings, as is also stated in Ref. [60]. The lesson is that one must always be very careful to be clear whether a frequentistic 95% CL result means "by chance", a 95% confidence on a given statement, or something else.

- It is impossible to construct a theory of uncertainty which is not affected by this "illness". Those methods which are advertised as being "objective" tend in reality to hide the hypotheses on which they are grounded. A typical example is the maximum likelihood method, of which we will discuss later.
- As the amount of information increases the dependence on initial prejudices diminishes.
- When the amount of information is very limited, or completely lacking, there is nothing to be ashamed of if the inference is dominated by *a priori* assumptions.

It is well known to all experienced physicists that conclusions drawn from an experimental result (and sometimes even the "result" itself!) often depend on prejudices about the phenomenon under study. Some examples:

- When doing quick checks on a device, a single measurement is usually performed if the value is "what it should be", but if it is not then many measurements tend to be made.
- Results are sometimes influenced by previous results or by theoretical predictions. See for example Fig. 3.5 taken from the Particle Data Book [51]. The interesting book *"How experiments end"* [61] discusses, among others, the issue of when experimentalists are "happy with the result" and stop "correcting for the systematics".
- Slight deviations from the background might be interpreted as a signal (e.g. as for the first claim of discovery of the top quark in spring '94), while larger 'signals' are viewed with suspicion if they are unwanted by the physics 'establishment'[11].
- Experiments are planned and financed according to the prejudices of the moment (for a delightful report see Ref. [63]).

These comments are not intended to justify unscrupulous behavior or sloppy analysis. They are intended, instead, to remind us — if need be — that scientific research is ruled by subjectivity much more than outsiders imagine. The transition from subjectivity to "objectivity" begins when there is a large consensus among the most influential people about how to interpret the results[12].

[11] A case, concerning the search for electron compositeness in e^+e^- collisions, is discussed in Ref. [62].

[12] *"A theory needs to be confirmed by experiments. But it is also true that an experimental result needs to be confirmed by a theory."* This sentence expresses clearly — though paradoxically — the idea that it is difficult to accept a result which is not

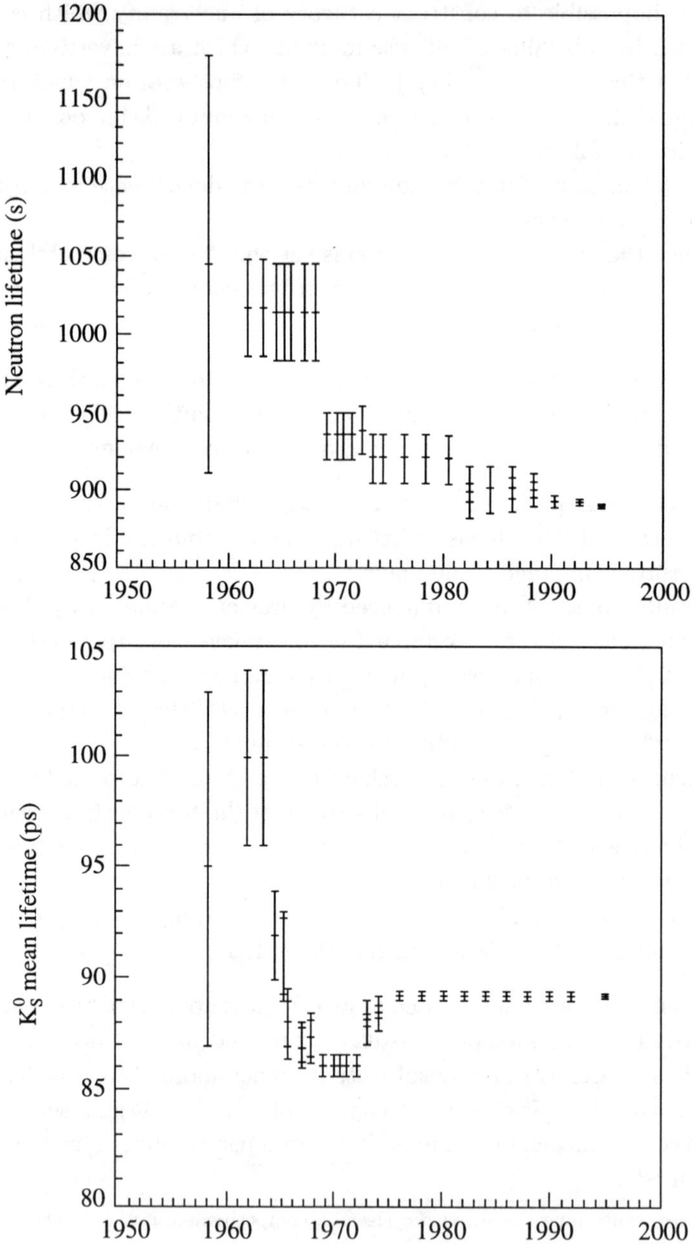

Fig. 3.5 Results on two physical quantities as a function of the publication date.

In this context, the subjective approach to statistical inference at least teaches us that every assumption must be stated clearly and all available information which could influence conclusions must be weighed with the maximum 'attempt at objectivity'. [13]

What are the rules for choosing the "right" initial probabilities? As one can imagine, this is an open and debated question among scientists and philosophers. My personal point of view is that one should avoid pedantic discussion of the matter, because the idea of universally true priors reminds me terribly of the famous "angels' sex" debates.

If I had to give recommendations, they would be the following.

- The *a priori* probability should be chosen in the same spirit as the rational person who places a bet, seeking to minimize the risk of losing.
- General principles — like those that we will discuss in a while — may help, but since it may be difficult to apply elegant theoretical ideas in all practical situations, in many circumstances the *guess* of the "expert" can be relied on for guidance.
- To avoid using as prior the results of other experiments dealing with the same open problem, otherwise correlations between the results would prevent all comparison between the experiments and thus the detection of any systematic errors.

3.11.2 *Insufficient reason and Maximum Entropy*

The first and most famous criterion for choosing initial probabilities is the simple *Principle of Insufficient Reason* (or *Indifference Principle*): If there is no reason to prefer one hypothesis over alternatives, simply attribute the same probability to all of them. The indifference principle applied to coin and die tossing, to card games or to other simple and symmetric problems, yields to the well-known rule of probability evaluation that we have called combinatorial. Since it is impossible not to agree with this point of view, in the cases for which one judges that it does apply, the combinatorial "definition" of probability is recovered in the Bayesian approach if the word "definition" is simply replaced by "evaluation rule". We have in fact already

rationally justified.

[13]It may look paradoxical, but, due to the normative role of the coherent bet, the subjective assessments are more objective about using, without direct responsibility, someone else's formulae. For example, even the knowledge that somebody else has a different evaluation of the probability is new information which must be taken into account.

used this reasoning in previous examples.

A modern and more sophisticated version of the Indifference Principle is the Maximum Entropy Principle. The information entropy function of n mutually exclusive events, to each of which a probability p_i is assigned, is defined as [64]

$$H(p_1, p_2, \ldots p_n) = -K \sum_{i=1}^{n} p_i \ln p_i, \tag{3.28}$$

with K a positive constant. The principle states that *"in making inferences on the basis of partial information we must use that probability distribution which has the maximum entropy subject to whatever is known,"* [44]. Note that, in this case, "entropy" is synonymous with "uncertainty"[14] [44]. One can show that, in the case of absolute ignorance about the events E_i, the maximization of the information uncertainty, with the constraint that $\sum_{i=1}^{n} p_i = 1$, yields the classical $p_i = 1/n$ (any other result would have been worrying ...).

Although this principle is sometimes used in combination with the Bayes formula for inferences (also applied to measurement uncertainty, see Ref. [36]), it will not be used for applications in this book.

3.12 Solution to some problems

Here are the solutions to some of the examples discussed earlier.

3.12.1 *AIDS test*

The AIDS test problem (Example 7 of Sec. 1.9) is a very standard one. Let us solve it using the Bayes factor:

$$\begin{aligned}
\frac{P(\text{HIV}\,|\,\text{Positive})}{P(\overline{\text{HIV}}\,|\,\text{Positive})} &= \frac{P(\text{Positive}\,|\,\text{HIV})}{P(\text{Positive}\,|\,\overline{\text{HIV}})} \cdot \frac{P_\circ(\text{HIV})}{P(\overline{\text{HIV}})} \\
&= \frac{\approx 1}{0.002} \times \frac{0.1/60}{\approx 1} = 500 \times \frac{1}{600} = \frac{1}{1.2} \\
P(\text{HIV}\,|\,\text{Positive}) &= 45.5\%\,.
\end{aligned}$$

Writing Bayes' theorem in this way helps a lot in understanding what is going on. Stated in terms of signal to noise and selectivity (see problem 1

[14]For one of the clearest illustrations about information entropy and uncertainty, see Myron Tribus' *"Rational descriptions, decisions and designs"*. [48]

in Sec. 3.5.4), we are in a situation in which the selectivity of the test is not enough for the noisy conditions. So in order to be practically sure that the patient declared 'positive' is infected, with this performance of the analysis, one needs independent tests, unless the patient belongs to high-risk classes. For example, a double independent analysis on an average person would yield

$$P(\text{HIV}\,|\,\text{Positive}_1 \cap \text{Positive}_2) = 99.76\%\,,$$

similar[15] to that obtained in the case where a physician had a 'severe doubt' [i.e. $P_\circ(\text{HIV}) \approx P_\circ(\overline{\text{HIV}})$] that the patient could be infected:

$$P(\text{HIV}\,|\,\text{Positive}, P_\circ(\text{HIV}) \approx 0.5) = 99.80\%\,.$$

We see then that, as discussed several times (see Sec. 10.8), the conclusion obtained by arbitrary probability inversion is equivalent to assuming uniform priors.

Another interesting question, which usually worries those who approach Bayesian methods for the first time, is the stability of the result. A variation of a factor of two of the prior makes $P(\text{HIV}\,|\,\text{Positive})$ vary between 29% and 63%.

3.12.2 *Gold/silver ring problem*

The three-box problem (Sec. 3.5.4) seems to be intuitive for some, but not for everybody. Let us label the three boxes: A, Golden-Golden; B, Golden-Silver; C, Silver-Silver. The initial probability (i.e. before having checked the first ring) of having chosen the box A, B, or C is, by symmetry, $P_\circ(A) = P_\circ(B) = P_\circ(C) = 1/3$. This probability is updated after the event $E =$ 'the first ring extracted is golden' by Bayes' theorem:

$$P(A\,|\,E) = \frac{P(E\,|\,A)\,P_\circ(A)}{P(E\,|\,A)\,P_\circ(A) + P(E\,|\,B)\,P_\circ(B) + P(E\,|\,C)\,P_\circ(C)} = 2/3$$

$$P(B\,|\,E) = \frac{P(E\,|\,B)\,P_\circ(B)}{P(E\,|\,A)\,P_\circ(A) + P(E\,|\,B)\,P_\circ(B) + P(E\,|\,C)\,P_\circ(C)} = 1/3$$

$$P(C\,|\,E) = \frac{P(E\,|\,C)\,P_\circ(C)}{P(E\,|\,A)\,P_\circ(A) + P(E\,|\,B)\,P_\circ(B) + P(E\,|\,C)\,P_\circ(C)} = 0\,,$$

where $P(E\,|\,A)$, $P(E\,|\,B)$ and $P(E\,|\,C)$ are, respectively, 1, 1/2 and 0.

[15]There is nothing profound in the fact that the two cases give very similar results. It is just due to the numbers of these examples (i.e. $500 \approx 600$).

Finally, calling F = 'the next ring will be golden if I extract it from the same box', we have, using the probability rules:

$$\begin{aligned} P(F\,|\,E) &= P(F|A,E)\,P(A|E) + P(F|B,E)\,P(B|E) \\ &\quad + P(F|C,E)\,P(C|E) \\ &= 1\times 2/3 + 0\times 1/3 + 0\times 0 = 2/3\,. \end{aligned}$$

3.12.3 *Regular or double-head coin?*

In Sec. 1.1 we discussed that, even after having observed a long series of heads when tossing a coin, we cannot be sure that that coin has two heads (assuming we cannot inspect the coin). The Bayesian solution of the problem is:

$$\begin{aligned} \frac{P(\text{Double-head}\,|\,n\,\text{Heads})}{P(\text{Fair}\,|\,n\,\text{Heads})} &= \frac{P(n\,\text{Heads}\,|\,\text{Double-head})}{P(n\,\text{Heads}\,|\,\text{Fair})}\cdot\frac{P_\circ(\text{Double-head})}{P_\circ(\text{Fair})} \\ &= \frac{1}{(1/2)^n}\cdot\frac{P_\circ(\text{Double-head})}{P_\circ(\text{Fair})} \\ &= 2^n\times\frac{P_\circ(\text{Double-head})}{P_\circ(\text{Fair})}\,. \end{aligned}$$

The odd ratio in favor of the double-headed coin grows as 2^n, but the absolute probability depends on how much we initially believe this hypothesis. To turn this problem into a standard 'non-subjective' text-book exercise, we can imagine that the tossed coin was chosen at random from a box which contained 100 coins, 99 of which were regular. The initial odd ratio is then 1/99: We need at least seven consecutive heads before we lose our initial conviction that the coin is most likely regular.

3.12.4 *Which random generator is responsible for the observed number?*

The solution of the random generator example met in Sec. 1.9 requires a limit to continuous variables and the use of p.d.f.'s (see Chapter 4), but it is conceptually analogous to the discrete case:

$$\frac{P(A \mid x)}{P(B \mid x)} = \frac{f(x \mid A)}{f(x \mid A)} \times \frac{P(A)}{P(B)} \tag{3.29}$$

$$= \frac{(1/\sqrt{2\pi}\,\sigma_A)\,\exp[-(x-\mu_A)^2/(2\sigma_A^2)]}{(1/\sqrt{2\pi}\,\sigma_B)\,\exp[-(x-\mu_B)^2/(2\sigma_B^2)]} \times \frac{P(A)}{P(B)} \tag{3.30}$$

$$= \frac{\sigma_B}{\sigma_A}\exp\left[\frac{(x-\mu_B)^2}{2\sigma_B^2} - \frac{(x-\mu_A)^2}{2\sigma_A^2}\right] \times \frac{P(A)}{P(B)}, \tag{3.31}$$

which, for the particular case of the example ($\mu_A = 0$, $\mu_B = 6.02$, $\sigma_A = \sigma_B = 1$, $P(A)/P(B) = 10$ and $x = 3.3$), yields $P(A \mid x)/P(B \mid x) = 1.75$, i.e. $P(A \mid x) = 0.64$ and $P(B \mid x) = 0.46$. In the case of equiprobable generator the solution would have been $P(A \mid x)/P(B \mid x) = 0.175$, i.e. $P(A \mid x) = 0.15$ and $P(B \mid x) = 0.85$.

3.13 Some further examples showing the crucial role of background knowledge

Many exercises in probability text books are nothing but exercises in combinatorics or in measure theory, because the equiprobability of the elements of the relevant space is usually implicit. It is obvious that in such exercises the Bayesian approach cannot produce a solution which is different from the standard solution, simply because we start from the same hypothesis of equiprobability and from the same probability rules. As a consequence, the Bayesian point of view might seem superficially a superfluous philosophical construct. However, a person trained in the Bayesian approach is always very careful to consider all hypotheses, both stated and unstated. A couple of illuminating examples will remind us why it is essential to be very careful about background knowledge and hidden hypotheses.

The three box problem(s). This is a well-known problem, which I have been told has caused a lot of discussion among fans of riddles. To better understand the logical traps involved, we will consider two variations on the problem. In both variations we have the presenter of a TV game-show and three identical boxes, only one of which contains a rich prize.

(1) In the first case, imagine two contestants, each of whom chooses one box at random. Contestant B opens his chosen box and finds it does not contain the prize. Then the presenter offers player A the oppor-

tunity to exchange his box, still un-opened, with the third box. Is the offer to his advantage, disadvantage, or indifferent?

Solution. It is "clearly" indifferent, as there is no reason for preferring either one or the other of the two remaining unopened boxes.

(2) In the second case there is only one contestant, A. After he has chosen one box the presenter tells him that, although the boxes are identical, he knows which one contains the prize. Then he says that, out of the two remaining boxes, he will open one that does not contain the prize. The presenter gets a box, opens it, and the box turns out to be empty. Then, as in the previous problem, he offers the contestant the opportunity to exchange his box with the remaining third box.

Solution. "Obviously", this time it is to the contestant's advantage to take the third box. In fact the probability of finding the prize in it amounts to the probability that the prize was in one of the two remaining boxes before the presenter chose one, opened it, and found it to be empty: 2/3.

These problems seem paradoxical to many people, because the physical action is exactly the same: one box was found to be empty while the other two boxes were still un-opened. But the status of information is quite different in the two cases: In the first, contestant B chose an empty box by (unlucky) chance. In the second, the presenter chose an empty box on purpose.

Formally, we can write the two conditions as $C^{(1)} = \overline{E}_B$ and $C^{(2)} = \overline{E}_B \cup \overline{E}_C$, where E_A, E_B and E_C are the events "prize in box A", "prize in box B", "prize in box C". Using Eq. (3.9) we have for case (1):

$$\frac{P(E_A\,|\,C^{(1)})}{P(E_A)} = \frac{P(E_A\,|\,\overline{E}_B)}{P(E_A)} = \frac{P(\overline{E}_B\,|\,E_A)}{P(\overline{E}_B)} = \frac{1}{2/3} = \frac{3}{2} \tag{3.32}$$

and, hence, $P(E_A\,|\,C^{(1)}) = 1/2$. Instead, in case (2) the condition $C^{(2)}$ corresponds to the certain event, as can be easily understood ($C^{(2)} = \overline{E}_B \cup \overline{E}_C = \overline{E_B \cap E_C} = \overline{\emptyset} = \Omega$). As a consequence, $P(E_A\,|\,C^{(2)})$ is not updated with respect to the initial value of $1/3$ (no real new information is available!).

There are further interesting variations on the game.

(3) Contestant A might not trust the presenter, believing that it was pure chance that he was able to predict what the box contained. Given this belief, we recover precisely the first case, and the prob-

ability we assign to the prize being in the third box goes back to 1/2.

(4) We can complicate the problem still further, introducing a "degree of mistrust" in the presenter, i.e a probability of bluff $p_b = P(\text{bluff})$. Using the probability rules, namely Eq. 3.10, and calling E the event "the prize is in the third box", we get

$$\begin{aligned} P(E\,|\,I) &= P(E\,|\,\text{bluff}, I)\,P(\text{bluff}\,|\,I) \\ &\quad + P(E\,|\,\overline{\text{bluff}}, I)\,P(\overline{\text{bluff}}\,|\,I) \qquad (3.33)\\ &= \frac{1}{2}\,p_b + \frac{2}{3}\,(1-p_b) \\ &= \frac{2}{3} - \frac{1}{6}\,p_b\,. \qquad (3.34) \end{aligned}$$

We have a smooth transition between the two solutions, depending on p_b. Nevertheless, accepting the offer to exchange boxes is never a bad decision.

The two envelope "paradox". Let us consider another problem which has no reasonable solution within standard probability, and hence is known as a paradox. As in the previous example, we will consider various ways of formulating the problem in order to highlight the logical mistake which gives rise to the "paradox".

(1) Imagine that someone shows you two envelopes and tells you that they each contain one check, but that the value of one of these checks is double the value of the other (though the exact amount of money involved is initially unknown). You choose one envelope, see the check, and then you are given the opportunity to exchange envelopes. Is it to your advantage to do so?
Standard solution: yes it is to your advantage, because you might find in the other envelope half or twice the amount of money you read on the first check — let us call it A. If it is half, you will lose $A/2$ $(= -A + A/2)$. Otherwise you will gain A $(= -A + 2\,A)$. 'Therefore' it is to your advantage to exchange envelopes. But this argument is already valid before the first envelope is opened, and can be repeated after the envelopes have been exchanged, and this certainly is a paradox.

(2) Now let us take the same problem, but with a couple of variations. First, imagine that the game is real and is played among normal students. Second, let us say that the ratio of values between the two

checks is 100. A student opens the first envelope and finds 100 dollars.[16] What should the student do? Yes, you are right, he should keep the check: he has no chance of finding \$10000 in the other envelope.

Solution of the paradox: the origin of the paradox lies in considering the two hypotheses equiprobable. If this was reasonable to do so, then the expected value of gain G in problem (1) would be

$$\mathrm{E}(G) = \frac{1}{2}\,(-A + A/2) + \frac{1}{2}\,(-A + 2\,A) = \frac{1}{4}\,A\,.$$

Extending the reasoning to the second problem, $\mathrm{E}(G)$ would be about $49A$. But now the situation becomes so extreme that the hidden hypothesis of the 'standard solution' becomes plain to everybody's eyes.

(3) Consider now a third case, again among normal students, but this time with a factor of 10 between the values of the two checks. A student, one who knows the person who has prepared the envelopes well, finds $A = \$10$ and has serious doubts about whether to exchange. How much does he believe he will find a check for \$100 in the other envelope?

Solution: since he finds himself in a status of indifference, the expected gain is about zero. It follows that he considers the chance of finding $1/10A$ much higher than the chance of finding $10A$, such that the ratio of probabilities (the odds) compensates the ratio of possible gains. This means that he unconsciously assigns a probability of about 9% to finding \$100 in the unopened envelope (the exact solution is $1/(1+r)$, where r is the prize ratio).

Moral. It is always very dangerous to calculate probabilities in a way which does not take into account a real or realistic situation and includes the full status of information about facts and/or persons involved. I do not know of any paradoxical problem which involves real people, in real situations, handling real money. As de Finetti used to say, *"either probability refers to real events or it is nothing"*.

[16]The real life envelopes that I prepare for students contain 1000 lire (about half a dollar) and 1 lira. When somebody finds 1000 lire and decides to exchange envelopes, I ask: *"Do you really think I have come here today prepared to losing one million lire with 50% probability?"*

Chapter 4

Probability distributions (a concise reminder)

In the following chapters it will be assumed that the reader is familiar with random variables, distributions, probability density functions, and expected values, as well as with the most frequently used distributions. This chapter is only intended as a summary of concepts and as a presentation of the notation used in the subsequent sections.

4.1 Discrete variables

Uncertain numbers are numbers in respect of which we are in a condition of uncertainty. They can be the number associated with the outcome of a die, to the number which will be read on a scale when a measurement is performed, or to the numerical value of a physics quantity. In the sequel, we will call uncertain numbers also "random variables", to come close to what physicists are used to, but one should not think, then, that "random variables" are only associated with the outcomes of repeated experiments, or to some idealistic, but of no practical relevance, definition of 'randomness' [54]. Stated simply, to define a random variable X means to find a rule which allows a real number to be related univocally (but not necessarily biunivocal) to an event (E). One could write this expression $X(E)$. Discrete variables assume a countable range, finite or not. We shall indicate the variable with X and its numerical realization x; and differently from other notations, the symbol x (in place of n or k) is also used for discrete variables.

Probability function

To each possible value of X we associate a degree of belief:

$$f(x) = P(X = x)\,. \tag{4.1}$$

$f(x)$, being a probability, must satisfy the following properties:

$$0 \leq f(x_i) \leq 1\,, \tag{4.2}$$

$$P(X = x_i \cup X = x_j) = f(x_i) + f(x_j)\,, \tag{4.3}$$

$$\sum_i f(x_i) = 1\,. \tag{4.4}$$

Cumulative distribution function

$$F(x_k) \equiv P(X \leq x_k) = \sum_{x_i \leq x_k} f(x_i)\,. \tag{4.5}$$

Properties:

$$F(-\infty) = 0, \tag{4.6}$$

$$F(+\infty) = 1, \tag{4.7}$$

$$F(x_i) - F(x_{i-1}) = f(x_i), \tag{4.8}$$

$$\lim_{\epsilon \to 0} F(x + \epsilon) = F(x) \qquad \text{(right side continuity)}\,. \tag{4.9}$$

Expected value (mean)

$$\mu \equiv \mathrm{E}(X) = \sum_i x_i f(x_i)\,. \tag{4.10}$$

In general, given a function $g(X)$ of X,

$$\mathrm{E}[g(X)] = \sum_i g(x_i) f(x_i)\,. \tag{4.11}$$

$\mathrm{E}(\cdot)$ is a linear operator:

$$\mathrm{E}(aX + b) = a\,\mathrm{E}(X) + b\,. \tag{4.12}$$

Variance and standard deviation

Variance:

$$\sigma^2 \equiv \mathrm{Var}(X) = \mathrm{E}[(X - \mu)^2] = \mathrm{E}(X^2) - \mu^2\,. \tag{4.13}$$

Standard deviation:

$$\sigma = \sqrt{\sigma^2}\,. \tag{4.14}$$

Transformation properties:

$$\mathrm{Var}(aX + b) = a^2\,\mathrm{Var}(X)\,, \tag{4.15}$$

$$\sigma(aX + b) = |a|\,\sigma(X)\,. \tag{4.16}$$

Moments

Expected value and variance are particular cases of *moments*. In general,

$$\mathrm{E}[(X - c)^r] \tag{4.17}$$

defines a *moment of order r about c* of a probability distribution. If c is not mentioned, $c = 0$ is implicit and one simply talks of moment of order r. Moments about $c = \mathrm{E}(X)$ are called *central moments.* Expected value is the first order moment, variance the second order central moment.

Binomial distribution

$X \sim \mathcal{B}_{n,p}$ (hereafter "$\sim$" stands for "follows"); $\mathcal{B}_{n,p}$ stands for *binomial* with parameters n and p:

$$f(x\,|\,\mathcal{B}_{n,p}) = \frac{n!}{(n-x)!\,x!}\,p^x\,(1-p)^{n-x}\,, \qquad \begin{cases} n = 1, 2, \ldots, \infty \\ 0 \le p \le 1 \\ x = 0, 1, \ldots, n \end{cases}. \tag{4.18}$$

Expected value, standard deviation and *variation coefficient*:

$$\mu = n\,p, \tag{4.19}$$

$$\sigma = \sqrt{n\,p\,(1-p)}, \tag{4.20}$$

$$v \equiv \frac{\sigma}{\mu} = \frac{\sqrt{n\,p\,(1-p)}}{n\,p} \propto \frac{1}{\sqrt{n}}\,. \tag{4.21}$$

$1 - p$ is often indicated by q.

Poisson distribution

$X \sim \mathcal{P}_\lambda$:

$$f(x \,|\, \mathcal{P}_\lambda) = \frac{\lambda^x}{x!}\, e^{-\lambda} \qquad \begin{cases} 0 < \lambda < \infty \\ x = 0, 1, \ldots, \infty \end{cases} . \tag{4.22}$$

Expected value, standard deviation and variation coefficient:

$$\mu = \lambda, \tag{4.23}$$

$$\sigma = \sqrt{\lambda}, \tag{4.24}$$

$$v = \frac{1}{\sqrt{\lambda}}. \tag{4.25}$$

Binomial → Poisson

$$\mathcal{B}_{n,p} \xrightarrow[\substack{n \to \infty \\ p \to 0 \\ (n\,p = \lambda)}]{} \mathcal{P}_\lambda \,.$$

4.2 Continuous variables: probability and probability density function

Moving from discrete to continuous variables there are the usual problems with infinite possibilities, similar to those found in Zeno's "Achilles and the tortoise" paradox. In both cases the answer is given by infinitesimal calculus. But some comments are needed:

- The probability of each of the realizations of X is zero ($P(X = x) = 0$); but this does not mean that each value is impossible, otherwise it would be impossible to get any result.
- Although all values x have zero probability, one usually assigns different degrees of belief to them, quantified by the probability density function (p.d.f.) $f(x)$. Writing $f(x_1) > f(x_2)$, for example, indicates that our degree of belief in x_1 is greater than that in x_2.
- The probability that a random variable lies inside a finite interval, for example $P(a \leq X \leq b)$, is instead finite. If the distance between a and b becomes infinitesimal, then the probability becomes infinitesimal too. If all the values of X have the same degree of belief (and not only equal numerical probability $P(x) = 0$) the infinitesimal probability is simply proportional to the infinitesimal interval $\mathrm{d}P = k\,\mathrm{d}x$. In the

general case the ratio between two infinitesimal probabilities around two different points will be equal to the ratio of the degrees of belief in the points (this argument implies the continuity of $f(x)$ on either side of the values). It follows that $\mathrm{d}P = f(x)\,\mathrm{d}x$ and then

$$P(a \le X \le b) = \int_a^b f(x)\,\mathrm{d}x\,. \tag{4.26}$$

- $f(x)$ has a dimension inverse to that of the random variable.

After this short introduction, here is a list of definitions, properties and notations:

Cumulative distribution function

$$F(x) = P(X \le x) = \int_{-\infty}^{x} f(x')\,\mathrm{d}x'\,, \tag{4.27}$$

or

$$f(x) = \frac{\mathrm{d}F(x)}{\mathrm{d}x} \tag{4.28}$$

Properties of $f(x)$ and $F(x)$

- $f(x) \ge 0$,
- $\int_{-\infty}^{+\infty} f(x)\,\mathrm{d}x = 1$,
- $0 \le F(x) \le 1$,
- $P(a \le X \le b) = \int_a^b f(x)\,\mathrm{d}x = \int_{-\infty}^b f(x)\,\mathrm{d}x - \int_{-\infty}^a f(x)\,\mathrm{d}x = F(b) - F(a)$,
- if $x_2 > x_1$ then $F(x_2) \ge F(x_1)$,
- $\lim_{x\to-\infty} F(x) = 0$,
  $\lim_{x\to+\infty} F(x) = 1$.

Expected value of continuous variable distributions

$$\mathrm{E}(X) = \int_{-\infty}^{+\infty} x\, f(x)\,\mathrm{d}x, \tag{4.29}$$

$$\mathrm{E}[g(X)] = \int_{-\infty}^{+\infty} g(x)\, f(x)\,\mathrm{d}x. \tag{4.30}$$

Uniform distribution

$X \sim \mathcal{K}(a,b)$: [1]

$$f(x\,|\,\mathcal{K}(a,b)) = \frac{1}{b-a} \qquad (a \le x \le b), \tag{4.31}$$

$$F(x\,|\,\mathcal{K}(a,b)) = \frac{x-a}{b-a}\,. \tag{4.32}$$

Expected value and standard deviation:

$$\mu = \frac{a+b}{2}, \tag{4.33}$$

$$\sigma = \frac{b-a}{\sqrt{12}} = \frac{(b-a)/2}{\sqrt{3}} = \frac{\Delta}{\sqrt{3}}\,. \tag{4.34}$$

Normal (Gaussian) distribution

$X \sim \mathcal{N}(\mu,\sigma)$:

$$f(x\,|\,\mathcal{N}(\mu,\sigma)) = \frac{1}{\sqrt{2\pi}\,\sigma}\exp\left[-\frac{(x-\mu)^2}{2\,\sigma^2}\right] \qquad \begin{cases} -\infty < \mu < +\infty \\ 0 < \sigma < \infty \\ -\infty < x < +\infty \end{cases}, \tag{4.35}$$

where μ and σ (both real) are the expected value and standard deviation,[2] respectively. In the normal distribution mean, mode and median concide. Moreover, it can be easily proved that the variance is related to the second derivative of the exponent by

$$-\left.\frac{\partial^2 \ln f(x)}{\partial x^2}\right|_{x=\mu} = \frac{1}{\sigma^2}\,. \tag{4.36}$$

These observations are very important in applications, when $f(x)$ can have a very complicated mathematical expression, but, nevertheless, has approximately a Gaussian shape: expected value and variance can be estimated from the mode and Eq. (4.36), respectively. That is, evaluations that involve integration are replaced by evaluations that involve differentiation, a usually easier task.

[1] The symbols of the following distributions have the parameters within parentheses to indicate that the variables are continuous.

[2] Mathematicians and statisticians prefer to take σ^2, instead of σ, as second parameter of the normal distribution. Here the standard deviation is preferred, since it is homogeneous to μ and it has a more immediate physical interpretation. So, one has to pay attention to be sure about the meaning of expressions like $\mathcal{N}(0.5, 0.8)$.

Standard normal distribution

This is the name given to the particular normal distribution of mean 0 and standard deviation 1, usually indicated by Z:

$$Z \sim \mathcal{N}(0,1)\,. \tag{4.37}$$

Exponential distribution

$T \sim \mathcal{E}(\tau)$:

$$f(t\,|\,\mathcal{E}(\tau)) = \frac{1}{\tau}\,e^{-t/\tau} \qquad \begin{cases} 0 \leq \tau < \infty, \\ 0 \leq t < \infty \end{cases} \tag{4.38}$$

$$F(t\,|\,\mathcal{E}(\tau)) = 1 - e^{-t/\tau}. \tag{4.39}$$

We use the symbol t instead of x because this distribution will be applied to the *time domain.*

Survival probability:

$$P(T > t) = 1 - F(t\,|\,\mathcal{E}(\tau)) = e^{-t/\tau}. \tag{4.40}$$

Expected value and standard deviation:

$$\mu = \tau \tag{4.41}$$

$$\sigma = \tau. \tag{4.42}$$

The parameter τ has the physical meaning of (expected) *lifetime.*

Poisson $\leftrightarrow$ Exponential

The Poisson and exponential distribution represent two aspects of the *Poisson process.* If X (= "number of counts during the time Δt") is Poisson distributed then T (= "interval of time to wait — starting from any instant! — before the first count is recorded") is exponentially distributed:

$$X \sim f(x\,|\,\mathcal{P}_\lambda) \quad \Longleftrightarrow \quad T \sim f(t\,|\,\mathcal{E}(\tau)) \tag{4.43}$$

$$(\tau = \tfrac{\Delta T}{\lambda})\,. \tag{4.44}$$

Also the gamma distribution is related to the Poisson process: for c integer it describes the waiting time before c counts are recorded.

Gamma distribution

$X \sim \text{Gamma}(c, r)$:

$$f(x\,|\,\text{Gamma}(c,r)) = \frac{r^c}{\Gamma(c)} x^{c-1} e^{-r\,x} \qquad \begin{cases} r,\, c > 0 \\ x \geq 0 \end{cases}\,, \tag{4.45}$$

where

$$\Gamma(c) = \int_0^\infty x^{c-1} e^{-x} dx$$

(for n integer, $\Gamma(n+1) = n!$). c is called *shape* parameter, while $1/r$ is the *scale* parameter. Expected value, variance and mode are

$$\mathrm{E}(X) = \frac{c}{r} \tag{4.46}$$

$$\mathrm{Var}(X) = \frac{c}{r^2} \tag{4.47}$$

$$\mathrm{mode}(X) = \begin{cases} 0 & \text{if } c < 1 \\ \frac{c-1}{r} & \text{if } c \geq 1 \end{cases} . \tag{4.48}$$

If c is integer, the distribution is also known as *Erlang*, describing the time to be waited before observing c events in a Poisson process of intensity r (events per unit of time). For $c = 1$ the Gamma distribution recovers the exponential.

Chi-square distribution

The well known χ^2 distribution with ν *degrees of freedom* is formally nothing but a Gamma distribution with $c = \nu/2$ and $r = 1/2$:

$$\begin{aligned} f(x \,|\, \chi^2_\nu) &= f(x \,|\, \mathrm{Gamma}(\nu/2, 1/2)) \\ &= \frac{2^{-\nu/2}}{\Gamma(\nu/2)} x^{\nu/2-1} e^{-x/2} . \end{aligned} \tag{4.49}$$

Expected value, variance and mode follow from Eqs. (4.46)–(4.48):

$$\mathrm{E}(X) = \nu \tag{4.50}$$

$$\mathrm{Var}(X) = 2\nu \tag{4.51}$$

$$\mathrm{mode}(X) = \begin{cases} 0 & \text{if } \nu < 2 \\ \nu - 2 & \text{if } \nu \geq 2 \end{cases} . \tag{4.52}$$

Note that the χ^2_ν distribution can be obtained as sum of ν independent standardized Gaussian quantities:

$$Z_i \sim \mathcal{N}(0,1) \implies \sum_{i=1}^{\nu} Z_i^2 \sim \chi^2_\nu . \tag{4.53}$$

In data analysis the Pearson χ^2 is also well-known, given by Eq. (4.92).

Beta distribution

$X \sim \text{Beta}(r, s)$):

$$f(x \,|\, \text{Beta}(r, s)) = \frac{1}{\beta(r, s)} x^{r-1}(1-x)^{s-1} \qquad \begin{cases} r,\, s > 0 \\ 0 \le x \le 1 \,. \end{cases} \tag{4.54}$$

The denominator is just for normalization, i.e.

$$\beta(r, s) = \int_0^1 x^{r-1}(1-x)^{s-1}\, \mathrm{d}x \,.$$

Indeed this integral defines the beta function, resulting in

$$\beta(r, s) = \frac{\Gamma(r)\,\Gamma(s)}{\Gamma(r+s)} \,.$$

Since the beta distribution is not very popular among physicists, but very interesting for inferential purposes as *conjugate* distribution of the binomial, we show in Fig. 4.1 the variety of shapes that it can assume depending on the parameters r and s. Expected value and variance are:

$$\mathrm{E}(X) = \frac{r}{r+s} \tag{4.55}$$

$$\mathrm{Var}(X) = \frac{rs}{(r+s+1)\,(r+s)^2} \,. \tag{4.56}$$

If $r > 1$ and $s > 1$ the mode is unique, equal to $(r-1)/(r+s-2)$.

Triangular distribution

A convenient distribution for a rough description of subjective uncertainty on the value of influence quantities ('systematic effects') is given by the *triangular* distribution. This distribution models beliefs which decrease linearly in either side of the maximum (x_0) up to $x_0 + \Delta x_+$ on the right side and $x_0 - \Delta x_-$ on the left side (see Fig. 8.1). Expected value and variance are given by

$$\mathrm{E}(X) = x_0 + \frac{\Delta x_+ - \Delta x_-}{3} \tag{4.57}$$

$$\sigma^2(X) = \frac{\Delta^2 x_+ + \Delta^2 x_- + \Delta x_+ \Delta x_-}{18} \,. \tag{4.58}$$

In the case of a *symmetric triangular* distribution ($\Delta x_+ = \Delta x_- = \Delta x$)

we get

$$\mathrm{E}(X) = x_0 \tag{4.59}$$

$$\sigma(X) = \frac{\Delta x}{\sqrt{6}}\,. \tag{4.60}$$

4.3 Distribution of several random variables

We only consider the case of two continuous variables (X and Y). The extension to more variables is straightforward. The infinitesimal element of probability is $\mathrm{d}F(x, y) = f(x, y)\,\mathrm{d}x\,\mathrm{d}y$, and the probability density function

$$f(x, y) = \frac{\partial^2 F(x, y)}{\partial x\, \partial y}\,. \tag{4.61}$$

The probability of finding the variable inside a certain area A is

$$\int_A f(x, y)\,\mathrm{d}x\,\mathrm{d}y\,. \tag{4.62}$$

Marginal distributions

$$f_X(x) = \int f(x, y)\,\mathrm{d}y, \tag{4.63}$$

$$f_Y(y) = \int f(x, y)\,\mathrm{d}x\,. \tag{4.64}$$

The subscripts X and Y indicate that $f_X(x)$ and $f_Y(y)$ are only functions of X and Y, respectively (to avoid fooling around with different symbols to indicate the generic function), but in most cases we will drop the subscripts if the context helps in resolving ambiguities.

Conditional distributions

$$f_X(x \,|\, y) = \frac{f(x, y)}{f_Y(y)} = \frac{f(x, y)}{\int f(x, y)\,\mathrm{d}x}, \tag{4.65}$$

$$f_Y(y \,|\, x) = \frac{f(x, y)}{f_X(x)}, \tag{4.66}$$

$$f(x, y) = f_X(x \,|\, y)\, f_Y(y) \tag{4.67}$$

$$= f_Y(y \,|\, x)\, f_X(x)\,. \tag{4.68}$$

Independent random variables

$$f(x, y) = f_X(x)\, f_Y(y) \tag{4.69}$$

(it implies $f_X(x\,|\,y) = f_X(x)$ and $f_Y(y\,|\,x) = f_Y(y)$).

Bayes' theorem for continuous random variables

$$\boxed{f(h\,|\,e) = \frac{f(e\,|\,h)\, f_h(h)}{\int f(e\,|\,h)\, f_h(h)\,\mathrm{d}h}\,.} \tag{4.70}$$

(See proof in Section 2.7.)

Expected value

$$\mu_X = \mathrm{E}(X) = \int x\, f(x,y)\,\mathrm{d}x\,\mathrm{d}y \tag{4.71}$$

$$= \int x\, f_X(x)\,\mathrm{d}x\,, \tag{4.72}$$

and analogously for Y. In general

$$\mathrm{E}[g(X,Y)] = \int g(x,y)\, f(x,y)\,\mathrm{d}x\,\mathrm{d}y\,. \tag{4.73}$$

Variance:

$$\sigma_X^2 = \mathrm{E}[((X-\mathrm{E}(X))^2] = \mathrm{E}(X^2) - \mathrm{E}^2(X)\,, \tag{4.74}$$

and analogously for Y. In practice, expected value and variance are equal to those calculated only considering the variable of interest.

Covariance

$$\mathrm{Cov}(X,Y) = \mathrm{E}\,[\,(X-\mathrm{E}(X))\cdot(Y-\mathrm{E}(Y))\,] \tag{4.75}$$

$$= \mathrm{E}(XY) - \mathrm{E}(X)\cdot\mathrm{E}(Y)\,. \tag{4.76}$$

If X and Y are independent, then $\mathrm{E}(XY) = \mathrm{E}(X)\cdot\mathrm{E}(Y)$ and hence $\mathrm{Cov}(X,Y) = 0$ (the opposite is true only if X, $Y \sim \mathcal{N}(\cdot)$). Note also that, if $Y = X$, then $\mathrm{Cov}(X,Y) = \mathrm{Var}(X)$.

Correlation coefficient

$$\rho(X,Y) = \frac{\mathrm{Cov}(X,Y)}{\sqrt{\mathrm{Var}(X)\,\mathrm{Var}(Y)}} \tag{4.77}$$

$$= \frac{\mathrm{Cov}(X,Y)}{\sigma_X\,\sigma_Y}\,. \tag{4.78}$$

$$(-1 \le \rho \le 1)$$

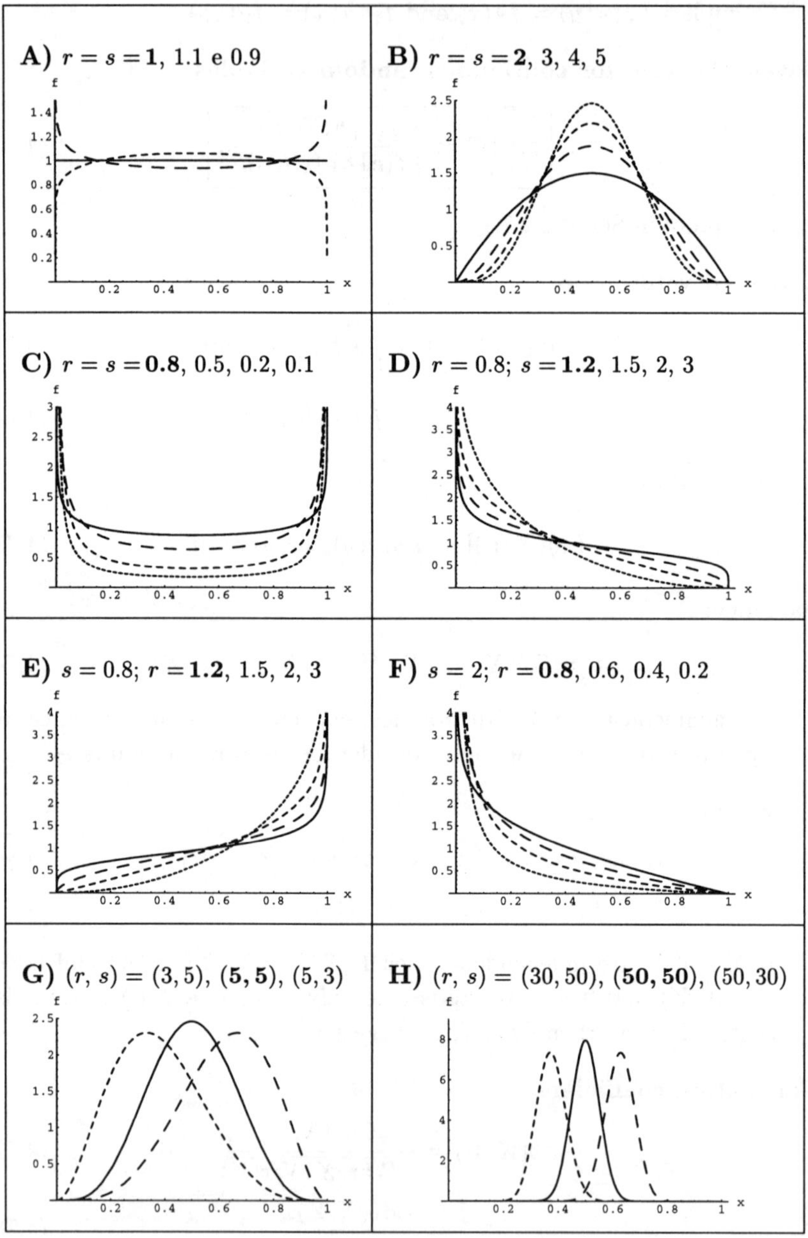

Fig. 4.1 Examples of Beta distributions for some values of r and s. The parameters in bold refer to continuous curves.

Covariance matrix and correlation matrix

Given n random quantities $\boldsymbol{X}$, the *covariance matrix* $\boldsymbol{V}$ is defined as

$$V_{ij} = \mathrm{E}\left[(X_i - \mathrm{E}(X_i)) \cdot (X_j - \mathrm{E}(X_j)) \right] , \tag{4.79}$$

where the diagonal terms are the variances and the off-diagonal ones are the covariances. It is also convenient to define a *correlation matrix*, given by $\rho(X_i, X_j)$. The diagonal terms of the correlation matrix are equal to unity.

Bivariate normal distribution

Joint probability density function of X and Y with correlation coefficient ρ (see Fig. 4.2):

$$\begin{aligned} f(x,y) = \frac{1}{2\,\pi\,\sigma_x\,\sigma_y\,\sqrt{1-\rho^2}} \exp\left\{ -\frac{1}{2\,(1-\rho^2)} \left[\frac{(x-\mu_x)^2}{\sigma_x^2} \right.\right. \\ \left.\left. -2\,\rho\,\frac{(x-\mu_x)(y-\mu_y)}{\sigma_x\,\sigma_y} + \frac{(y-\mu_y)^2}{\sigma_y^2} \right] \right\} . \end{aligned} \tag{4.80}$$

Marginal distributions:

$$X \sim \mathcal{N}(\mu_x, \sigma_x), \tag{4.81}$$

$$Y \sim \mathcal{N}(\mu_y, \sigma_y) . \tag{4.82}$$

Conditional distribution:

$$f(y \,|\, x_\circ) = \frac{1}{\sqrt{2\,\pi}\,\sigma_y\,\sqrt{1-\rho^2}} \exp\left[-\frac{\left(y - \left[\mu_y + \rho\,\frac{\sigma_y}{\sigma_x}\,(x_\circ - \mu_x) \right] \right)^2}{2\,\sigma_y^2\,(1-\rho^2)} \right], \tag{4.83}$$

i.e.

$$Y_{|x_\circ} \sim \mathcal{N}\left(\mu_y + \rho\,\frac{\sigma_y}{\sigma_x}\,(x_\circ - \mu_x)\, ,\, \sigma_y\sqrt{1-\rho^2} \right) . \tag{4.84}$$

The condition $X = x_\circ$ squeezes the standard deviation and shifts the mean of Y.

Multi-variate normal distribution

The extension to n random variables $\boldsymbol{X}$ is given by

$$f(\boldsymbol{x}) = (2\pi)^{-n/2} |\mathbf{V}|^{-1/2} \exp\left[-\frac{1}{2} \boldsymbol{\Delta}^T \mathbf{V}^{-1} \boldsymbol{\Delta} \right] , \tag{4.85}$$

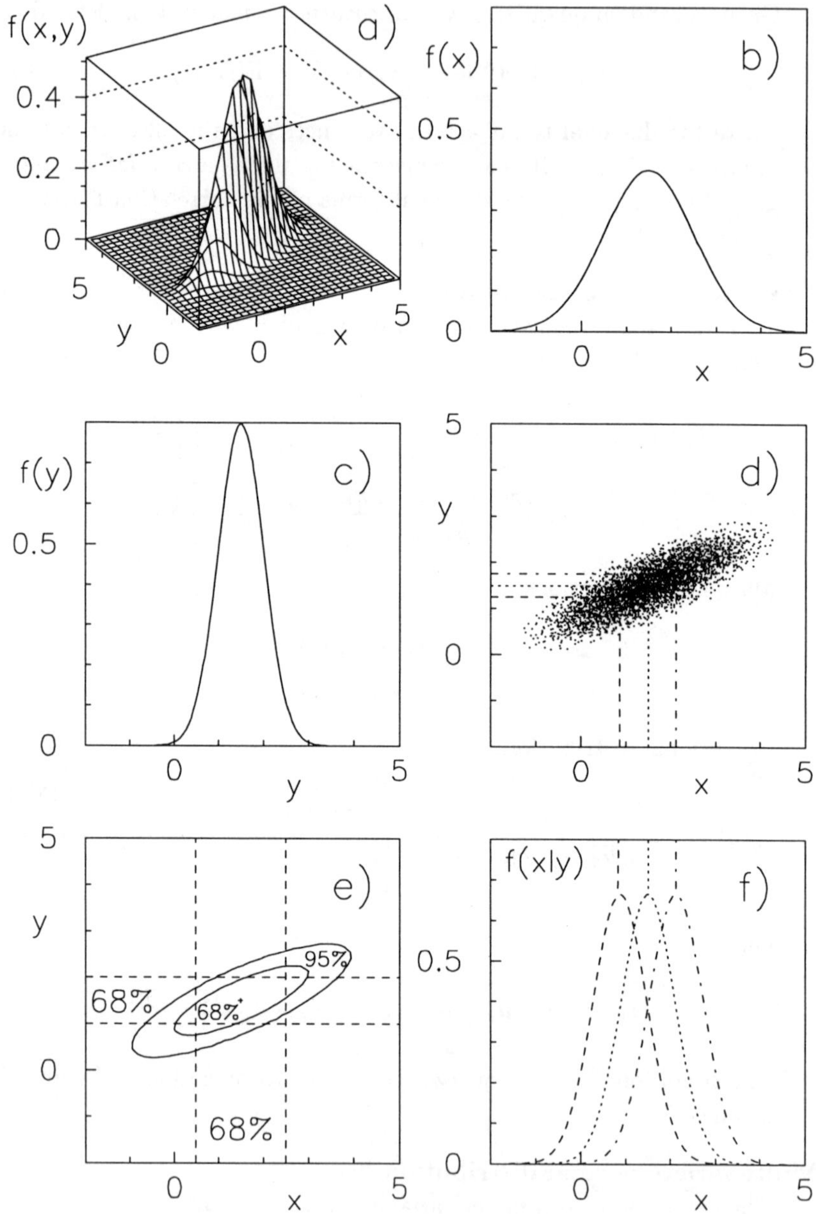

Fig. 4.2 Example of bivariate normal distribution.

where $\mathbf{\Delta}$ stands for the vector of differences $x_i - \mu_i$ and $|\mathbf{V}|$ is the determinant of the covariance matrix $\mathbf{V}$. It is easy to show that the *Hessian* of the logarithm of $f(\boldsymbol{x})$ is related to the inverse of the covariance matrix:

$$- \left. \frac{\partial^2 \ln f(\boldsymbol{x})}{\partial x_i \partial x_j} \right|_{\boldsymbol{\mu}} = (V^{-1})_{ij} \,. \tag{4.86}$$

Multinomial distribution

It is the extension of the binomial to the case of many possibilities (m), each with probability p_i:

$$f(\boldsymbol{x} \,|\, \mathcal{M}^m_{n\,\boldsymbol{p}}) = \frac{n!}{x_1!\, x_2! \cdots x_m!}\, p_1^{x_1} p_2^{x_2} \cdots p_m^{x_m} \,, \tag{4.87}$$

where $\boldsymbol{x}$ is the set of variables $\{x_1, x_2, \ldots, x_m\}$ and $\boldsymbol{p}$ the set of probabilities $\{p_1, p_2, \ldots, p_m\}$. For $m = 2$ the binomial distribution is recovered. Expected value and variance are given by

$$\mathrm{E}(X_i) = n\, p_i \tag{4.88}$$

$$\mathrm{Var}(X_i) = n\, p_i\, (1 - p_i) \,. \tag{4.89}$$

All variables are correlated. Covariances and correlation coefficients are given by

$$\mathrm{Cov}(X_i, X_j) = -n\, p_i\, p_j \tag{4.90}$$

$$\rho(X_i, X_j) = -\sqrt{\frac{p_i\, p_j}{(1 - p_i)\,(1 - p_j)}} \,. \tag{4.91}$$

In the binomial case x_1 and x_2 are 100% anticorrelated, due the constraint $x_2 = n - x_1$. Note that the covariance matrix is singular, as anybody who has tried to calculate the χ^2 as $\mathbf{\Delta}^T \mathbf{V}^{-1} \mathbf{\Delta}$ will have realized with disappointment. Nevertheless, Pearson proved that the quantity

$$\sum_{i=1}^{m} \frac{(x_i - n\, p_i)^2}{n\, p_i} \tag{4.92}$$

behaves like a χ^2_ν variable [Eq. (4.49)] with $\nu = m - 1$ if all $n\, p_i$ are 'large enough' (see Ref. [37] for details). The summation (4.92) is called Pearson-χ^2.

4.4 Propagation of uncertainty

The general problem is, given many (*final*) variables Y_j which depend on other (*initial*) variables X_i, to calculate $f(\boldsymbol{y})$ from the knowledge of $f(\boldsymbol{x})$, according to the following scheme:

$$f(x_1, x_2, \ldots, x_n) \xrightarrow[Y_j = Y_j(X_1, X_2, \ldots, X_n)]{} f(y_1, y_2, \ldots, y_m)\,. \tag{4.93}$$

This calculation can be quite challenging, and it is often performed by Monte Carlo techniques.

General solution for discrete variables

The probability of a given $Y = y$ is equal to the sum of the probability of each X_i such that $Y(x) = y$, where $Y()$ stands for the mathematical function relating X and Y. The extension to many variables is also straightforward. For example, if the uncertainty about X_1 and X_2 is modelled by the two-dimensional probability function $f(x_1, x_2)$ and the quantities Y_1 and Y_2 are related to them by the functions $Y_1 = Y_1(X_1, X_2)$ and $Y_2 = Y_2(X_1, X_2)$, the probability function of Y_1 and Y_2 will be

$$f(y_1, y_2) = \sum_{\substack{x_1, x_2 \\ \left\{\begin{array}{l} Y_1(x_1, x_2) = y_1 \\ Y_2(x_1, x_2) = y_2\,. \end{array}\right.}} f(x_1, x_2) \tag{4.94}$$

General solution for continuous variables

To deal with continuous variables we need to replace sums by integrals, and the constraints by suitable Dirac delta functions. Equation (4.94) will be replaced, then, by

$$f(y_1, y_2) = \int \delta(y_1 - Y_1(x_1, x_2))\, \delta(y_2 - Y_2(x_1, y_2))\, f(x_1, x_2)\, \mathrm{d}x_1 \mathrm{d}x_2\,. \tag{4.95}$$

A simple example of application of this formula is given in Fig. 4.3, where $Y = X_1 + X_2$, with X_1 and X_2 independent variables distributed according to an asymmetric triangular distribution (self-defined in Fig. 4.3). Note that the distribution parameters which matter in the propagation are expected value and standard deviation, ruled by Eqs. (4.98)–(4.99). There is, instead, no probability theory theorem which gives a simple propagation rule of mode, median and probability

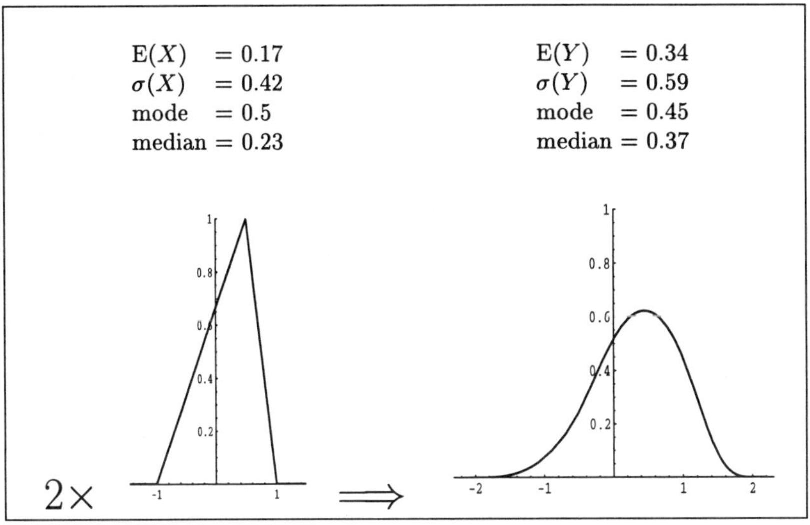

Fig. 4.3 Distribution of the sum of two independent quantities, each described by an asymmetric triangular p.d.f. with $x_0 = 0.5$, $\Delta x_+ = 0.5$ and $\Delta x_- = 1.5$ (see Fig. 8.1 for the parameters of triangular distributions). The p.d.f. of $Y = X_1 + X_2$ was calculated using Eq. (4.95). Note that $\mathrm{E}(X_1 + X_2) = \mathrm{E}(X_1) + \mathrm{E}(X_2)$ and $\sigma^2(X_1 + X_2) = \sigma^2(X_1) + \sigma^2(X_2)$, while, in general, $\mathrm{mode}(X_1 + X_2) \neq \mathrm{mode}(X_1) + \mathrm{mode}(X_2)$.

intervals. The simpler, better known, text book transformation formulae using the Jacobian are recovered using the properties of the delta function. For example, in the case of only one initial variable, and one final variable Eq. (4.95) reduces to

$$f_Y(y) = \frac{f_X(x)}{|\mathrm{d}y/\mathrm{d}x|}\,. \tag{4.96}$$

Some simple examples, starting from a uniform $f_X(x)$ are shown in Fig. 4.4. In the practical cases Eqs. (4.94–4.95) can be difficult to solve, and in many cases Monte Carlo methods are used, as suggested by the structure of the formulae (see also Sec. 2.10.2). For an example of application see Fig. 4.5.

Approximate solution

The solution becomes quite easy under the following conditions: there is a linear relation between $\boldsymbol{X}$ and $\boldsymbol{Y}$; we are interested only in expected values, variances and covariances. This situation can be sketched in the

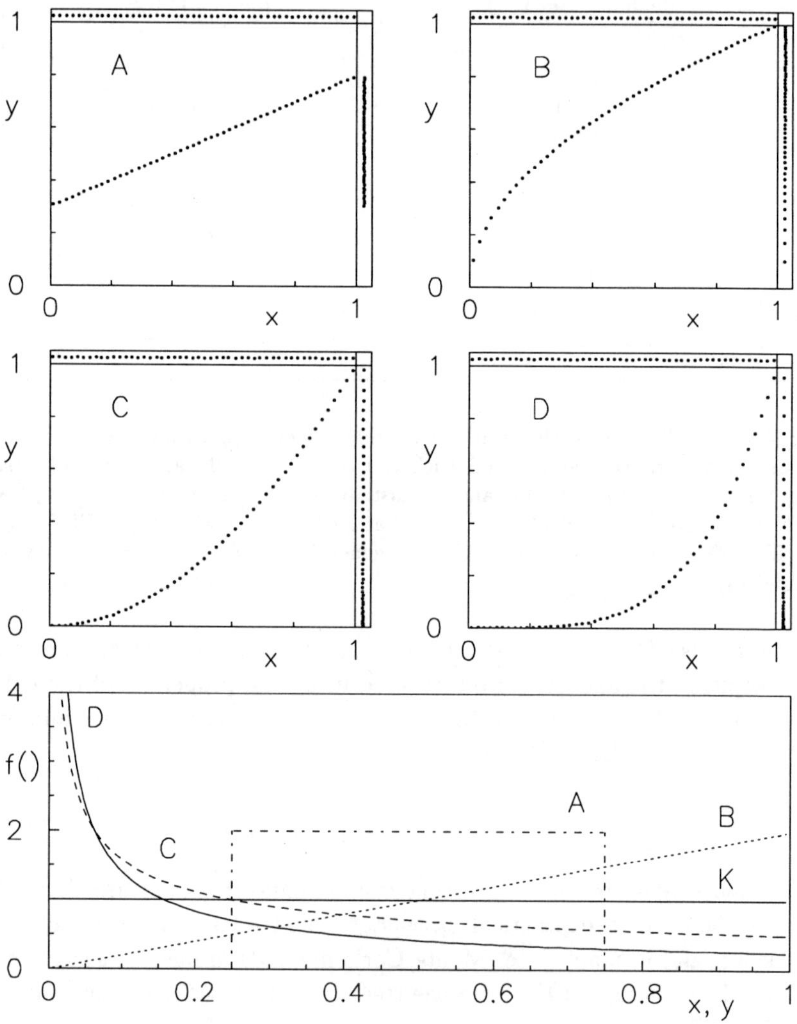

Fig. 4.4 Examples of variable changes starting from a uniform distribution ("K"): A) $Y = 0.5\,X + 0.25$; B) $Y = \sqrt{X}$; C) $Y = X^2$; D) $Y = X^4$. The dots projected on the two axes of the frames give a visual idea of the probability density functions inversely proportional to the slope of the function.

following way:

$$\begin{cases} \mathrm{E}(X_i) \\ \sigma(X_i) \\ \rho(X_i, X_{i'}) \end{cases} \xrightarrow[Y_j = c_{j0} + c_{j1}X_1 + c_{j2}X_2 + \cdots + c_{jn}X_n]{} \begin{cases} \mathrm{E}(Y_j) \\ \sigma(Y_j) \\ \rho(Y_j, Y_{j'}) \end{cases} . \tag{4.97}$$

Linear combinations of random variables

If $Y = \sum_i c_i X_i$, with c_i real, then:

$$\mu_Y = \mathrm{E}(Y) = \sum_i c_i\, \mathrm{E}(X_i) = \sum_i c_i\, \mu_i, \tag{4.98}$$

$$\begin{aligned} \sigma_Y^2 = \mathrm{Var}(Y) &= \sum_i c_i^2\, \mathrm{Var}(X_i) + 2\sum_{i<j} c_i\, c_j\, \mathrm{Cov}(X_i, X_j) \\ &= \sum_i c_i^2\, \mathrm{Var}(X_i) + \sum_{i\neq j} c_i\, c_j\, \mathrm{Cov}(X_i, X_j) \\ &= \sum_i c_i^2\, \sigma_i^2 + \sum_{i\neq j} \rho_{ij}\, c_i\, c_j\, \sigma_i\, \sigma_j \\ &= \sum_{ij} \rho_{ij}\, c_i\, c_j\, \sigma_i\, \sigma_j \\ &= \sum_{ij} c_i\, c_j\, \sigma_{ij}\, . \end{aligned} \tag{4.99}$$

σ_Y^2 has been written in different ways, with increasing levels of compactness, which can be found in the literature. In particular, we use the notations $\sigma_{ij} \equiv \mathrm{Cov}(X_i, X_j) = \rho_{ij}\, \sigma_i\, \sigma_j$ and $\sigma_{ii} = \sigma_i^2$.

The above results are easily extended to many final variables, but we need to also evaluate correlation coefficients among the Y_l. Given $Y_k = \sum_i c_{ki}\, X_i$, it follows

$$\begin{aligned} \sigma_{Y_{kl}} = \mathrm{Cov}(Y_k, Y_l) &= \sum_i c_{ki}\, c_{li}\, \mathrm{Var}(X_i) \\ &\quad + \sum_{i<j} (c_{ki}\, c_{lj} + c_{kj}\, c_{li})\, \mathrm{Cov}(X_i, X_j) \end{aligned} \tag{4.100}$$

$$\begin{aligned} &= \sum_i c_{ki}\, c_{li}\, \sigma_i^2 + \sum_{i<j} (c_{ki}\, c_{lj} + c_{kj}\, c_{li})\, \sigma_{ij} \\ &= \sum_i c_{ki}\, c_{li}\, \sigma_i^2 + \sum_{i\neq j} c_{ki}\, c_{lj}\, \sigma_{ij} \\ &= \sum_{ij} c_{ki}\, c_{lj}\, \sigma_{ij}\, . \end{aligned} \tag{4.101}$$

Indeed, this result also contains Eq. (4.99), as a special case in which $k = l$. Equation (4.100) shows that, even if the input quantities X_i are independent, several Y_k become correlated if they depend on the same X_i. Note that signs are important, and compensations might occur. Therefore it is important to take correlations with care. Equation (4.101) can be rewritten as $\sigma_{Y_{kl}} = \sum_{ij} c_{ki}\, \sigma_{ij}\, c_{lj}$, in order to stress its matrix form:

$$\mathbf{V}_Y = \mathbf{C}\,\mathbf{V}_X\,\mathbf{C}^T\,, \tag{4.102}$$

where $\mathbf{V}_X$ and $\mathbf{V}_Y$ are the covariance matrices of $\boldsymbol{X}$ and $\boldsymbol{Y}$, respectively, and $\mathbf{C}$ is the coefficient matrix.

Linearization

Many functions can be linearized,[3] and hence all previous results are recovered, if the derivatives $\partial Y_k/\partial X_i|_{\mathrm{E}(\boldsymbol{X})}$ $(\equiv c_{ki})$ are approximately constant in a range of a few standard deviations around $\mathrm{E}(\boldsymbol{X})$. The physics meaning of the derivatives c_{ki} is that of *sensitivity coefficients*.

The properties seen in this section are general properties of probability theory, and do not depend on the p.d.f. of $\boldsymbol{X}$. However, these properties say little about the probability function of $\boldsymbol{Y}$. Fortunately, the central limit theorem helps us in many cases of practical interest.

4.5 Central limit theorem

4.5.1 *Terms and role*

The well-known central limit theorem plays a crucial role in statistics and justifies the enormous importance that the Gaussian distribution has in many practical applications (this is why it appeared on 10 DM notes).

We have reminded ourselves in Eqs. (4.98)–(4.99) of the expression of the mean and variance of a linear combination of random variables,

$$Y = \sum_{i=1}^{n} c_i X_i\,,$$

in the most general case, which includes correlated variables ($\rho_{ij} \neq 0$). In the case of independent variables the variance is given by the simpler, and

[3] Next to linear order approximation is discussed in Chapter 12.

better known, expression

$$\sigma_Y^2 = \sum_{i=1}^{n} c_i^2\,\sigma_i^2 \qquad (\rho_{ij} = 0,\ i \neq j)\,. \tag{4.103}$$

This is a very general statement, valid for any number and kind of variables (with the obvious clause that all σ_i must be finite), but it does not give any information about the probability distribution of Y. Even if all X_i follow the same distributions $f(x)$, $f(y)$ is different from $f(x)$, with some exceptions, one of these being the normal.

The central limit theorem states that the distribution of a linear combination Y will be *approximately normal* if the variables X_i are independent and σ_Y^2 is much larger than any single component $c_i^2\sigma_i^2$ from a non-normally distributed X_i. The last condition is just to guarantee that there is no single random variable which dominates the fluctuations. The accuracy of the approximation improves as the number of variables n increases (the theorem says "when $n \to \infty$"):

$$n \to \infty \Longrightarrow Y \sim \mathcal{N}\left(\sum_{i=1}^{n} c_i\,\mathrm{E}(X_i),\ \left(\sum_{i=1}^{n} c_i^2\,\sigma_i^2\right)^{\frac{1}{2}}\right). \tag{4.104}$$

The proof of the theorem can be found in standard textbooks. For practical purposes, and if one is not very interested in the detailed behavior of the tails, n equal to 2 or 3 may already give a satisfactory approximation, especially if the X_i exhibits a Gaussian-like shape. See for example, Fig. 4.5, where samples of 10 000 events have been simulated,[4] starting from a uniform distribution and from a crazy square-wave distribution. The latter, depicting a kind of "worst practical case", shows that, already for $n = 20$ the distribution of the sum is practically normal. In the case of the uniform distribution $n = 3$ already gives an acceptable approximation as far as probability intervals of one or two standard deviations from the mean value are concerned. The figure also shows that, starting from a triangular distribution (obtained in the example from the sum of two uniform distributed variables), $n = 2$ is already sufficient. (The sum of two triangular distributed variables is equivalent to the sum of four uniform distributed variables.) For another example of central limit theorem at work see Fig. 12.3.

[4]Note that the Monte Carlo simulation does nothing but a numerical integration of Eq. (4.95).

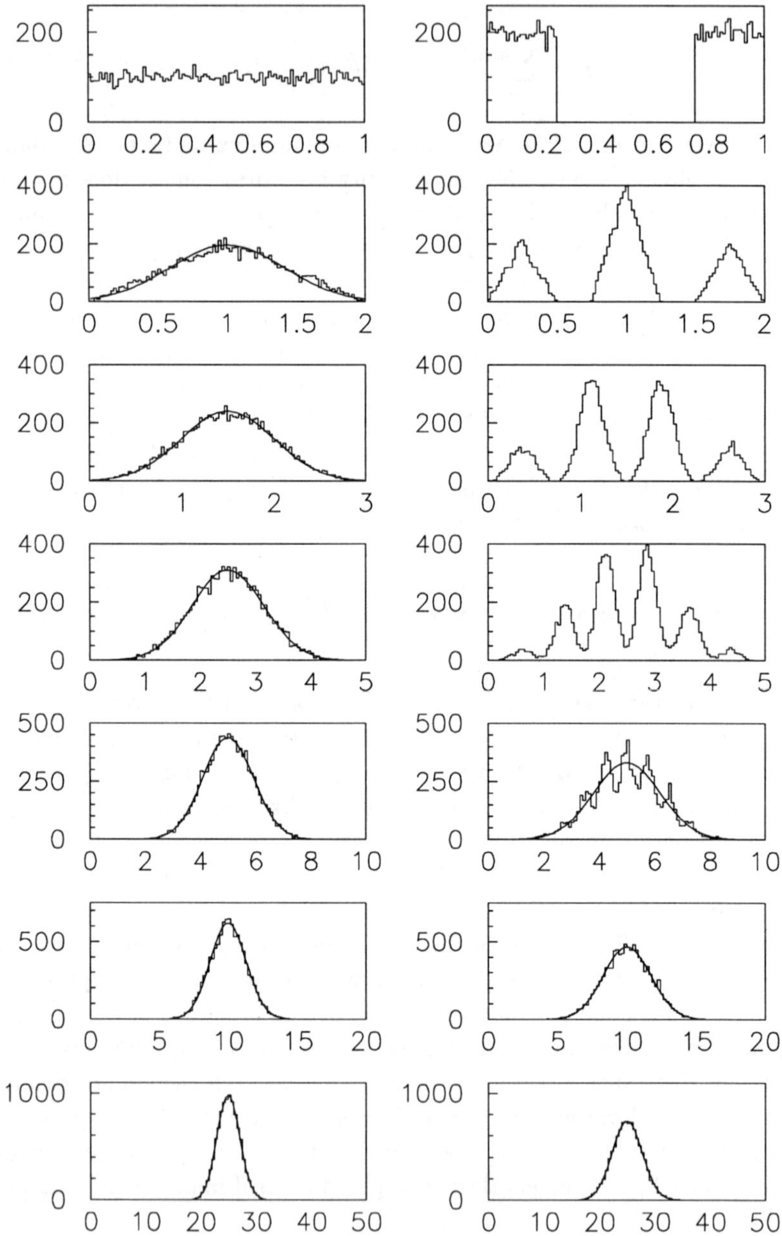

Fig. 4.5 Central limit theorem at work: The sum of n *iid* variables, for two different distributions, is shown. The values of n (top bottom) are 1, 2, 3, 5, 10, 20, 50.

4.5.2 *Distribution of a sample average*

As first application of the theorem, let us remind ourselves that a sample average $\overline{X}_n$ of n *independent identical distributed* ("*iid*") variables,

$$\overline{X}_n = \sum_{i=1}^{n} \frac{1}{n} X_i, \tag{4.105}$$

is normally distributed, since it is a linear combination of n variables X_i, with $c_i = 1/n$. Then,

$$\overline{X}_n \sim \mathcal{N}(\mu_{\overline{X}_n}, \sigma_{\overline{X}_n}), \tag{4.106}$$

$$\mu_{\overline{X}_n} = \sum_{i=1}^{n} \frac{1}{n}\, \mu = \mu, \tag{4.107}$$

$$\sigma^2_{\overline{X}_n} = \sum_{i=1}^{n} \left(\frac{1}{n}\right)^2 \sigma^2 = \frac{\sigma^2}{n}, \tag{4.108}$$

$$\sigma_{\overline{X}_n} = \frac{\sigma}{\sqrt{n}}. \tag{4.109}$$

This result, we repeat, is independent of the distribution of X and is already *approximately valid* for small values of n.

4.5.3 *Normal approximation of the binomial and of the Poisson distribution*

Another important application of the theorem is that the binomial and the Poisson distribution can be approximated, for "large numbers", by a Gaussian distribution. This is a general result, valid for all distributions which have the *reproductive property under the sum.* Distributions of this kind are the binomial, the Poisson and the χ^2. Let us go into more detail:

$\boxed{\mathcal{B}_{n,p} \rightarrow \mathcal{N}\left(n\,p, \sqrt{n\,p\,(1-p)}\right)}$ The reproductive property of the binomial states that if X_1, X_2, ..., X_m are m independent variables, each following a binomial distribution of parameter n_i and p, then their sum $Y = \sum_i X_i$ also follows a binomial distribution with parameters $n = \sum_i n_i$ and p. It is easy to be convinced of this property without any mathematics. Just think of what happens if one tosses bunches of three, of five and of ten coins, and then one considers the global result: a binomial with a large n can then always be seen as a sum of many binomials with smaller n_i. The application of the central limit theorem is straightforward, apart from deciding when the convergence is acceptable. The parameters on which one has to base a judgment are in this case $\mu = n\,p$ and the complementary quantity $\mu^c = n\,(1-p) = n - \mu$. If they are both $\gtrsim 10$ then the approximation starts to be reasonable.

$\mathcal{P}_\lambda \to \mathcal{N}\left(\lambda, \sqrt{\lambda}\right)$ The same argument holds for the Poisson distribution. In this case the approximation starts to be reasonable when $\mu = \lambda \gtrsim 10$.

4.5.4 *Normal distribution of measurement errors*

The central limit theorem is also important to *justify* why in many cases the distribution followed by the measured values around their average is approximately normal. Often, in fact, the random experimental error e, which causes the fluctuations of the measured values around the unknown true value of the physical quantity, can be seen as an incoherent sum of smaller contributions e_i

$$e = \sum_i e_i\,, \tag{4.110}$$

each contribution having a distribution which satisfies the conditions of the central limit theorem.

4.5.5 *Caution*

Following this commercial in favor of the miraculous properties of the central limit theorem, some words of caution are in order:

- Although I have tried to convince the reader that the convergence is rather fast in the cases of practical interest, the theorem only states that the asymptotic Gaussian distribution is reached for $n \to \infty$. As an example of very slow convergence, let us imagine 10^9 independent variables described by a Poisson distribution of $\lambda_i = 10^{-9}$: their sum is still far from a Gaussian.
- Sometimes the conditions of the theorem are not satisfied.
  - A single component dominates the fluctuation of the sum: a typical case is the well-known Landau ionization distribution; systematic errors may also have the same effect on the global error.
  - The condition of independence is lost if systematic errors affect a set of measurements, or if there is coherent noise.

- The tails of the distributions do exist and they are not always Gaussian! Moreover, random variables might take values several standard deviations away from the mean. And fluctuations show up without notice!

4.6 Laws of large numbers

The *convergence in probability* of the relative frequency to the probability ("frequency tends to probability" is the often-heard, simplified statement) is one of the best known *laws of large numbers.* However, it is a matter of fact that these laws are often misunderstood and misused. We shall not enter into technical details here, but we will discuss in Sec. 7.3 the particular 'law' expressed by Bernoulli's theorem. My aim in Sec. 7.3 will be to clarify two things which tend to be confused: "evaluating probability from frequency" as against "predicting frequency from probability" (and, hence, predicting future frequencies from past frequencies).

Chapter 5

Bayesian inference of continuous quantities

"... these problems are classified as probability of the causes, and are the most interesting of all from their scientific applications".

"An effect may be produced by the cause a or by the cause b. The effect has just been observed. We ask the probability that it is due to the cause a. This is an à posteriori probability of cause. But I could not calculate it, if a convention more or less justified did not tell me in advance what is the à priori probability for the cause a to come into play. I mean the probability of this event to some one who had not observed the effect."

(Henri Poincaré)

5.1 Measurement error and measurement uncertainty

One might assume that the concepts of error and uncertainty are well known to be not worth discussing. Nevertheless a few comments are needed (although for more details the DIN [3] and ISO [5, 6] recommendations should be consulted).

- The first concerns the terminology. In fact the words *error* and *uncertainty* are currently used almost as synonyms:
  - "error" to mean both error and uncertainty (but nobody says "Heisenberg Error Principle");
  - "uncertainty" only for the uncertainty.

  "Usually" we understand what each is talking about, but a more precise use of these nouns would really help. This is strongly called for by the DIN [3] and ISO [5, 6] recommendations. They state in fact that

  - error is *"the result of a measurement minus a true value of the measurand"* - it follows that the error is usually unknown;
  - uncertainty is a *"parameter, associated with the result of a measurement, that characterizes the dispersion of the values that could reasonably be attributed to the measurand"*;
- Within the physics community there is an established practice for reporting the final uncertainty of a measurement in the form of standard deviation. This is also recommended by the mentioned standards. However, this should be done at each step of the analysis, instead of estimating "maximum error bounds" and using them as standard deviation in the "error propagation".
- The process of measurement is a complex one and it is difficult to disentangle the different contributions which cause the total error. In particular, the active role of the experimentalist is sometimes overlooked. For this reason it is often incorrect to quote the ("nominal") uncertainty due to the instrument as if it were the uncertainty of the measurement.

5.1.1 *General form of Bayesian inference*

In the Bayesian framework the inference is performed by calculating the final distribution of the random variable associated with the true values of the physical quantities from all available information. Let us call $\boldsymbol{x} = \{x_1, x_2, \ldots, x_n\}$ the *n-tuple* ("vector") of observables, $\boldsymbol{\mu} = \{\mu_1, \mu_2, \ldots, \mu_n\}$ the n-tuple of the true values of the physical quantities of interest, and $\boldsymbol{h} = \{h_1, h_2, \ldots, h_n\}$ the n-tuple of all the possible realizations of the *influence quantity* H_i (or variable); The term 'influence quantity' is used here with an extended meaning, to indicate not only external factors which could influence the result (temperature, atmospheric pressure, and so on) but also any possible calibration constant and any source of systematic errors. In fact the distinction between $\boldsymbol{\mu}$ and $\boldsymbol{h}$ is artificial, since they are all conditional hypotheses. We separate them simply because at the end we will "marginalize" the final joint distribution functions with respect to $\boldsymbol{\mu}$, integrating the joint distribution with respect to the other hypotheses considered as influence variables.

The likelihood of the *sample* $\boldsymbol{x}$ being produced from $\boldsymbol{h}$ and $\boldsymbol{\mu}$ and the initial probability are

$$f(\boldsymbol{x} \,|\, \boldsymbol{\mu}, \boldsymbol{h}, H_\circ)$$

and

$$f_\circ(\boldsymbol{\mu}, \boldsymbol{h}) = f(\boldsymbol{\mu}, \boldsymbol{h} \,|\, H_\circ)\,, \tag{5.1}$$

respectively. $H_\circ$ is intended to remind us, yet again, that likelihoods and priors — and hence conclusions — depend on all explicit and implicit assumptions within the problem, and in particular on the parametric functions used to model priors and likelihoods. To simplify the formulae, $H_\circ$ will no longer be written explicitly.

Using the Bayes formula for multidimensional continuous distributions [an extension of Eq. (4.70)] we obtain the most general formula of inference,

$$f(\boldsymbol{\mu}, \boldsymbol{h} \,|\, \boldsymbol{x}) = \frac{f(\boldsymbol{x} \,|\, \boldsymbol{\mu}, \boldsymbol{h})\, f_\circ(\boldsymbol{\mu}, \boldsymbol{h})}{\int f(\boldsymbol{x} \,|\, \boldsymbol{\mu}, \boldsymbol{h})\, f_\circ(\boldsymbol{\mu}, \boldsymbol{h})\, \mathrm{d}\boldsymbol{\mu}\, \mathrm{d}\boldsymbol{h}}\,, \tag{5.2}$$

yielding the joint distribution of all conditional variables $\boldsymbol{\mu}$ and $\boldsymbol{h}$ which are responsible for the observed sample $\boldsymbol{x}$. To obtain the final distribution of $\boldsymbol{\mu}$ one has to integrate Eq. (5.2) over all possible values of $\boldsymbol{h}$, obtaining

$$\boxed{f(\boldsymbol{\mu} \,|\, \boldsymbol{x}) = \frac{\int f(\boldsymbol{x} \,|\, \boldsymbol{\mu}, \boldsymbol{h})\, f_\circ(\boldsymbol{\mu}, \boldsymbol{h})\, \mathrm{d}\boldsymbol{h}}{\int f(\boldsymbol{x} \,|\, \boldsymbol{\mu}, \boldsymbol{h})\, f_\circ(\boldsymbol{\mu}, \boldsymbol{h})\, \mathrm{d}\boldsymbol{\mu}\, \mathrm{d}\boldsymbol{h}}\,.} \tag{5.3}$$

Apart from the technical problem of evaluating the integrals, if need be numerically or using Monte Carlo methods[1], Eq. (5.3) represents the most general form of *hypothetical inductive inference.* The word "hypothetical" reminds us of $H_\circ$.

When all the sources of influence are under control, i.e. they can be assumed to take a precise value, the initial distribution can be factorized by a $f_\circ(\boldsymbol{\mu})$ and a Dirac $\delta(\boldsymbol{h} - \boldsymbol{h}_\circ)$, obtaining the much simpler formula

$$\begin{aligned} f(\boldsymbol{\mu} \,|\, \boldsymbol{x}) &= \frac{\int f(\boldsymbol{x} \,|\, \boldsymbol{\mu}, \boldsymbol{h})\, f_\circ(\boldsymbol{\mu})\, \delta(\boldsymbol{h} - \boldsymbol{h}_\circ)\, \mathrm{d}\boldsymbol{h}}{\int f(\boldsymbol{x} \,|\, \boldsymbol{\mu}, \boldsymbol{h})\, f_\circ(\boldsymbol{\mu})\, \delta(\boldsymbol{h} - \boldsymbol{h}_\circ)\, \mathrm{d}\boldsymbol{\mu}\, \mathrm{d}\boldsymbol{h}} \\ &= \frac{f(\boldsymbol{x} \,|\, \boldsymbol{\mu}, \boldsymbol{h}_\circ)\, f_\circ(\boldsymbol{\mu})}{\int f(\boldsymbol{x} \,|\, \boldsymbol{\mu}, \boldsymbol{h}_\circ)\, f_\circ(\boldsymbol{\mu})\, \mathrm{d}\boldsymbol{\mu}}\,. \end{aligned} \tag{5.4}$$

Even if formulae (5.3)–(5.4) look complicated because of the multidimensional integration and of the continuous nature of $\boldsymbol{\mu}$, conceptually they are identical to the example of the $\mathrm{d}E/\mathrm{d}x$ measurement discussed in Sec. 3.5.3.

The final probability density function provides the most complete and detailed information about the unknown quantities, but sometimes (almost

[1]This is conceptually what experimentalists do when they change all the parameters of the Monte Carlo simulation in order to estimate the "systematic error".

always ...) one is not interested in full knowledge of $f(\boldsymbol{\mu})$, but just in a few numbers which summarize at best the position and the width of the distribution (for example when publishing the result in a journal in the most compact way). The most natural quantities for this purpose are the expectation value and the variance, or the standard deviation. Then the Bayesian best estimate of a physical quantity is:

$$\widehat{\mu}_i = \mathrm{E}[\mu_i] = \int \mu_i\, f(\boldsymbol{\mu}\,|\,\boldsymbol{x})\, \mathrm{d}\boldsymbol{\mu}, \tag{5.5}$$

$$\sigma^2_{\mu_i} \equiv \mathrm{Var}(\mu_i) = \mathrm{E}[\mu_i^2] - \mathrm{E}^2[\mu_i]. \tag{5.6}$$

When many true values are inferred from the same data the numbers which synthesize the result are not only the expected values and variances. Also the covariances (or the correlation coefficients) should be reported.

In the following sections we will deal in most cases with only one value to infer:

$$f(\mu\,|\,\boldsymbol{x}) = \ldots . \tag{5.7}$$

5.2 Bayesian inference and maximum likelihood

We have already said that the dependence of the final probabilities on the initial ones gets weaker as the amount of experimental information increases. Without going into mathematical complications (the proof of this statement can be found for example in Ref. [49]) this simply means that, asymptotically, whatever $f_\circ(\mu)$ one puts in Eq. (5.4), $f(\mu\,|\,\boldsymbol{x})$ is unaffected. This happens when the "width" of $f_\circ(\mu)$ is much larger than that of the likelihood, when the latter is considered as a mathematical function of μ. Therefore $f_\circ(\mu)$ acts as a constant in the region of μ where the likelihood is significantly different from 0. This is "equivalent" to dropping $f_\circ(\mu)$ from Eq. (5.4). This results in

$$f(\mu\,|\,\boldsymbol{x}) \approx \frac{f(\boldsymbol{x}\,|\,\mu, \boldsymbol{h}_\circ)}{\int f(\boldsymbol{x}\,|\,\mu, \boldsymbol{h}_\circ)\, \mathrm{d}\mu} . \tag{5.8}$$

Since the denominator of the Bayes formula has the technical role of properly normalizing the probability density function, the result can be written in the simple form

$$f(\mu\,|\,\boldsymbol{x}) \propto f(\boldsymbol{x}\,|\,\mu, \boldsymbol{h}_\circ) \equiv \text{“}\mathcal{L}(\mu; \boldsymbol{x}, \boldsymbol{h}_\circ)\text{”} . \tag{5.9}$$

Asymptotically the final probability is just the (normalized) likelihood! The notation $\mathcal{L}$ is that used in the maximum likelihood literature (note that, not only does f become $\mathcal{L}$, but also "$\,|\,$" has been replaced by ";": $\mathcal{L}$ has no probabilistic interpretation, when referring to μ, in conventional statistics).

If the mean value of $f(\mu\,|\,\boldsymbol{x})$ coincides with the value for which $f(\mu\,|\,\boldsymbol{x})$ has a maximum, we obtain the maximum likelihood method. This does not mean that the Bayesian methods are "blessed" because of this achievement, and hence they can be used only in those cases where they provide the same results. It is the other way round: The maximum likelihood method gets justified when all the limiting conditions of the approach ($\rightarrow$ insensitivity of the result from the initial probability $\rightarrow$ large number of events) are satisfied.

Even if in this asymptotic limit the two approaches yield the same numerical results, there are differences in their interpretation:

- The likelihood, after proper normalization, has a probabilistic meaning for Bayesians but not for frequentists; so Bayesians can say that the probability that μ is in a certain interval is, for example, 68 %, while this statement is blasphemous for a frequentist ("the true value is a constant" from his point of view).
- Frequentists prefer to choose $\widehat{\mu}_L$, the value which maximizes the likelihood, as estimator. For Bayesians, on the other hand, the expectation value $\widehat{\mu}_B = \mathrm{E}[\mu]$ (also called the *prevision*) is more appropriate. This is justified by the fact that the assumption of the $\mathrm{E}[\mu]$ as best estimate of μ minimizes the risk of a bet (always keep the bet in mind!). For example, if the final distribution is exponential with parameter τ (let us think for a moment of particle decays) the maximum likelihood method would recommend betting on the value $t = 0$, whereas the Bayesian approach suggests the value $t = \tau$. If the terms of the bet are "whoever gets <u>closest</u> wins", what is the best strategy? And then, what is the best strategy if the terms are "whoever gets the <u>exact</u> value wins"? But now think of the probability of getting the exact value and of the probability of getting closest.

5.3 The dog, the hunter and the biased Bayesian estimators

One of the most important tests to judge the quality of an estimator, is whether or not it is *correct* (not biased). Maximum likelihood estimators are usually correct, while Bayesian estimators — analyzed within the maximum likelihood framework — are often not. This could be considered a weak point; however the Bayes estimators are simply naturally consistent with the state of information before new data become available. In the maximum likelihood method, on the other hand, it is not clear what the assumptions are.

Let us take an example which shows the logic of frequentistic inference and why the use of reasonable prior distributions yields results which that frame classifies as distorted. Imagine meeting a hunting dog in the country. Let us assume we know that there is a 50 % probability of finding the dog within a radius of 100 m centred on the position of the hunter (this is our likelihood). Where is the hunter? He is with 50 % probability within a radius of 100 m around the position of the dog, with equal probability in all directions. "Obvious". This is exactly the logic scheme used in the frequentistic approach to build confidence regions from the estimator (the dog in this example). This however assumes that the hunter can be anywhere in the country. But now let us change the state of information: "the dog is by a river"; "the dog has collected a duck and runs in a certain direction"; "the dog is sleeping"; "the dog is in a field surrounded by a fence through which he can pass without problems, but the hunter cannot". Given any new condition the conclusion changes. Some of the new conditions change our likelihood, but some others only influence the initial distribution. For example, the case of the dog in an enclosure inaccessible to the hunter is exactly the problem encountered when measuring a quantity close to the edge of its physical region, which is quite common in frontier research.

5.4 Choice of the initial probability density function

The title of this section is similar to that of Sec. 3.11, but the problem and the conclusions will be different.

5.4.1 *Difference with respect to the discrete case*

In Sec. 3.11 we said that the Indifference Principle (or, in its refined modern version, the Maximum Entropy Principle) was a good choice. Here there are problems with *infinities* and with the fact that it is possible to map an infinite number of points contained in a finite region onto an infinite number of points contained in a larger or smaller finite region. This changes the probability density function. If, moreover, the transformation from one set of variables to the other is not linear (see, e.g., Fig. 4.4), what is uniform in one variable (X) is not uniform in another variable (e.g. $Y = X^2$). This problem does not exist in the case of discrete variables, since if $X = x_i$ has a probability $f(x_i)$ then $Y = x_i^2$ has the same probability. A different way of stating the problem is that the *Jacobian* of the transformation squeezes or stretches the metrics, changing the probability density function.

We will not enter into open discussion about the optimal choice of the distribution. Essentially we shall use the uniform distribution, being careful to employ the variable which "seems" most appropriate for the problem, but You may disagree — surely with good reason — if You have a different kind of experiment in mind.

The same problem is also present, but well hidden, in the maximum likelihood method. For example, it is possible to demonstrate that, in the case of normally distributed likelihoods, a uniform distribution of the mean μ is implicitly assumed (see Sec. 6.2). There is nothing wrong with this, but one should be aware of it.

5.4.2 *Bertrand paradox and angels' sex*

A good example to help understand the problems outlined in the previous section is the so-called Bertrand paradox.

Problem: Given a circle of radius R and a chord drawn randomly on it, what is the probability that the length L of the chord is smaller than R?

Solution 1: Choose "randomly" two points on the circumference and draw a chord between them: $\Rightarrow P(L < R) = 1/3 = 0.33$.

Solution 2: Choose a straight line passing through the centre of the circle; then draw a second line, orthogonal to the first, and which intersects it inside the circle at a "random" distance from the centre: $\Rightarrow P(L < R) = 1 - \sqrt{3}/2 = 0.13$.

Solution 3: Choose "randomly" a point inside the circle and draw a straight line orthogonal to the radius that passes through the chosen point $\Rightarrow P(L < R) = 1/4 = 0.25$.

Your solution: ....?

Question: What is the origin of the paradox?

Answer: The problem does not specify how to "randomly" choose the chord. The three solutions take a uniform distribution: along the circumference; along the radius; inside the circle. What is uniform in one variable is not uniform in the others!

Question: Which is the right solution?

In principle you may imagine an infinite number of different solutions. From a physicist's viewpoint any attempt to answer this question is a waste of time. The reason why the paradox has been compared to the Byzantine discussions about the sex of angels is that there are indeed people arguing about it. For example, there is a school of thought which insists that Solution 2 is the right one.

In fact this kind of paradox, together with abuse of the Indifference Principle for problems like "what is the probability that the sun will rise tomorrow morning" threw a shadow over Bayesian methods at the end of the last century. The Maximum Likelihood method, which does not make explicit use of prior distributions, was then seen as a valid solution to the problem. But in reality the ambiguity of the proper metrics on which the initial distribution is uniform has an equivalent in the arbitrariness of the variable used in the likelihood function (usually the 'best estimate' has the nice property of being invariant, but the interpretation of the 'error analysis' does not! — See Sec. 12.2). In the end, what was criticized when it was stated explicitly in the Bayes formula is accepted passively when it is hidden in the maximum likelihood method.

Chapter 6

Gaussian likelihood

"Functio nostra fiet $\varphi(z) = \frac{h}{\sqrt{\pi}}\, e^{-h^2 z^2}$"

(Carl F. Gauss)

6.1 Normally distributed observables

The first application of the Bayesian inference will be that of a normally distributed quantity. Let us take a data sample $\boldsymbol{q}$ of n_1 measurements, of which we calculate the average $\overline{q}_{n_1}$. In our formalism $\overline{q}_{n_1}$ is a value that the random variable $\overline{Q}_{n_1}$ can assume. Let us assume we know the standard deviation σ of the variable Q, either because n_1 is very large and can be estimated accurately from the sample or because it was known *a priori*. (We are not going to discuss in this primer the case of small samples and unknown variance — for criticisms about the standard treatment of the small-sample problem see Ref. [33].) The property of the average (see Sec. 4.5.2) tells us that the likelihood $f(\overline{q}_{n_1} \,|\, \mu, \sigma)$ is Gaussian:

$$\overline{Q}_{n_1} \sim \mathcal{N}(\mu, \sigma/\sqrt{n_1}). \tag{6.1}$$

To simplify the following notation, let us call x_1 this average and σ_1 the standard deviation of the average:

$$x_1 = \overline{q}_{n_1}, \tag{6.2}$$

$$\sigma_1 = \sigma/\sqrt{n_1}\,. \tag{6.3}$$

6.2 Final distribution, prevision and credibility intervals of the true value

We then apply Eq. (5.4) and get

$$f(\mu\,|\,x_1, \mathcal{N}(\cdot, \sigma_1)) = \frac{\frac{1}{\sqrt{2\,\pi}\,\sigma_1}\,\exp\left[-\frac{(x_1-\mu)^2}{2\,\sigma_1^2}\right] f_\circ(\mu)}{\int \frac{1}{\sqrt{2\,\pi}\,\sigma_1}\,\exp\left[-\frac{(x_1-\mu)^2}{2\,\sigma_1^2}\right] f_\circ(\mu)\,\mathrm{d}\mu}\,. \tag{6.4}$$

At this point we have to make a choice for $f_\circ(\mu)$. A reasonable choice is to take, as a first guess, a uniform distribution defined over a "large" interval which includes x_1. It is not really important how large the interval is, for a few σ_1 away from x_1 the integrand at the denominator tends to zero because of the Gaussian function. What is important is that a constant $f_\circ(\mu)$ can be simplified in Eq. (6.4), obtaining

$$f(\mu\,|\,x_1, \mathcal{N}(\cdot, \sigma_1)) = \frac{\frac{1}{\sqrt{2\,\pi}\,\sigma_1}\,\exp\left[-\frac{(x_1-\mu)^2}{2\,\sigma_1^2}\right]}{\int_{-\infty}^{\infty} \frac{1}{\sqrt{2\,\pi}\,\sigma_1}\,\exp\left[-\frac{(x_1-\mu)^2}{2\,\sigma_1^2}\right] \mathrm{d}\mu}\,. \tag{6.5}$$

The integral in the denominator is equal to unity, since integrating with respect to μ is equivalent to integrating with respect to x_1. The final result is then

$$f(\mu) = f(\mu\,|\,x_1, \mathcal{N}(\cdot, \sigma_1)) = \frac{1}{\sqrt{2\,\pi}\,\sigma_1}\,\exp\left[-\frac{(\mu - x_1)^2}{2\,\sigma_1^2}\right] : \tag{6.6}$$

- the true value is normally distributed around x_1;
- its best estimate (*prevision*) is $\mathrm{E}[\mu] = x_1$;
- its variance is $\sigma_\mu = \sigma_1$;
- the "confidence intervals", or *credibility intervals*, in which there is a certain probability of finding the true value are easily calculable:

Probability level (confidence level) (%)	Credibility interval (confidence interval)
68.3	$x_1 \pm \sigma_1$
90.0	$x_1 \pm 1.65\,\sigma_1$
95.0	$x_1 \pm 1.96\,\sigma_1$
99.0	$x_1 \pm 2.58\,\sigma_1$
99.73	$x_1 \pm 3\,\sigma_1$

6.3 Combination of several measurements – Role of priors

Let us imagine making a second set of measurements of the physical quantity, which we assume unchanged from the previous set of measurements. How will our knowledge of μ change after this new information? Let us call $x_2 = \overline{q}_{n_2}$ and $\sigma_2 = \sigma'/\sqrt{n_2}$ the new average and standard deviation of the average (σ' may be different from σ of the sample of n_1 measurements), respectively. Applying Bayes' theorem a second time we now have to use *as initial distribution the final probability of the previous inference*:

$$f(\mu \,|\, x_1, \sigma_1, x_2, \sigma_2, \mathcal{N}) = \frac{\frac{1}{\sqrt{2\,\pi}\,\sigma_2}\,\exp\left[-\frac{(x_2-\mu)^2}{2\,\sigma_2^2}\right] f(\mu \,|\, x_1, \mathcal{N}(\cdot, \sigma_1))}{\int \frac{1}{\sqrt{2\,\pi}\,\sigma_2}\,\exp\left[-\frac{(x_2-\mu)^2}{2\,\sigma_2^2}\right] f(\mu \,|\, x_1, \mathcal{N}(\cdot, \sigma_1))\,\mathrm{d}\mu}\,. \tag{6.7}$$

The integral is not as simple as the previous one, but still feasible analytically. The final result is

$$f(\mu \,|\, x_1, \sigma_1, x_2, \sigma_2, \mathcal{N}) = \frac{1}{\sqrt{2\,\pi}\,\sigma_A}\,\exp\left[-\frac{(\mu - x_A)^2}{2\,\sigma_A^2}\right], \tag{6.8}$$

where

$$x_A = \frac{x_1/\sigma_1^2 + x_2/\sigma_2^2}{1/\sigma_1^2 + 1/\sigma_2^2}\,, \tag{6.9}$$

$$\frac{1}{\sigma_A^2} = \frac{1}{\sigma_1^2} + \frac{1}{\sigma_2^2}\,. \tag{6.10}$$

One recognizes the famous formula of the weighted average with the inverse of the variances, usually obtained from maximum likelihood. There are some comments to be made.

- Bayes' theorem updates the knowledge about μ in an automatic and natural way.
- If $\sigma_1 \gg \sigma_2$ (and x_1 is not "too far" from x_2) the final result is only determined by the second sample of measurements. This suggests that an alternative *vague a priori* distribution can be, instead of uniform, a Gaussian with a *large enough* variance and a *reasonable* mean.
- The combination of the samples requires a subjective judgment that the two samples are really coming from the same true value μ. We will not discuss here this point, but a hint on how to proceed is to: take the inference on the difference of two measurements, D, as explained at the end of Sec. 6.8 and judge yourself whether $D = 0$ is consistent with the

probability density function of D. As is easy to imagine, the problem of the "outliers" should be treated with care, and surely avoiding automatic prescriptions. An example of solution will be discussed in detail in Chapter 11.

6.3.1 *Update of estimates in terms of Kalman filter*

An interesting way of writing Eq. (6.9) is considering x_1 and x_A the estimates of μ at times t_1 and t_2, respectively before and after the observation x_2 happened at time t_2. The uncertainties about μ at t_1 and t_2 are $\sigma_\mu(t_1) = \sigma_1$ and $\sigma_\mu(t_2) = \sigma_A$, respectively. Indicating the *estimates* at different times by $\hat{\mu}(t)$, we can rewrite Eq. (6.9) as

$$\begin{aligned}
\hat{\mu}(t_2) &= \frac{\sigma_\mu^2(t_1)}{\sigma_x^2(t_2)+\sigma_\mu^2(t_1)}\, x(t_2) + \frac{\sigma_x^2(t_2)}{\sigma_x^2(t_2)+\sigma_\mu^2(t_1)}\, \hat{\mu}(t_1) \\
&= \hat{\mu}(t_1) + \frac{\sigma_\mu^2(t_1)}{\sigma_x^2(t_2)+\sigma_\mu^2(t_2)}\, [x(t_2) - \hat{\mu}(t_1)] \qquad (6.11) \\
&= \hat{\mu}(t_1) + K(t_2)\, [x(t_2) - \hat{\mu}(t_1)] \qquad (6.12)
\end{aligned}$$

$$\sigma_\mu^2(t_2) = \sigma_\mu^2(t_1) - K(t_2)\, \sigma_\mu^2(t_1)\,, \qquad (6.13)$$

where

$$K(t_2) = \frac{\sigma_\mu^2(t_1)}{\sigma_x^2(t_2)+\sigma_\mu^2(t_1)}\,. \qquad (6.14)$$

Indeed, we have given Eq. (6.9) the structure of a *Kalman filter* [65]. The new observation 'corrects' the estimate by a quantity given by the *innovation* (or *residual*) $[x(t_2) - \hat{\mu}(t_1)]$ times the *blending factor* (or *gain*) $K(t_2)$. For an introduction about Kalman filter and its probabilistic origin, see Refs. [66] and [67].

6.4 Conjugate priors

In Sec. 6.3 we introduced with a practical example the concept of *conjugate priors*: a prior such that the product likelihood $\times$ prior, i.e. the posterior, belongs to the same family as the prior. This is a well known technique to simplify the calculations.

The prior conjugate of a Gaussian likelihood is a Gaussian, due to the well known property that products of Gaussians are still Gaussians. Less trivial conjugate priors will be shown in the next chapter.

6.5 Improper priors — never take models literally!

There is another important concept we introduced in a practical way in Sec. 6.2: a prior uniform from $-\infty$ to $+\infty$. Obviously, this p.d.f. is not normalizable, all moments of the distribution being infinite. Not normalizable priors are called *improper*. They must be considered as mathematical commodities to model our uncertainty over a wide range. For instance, if we are interested in a length or a mass, 'infinite values' make no sense. Similarly, the value of these quantities cannot be negative, and even less they can be '$-\infty$'. Nevertheless, if we make a measurement with a 'precise' instrument having a Gaussian response, the product likelihood $\times$ prior is immediately dumped by the Gaussian, when $|\mu - x|$ is large enough. It follows that the result does not change, if we extend the range of the vague prior to infinity. This teaches us that we can make an easy use of improper priors as long as the likelihood is Gaussian or, more in general, 'closed', in the sense that will be discussed at length in Chapter 13. Otherwise, much more care and deep thought is needed.

I take the opportunity here to remark that we should be careful about taking models too seriously. If we take once more the Gaussian case, a Gaussian response of a detector should be considered as a practical model, but should never be treated literally. Extreme values will never be observed, not "because they are very improbable, as the Gaussian shows", but simply because all instrument scales are by construction finite! There are several problems in 'error propagation' (think for example of the p.d.f. $Y = 1/X$, starting from a Gaussian distributed X) in which presumed paradoxes vanish as soon as we try to model our knowledge in the most realistic way (an effect similar to that discussed in Sec. 3.13).

6.6 Predictive distribution

We have seen the importance of $f(x \mid \mu)$ for making inferences. It is important to stress that this p.d.f. does not describe 'probabilities of (future) observations', but only 'probabilities of (future) observations under the hypothesis that the true value is precisely μ'. However, we do not know the exact value of μ, since our knowledges is described by $f(\mu \mid I)$. Again, we use the rules of probability, and weigh all infinite p.d.f.'s $f(x \mid \mu)$ with how much we believe in each value of μ:

$$f(x \mid I) = \int f(x \mid \mu, I)\, f(\mu \mid I)\, \mathrm{d}\mu\,. \tag{6.15}$$

In the case where our knowledge about μ comes from a measurement modelled by a Gaussian likelihood with standard deviation σ_p (p stands for 'past'), and the response of the future experiment follows the same model, but with σ_f, we have:

$$\begin{aligned} f(x_f \mid x_p) &= \int \frac{1}{\sqrt{2\pi}\,\sigma_f} \exp\left[-\frac{(x_f-\mu)^2}{2\,\sigma_f^2}\right] \frac{1}{\sqrt{2\pi}\,\sigma_p} \exp\left[-\frac{(\mu-x_p)^2}{2\,\sigma_p^2}\right] \mathrm{d}\mu \\ &= \frac{1}{\sqrt{2\pi}\sqrt{\sigma_p^2+\sigma_f^2}} \exp\left[-\frac{(x_f-x_p)^2}{2\,(\sigma_p^2+\sigma_f^2)}\right], \end{aligned} \tag{6.16}$$

resulting in

$$\mathrm{E}[X_f] = x_p \tag{6.17}$$

$$\sigma(X_f) = \sqrt{\sigma_p^2+\sigma_f^2}\,. \tag{6.18}$$

Note that the *predictive* distribution (6.16) describes the uncertainty about the not-yet-known value x_f, conditioned by the previous observation x_p, while the kind of 'metaphysical' object μ (i.e. about which we can have no direct experience) disappears. Figure 6.1 shows the inferential scheme for predicting the future observation x_f given the past observation x_p.

It is worth noting the particular case $\sigma_f = \sigma_p = \sigma_\circ/\sqrt{n}$ (i.e. the n measurements can be considered as a single 'equivalent' measurement — a schematization related to what statisticians call *sufficiency*): there is a 52% probability that the new measurement will fall within $\pm\sigma_\circ/\sqrt{n}$ of the previous one. It is not uncommon to hear people saying that such a probability is 68% and the kind of logical mistake they are making is clear.

6.7 Measurements close to the edge of the physical region

A case which has essentially no solution in the maximum likelihood approach is when a measurement is performed at the edge of the physical region and the measured value comes out very close to it, or even on the unphysical region. Let us take a numerical example.

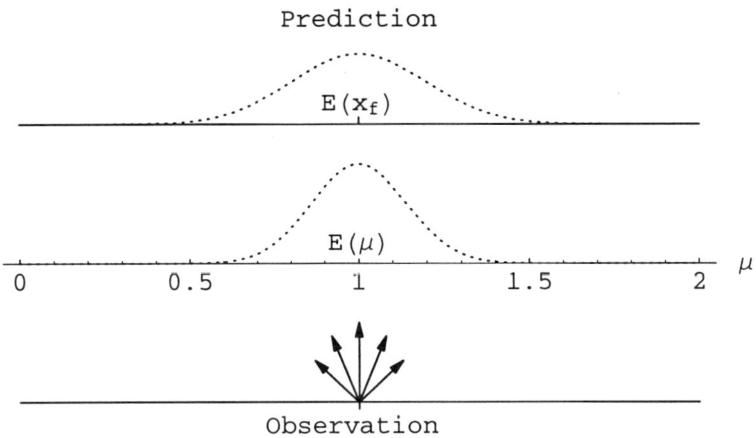

Fig. 6.1 Scheme of predictive inference which skips the intermediate 'metaphysical' step μ.

Problem: An experiment is planned to measure a neutrino mass. The simulations show that the mass resolution is $3.3\,\mathrm{eV}/c^2$, largely independent of the mass value, and that the measured mass is normally distributed around the true mass[1]. The mass value which results from the analysis procedure,[2] and corrected for all known systematic effects, is $x = -5.41\,\mathrm{eV}/c^2$. What have we learned about the neutrino mass?

Solution: Our *a priori* value of the mass is that it is positive and not too large (otherwise it would already have been measured in other experiments). One can take any vague distribution which assigns a probability density function between 0 and 20 or 30 eV/c^2. In fact, if an experiment having a resolution of $\sigma = 3.3\,\mathrm{eV}/c^2$ has been planned and financed by rational people, with the *hope* of finding evidence of non-negligible mass, it means that the mass was thought to be in that range. If there is no reason to prefer one of the values in that interval a uniform distribution can be used, for example

$$f_{\circ K}(m) = k = 1/30 \qquad (0 \le m \le 30)\,. \tag{6.19}$$

[1]In reality, often m^2 rather than m is normally distributed. In this case the terms of the problem change and a new solution should be worked out, following the trace indicated in this example.

[2]We consider detector and analysis machinery as a black box, no matter how complicated it is, and treat the numerical outcome as a result of a direct measurement [3].

Otherwise, if one thinks there is a greater chance of the mass having small rather than high values, a prior which reflects such an assumption could be chosen, for example a half normal with $\sigma_\circ = 10\,\text{eV}$

$$f_{\circ N}(m) = \frac{2}{\sqrt{2\,\pi}\,\sigma_\circ}\,\exp\left[-\frac{m^2}{2\,\sigma_\circ^2}\right] \qquad (m \geq 0)\,, \tag{6.20}$$

or a triangular distribution

$$f_{\circ T}(m) = \frac{1}{450}\,(30 - x) \qquad (0 \leq m \leq 30)\,. \tag{6.21}$$

Let us consider for simplicity the uniform distribution

$$f(m\,|\,x, f_{\circ K}) = \frac{\frac{1}{\sqrt{2\,\pi}\,\sigma}\,\exp\left[-\frac{(m-x)^2}{2\,\sigma^2}\right]k}{\int_0^{30}\frac{1}{\sqrt{2\,\pi}\,\sigma}\,\exp\left[-\frac{(m-x)^2}{2\,\sigma^2}\right]k\,\mathrm{d}\mu} \tag{6.22}$$

$$= \frac{19.8}{\sqrt{2\pi}\,\sigma}\,\exp\left[-\frac{(m-x)^2}{2\sigma^2}\right] \qquad (0 \leq m \leq 30)\,. \tag{6.23}$$

The value which has the highest degree of belief is $m = 0$, but $f(m)$ is nonvanishing up to $30\,\text{eV}/c^2$ (even if very small). We can define an interval, starting from $m = 0$, in which we believe that m should have a certain probability. For example this level of probability can be 95 %. One has to find the value $m_\circ$ for which the cumulative function $F(m_\circ)$ equals 0.95. This value of m is called the *upper limit* (or *upper bound*). The result is

$$m < 3.9\,\text{eV}/c^2 \qquad \text{at } 0.95\,\%\text{ probability}\,. \tag{6.24}$$

If we had assumed the other initial distributions the limit would have been in both cases

$$m < 3.7\,\text{eV}/c^2 \qquad \text{at } 0.95\,\%\text{ probability}\,, \tag{6.25}$$

practically the same (especially if compared with the experimental resolution of $3.3\,\text{eV}/c^2$).

Comment: Let us assume an *a priori* function sharply peaked at zero and see what happens. For example it could be of the kind[3]

$$f_{\circ S}(m) \propto \frac{1}{m}\,. \tag{6.26}$$

[3]For a deeper discussion about meaning and use of this prior, see Chapter 13.

To avoid singularities in the integral, let us take a power of m slightly greater than -1, for example -0.99, and let us limit its domain to 30, getting

$$f_\circ S(m) = \frac{0.01 \cdot 30^{0.01}}{m^{0.99}} \,. \tag{6.27}$$

The upper limit becomes

$$m < 0.006\,\mathrm{eV}/c^2 \quad \text{at} \;\; 0.95\,\%\ \text{probability}\,. \tag{6.28}$$

Any experienced physicist would find this result ridiculous. The upper limit is about 0.2 % of the experimental resolution; rather like expecting to resolve objects having dimensions smaller than a micron with a design ruler! Note instead that in the previous examples the limit was always of the order of magnitude of the experimental resolution σ. As $f_\circ S(m)$ becomes more and more peaked at zero (power of $x \to 1$) the limit gets smaller and smaller. This means that, asymptotically, the degree of belief that $m = 0$ is so high that whatever you measure you will conclude that $m = 0$: you could use the measurement to calibrate the apparatus! This means that this choice of initial distribution was unreasonable.

Instead, priors motivated by the ***positive attitude*** of the researchers are much more stable, and even when the observation is "very negative" the result is stable, and one always gets a limit of the order of the experimental resolution. Anyhow, it is also clear that when x is several σ below zero one starts to suspect that "something is wrong with the experiment", which formally corresponds to doubts about the likelihood itself. In this case one needs to change analysis model. An example of remodelling the likelihood is shown in Chapter 11.

We shall come back to this delicate issue in Chapter 13.

6.8 Uncertainty of the instrument scale offset

In our scheme any quantity of influence of which we do not know the exact value is a source of systematic error. It will change the final distribution of μ and hence its uncertainty. We have already discussed the most general case in Sec. 5.1.1. Let us make a simple application making a small variation to the example in Sec. 6.2: the "zero" of the instrument is not known exactly, owing to calibration uncertainty. This can be parametrized assuming that its true value Z is normally distributed around 0 (i.e. the calibration was properly done!) with a standard deviation σ_Z. Since, most probably, the true value of μ is independent of the true value of Z, the initial joint probability density function can be written as the product of the marginal ones:

$$f_\circ(\mu, z) = f_\circ(\mu)\, f_\circ(z) = k\, \frac{1}{\sqrt{2\,\pi}\,\sigma_Z} \exp\left[-\frac{z^2}{2\,\sigma_Z^2}\right]. \tag{6.29}$$

Also the likelihood changes with respect to Eq. (6.1):

$$f(x_1 \,|\, \mu, z) = \frac{1}{\sqrt{2\,\pi}\,\sigma_1} \exp\left[-\frac{(x_1 - \mu - z)^2}{2\,\sigma_1^2}\right]. \tag{6.30}$$

Putting all the pieces together and making use of Eq. (5.3) we finally get

$$f(\mu \,|\, x_1, \ldots, f_\circ(z)) = \frac{\int \frac{1}{\sqrt{2\,\pi}\,\sigma_1} \exp\left[-\frac{(x_1-\mu-z)^2}{2\,\sigma_1^2}\right] \frac{1}{\sqrt{2\,\pi}\,\sigma_Z} \exp\left[-\frac{z^2}{2\,\sigma_Z^2}\right] \mathrm{d}z}{\iint \frac{1}{\sqrt{2\,\pi}\,\sigma_1} \exp\left[-\frac{(x_1-\mu-z)^2}{2\,\sigma_1^2}\right] \frac{1}{\sqrt{2\,\pi}\,\sigma_Z} \exp\left[-\frac{z^2}{2\,\sigma_Z^2}\right] \mathrm{d}\mu\, \mathrm{d}z}.$$

Integrating we get

$$f(\mu) = f(\mu \,|\, x_1, \ldots, f_\circ(z)) = \frac{1}{\sqrt{2\,\pi}\,\sqrt{\sigma_1^2 + \sigma_Z^2}} \exp\left[-\frac{(\mu - x_1)^2}{2\,(\sigma_1^2 + \sigma_Z^2)}\right]. \tag{6.31}$$

(It may help to know that

$$\int_{-\infty}^{+\infty} \exp\left[b\,x - \frac{x^2}{a^2}\right] \mathrm{d}x = \sqrt{a^2\,\pi}\, \exp\left[\frac{a^2\,b^2}{4}\right] .)$$

For an introduction to Bayesian methods, where Gaussian integrals are also discussed, see e.g. Ref. [46]. The result is that $f(\mu)$ is still a Gaussian, but with a larger variance. The global standard uncertainty is the quadratic combination of that due to the statistical fluctuation of the data sample and the uncertainty due to the imperfect knowledge of the *systematic effect*:

$$\sigma_{tot}^2 = \sigma_1^2 + \sigma_Z^2\,. \tag{6.32}$$

This result (a theorem under well stated conditions!) is often used as a 'prescription', although there are still some "old-fashioned" recipes which require different combinations of the contributions to be performed.

It must be noted that in this framework it makes no sense to speak of "statistical" and "systematical" uncertainties, as if they were of a different nature. They have the same probabilistic nature: $\overline{Q}_{n_1}$ is around μ with a standard deviation σ_1, and Z is around 0 with standard deviation σ_Z. What distinguishes the two components is how the knowledge of the uncertainty is gained: in one case (σ_1) from repeated measurements; in the second case (σ_Z) the evaluation was done by someone else (the constructor of the instrument), or in a previous experiment, or guessed from the knowledge of the detector, or by simulation, etc. This is the reason why the ISO Guide [5] prefers the generic names *Type A* and *Type B* for the two kinds of contribution to global uncertainty (see Sec. 8.7). In particular, the name "systematic uncertainty" should be avoided, while it is correct to speak about "uncertainty due to a systematic effect".

6.9 Correction for known systematic errors

It is easy to be convinced that if our prior knowledge about Z was of the kind

$$Z \sim \mathcal{N}(z_\circ, \sigma_Z) \tag{6.33}$$

the result would have been

$$\mu \sim \mathcal{N}\left(x_1 - z_\circ, \sqrt{\sigma_1^2 + \sigma_Z^2}\right) , \tag{6.34}$$

i.e. one has first to correct the result for the best value of the systematic error and then include in the global uncertainty a term due to imperfect knowledge about it. This is a well known and practised procedure, although there are still people who confuse $z_\circ$ with its uncertainty.

6.10 Measuring two quantities with the same instrument having an uncertainty of the scale offset

Let us take an example which is a little more complicated (at least from the mathematical point of view) but conceptually very simple and also very common in laboratory practice. We measure two physical quantities with the same instrument, assumed to have an uncertainty on the "zero", modelled with a normal distribution as in the previous sections. For each of the quantities we collect a sample of data under the same conditions, which means that the unknown offset error does not change from one set of measurements to the other. Calling μ_1 and μ_2 the true values, x_1 and x_2 the sample averages, σ_1 and σ_2 the average's standard deviations, and Z the true value of the "zero", the initial probability density and the likelihood are

$$
\begin{aligned}
f_\circ(\mu_1,\mu_2,z) &= f_\circ(\mu_1)\,f_\circ(\mu_2)\,f_\circ(z) = k\,\frac{1}{\sqrt{2\,\pi}\,\sigma_Z}\,\exp\left[-\frac{z^2}{2\,\sigma_Z^2}\right]\\
f(x_1,x_2\,|\,\mu_1,\mu_2,z) &= \frac{1}{\sqrt{2\,\pi}\,\sigma_1}\,\exp\left[-\frac{(x_1-\mu_1-z)^2}{2\,\sigma_1^2}\right]\\
&\quad\times\frac{1}{\sqrt{2\,\pi}\,\sigma_2}\,\exp\left[-\frac{(x_2-\mu_2-z)^2}{2\,\sigma_2^2}\right]\\
&= \frac{1}{2\,\pi\,\sigma_1\sigma_2}\,\exp\left[-\frac{1}{2}\left(\frac{(x_1-\mu_1-z)^2}{\sigma_1^2}\right.\right.\\
&\quad\left.\left.+\frac{(x_2-\mu_2-z)^2}{\sigma_2^2}\right)\right]\,. \qquad (6.35)
\end{aligned}
$$

The result of the inference is now the joint probability density function of μ_1 and μ_2:

$$
f(\mu_1,\mu_2\,|\,x_1,x_2,\sigma_1,\sigma_2,f_\circ(z)) = \frac{\int f(x_1,x_2\,|\,\mu_1,\mu_2,z)\,f_\circ(\mu_1,\mu_2,z)\mathrm{d}z}{\int\ldots\,\mathrm{d}\mu_1\,\mathrm{d}\mu_2\,\mathrm{d}z}\,, \qquad (6.36)
$$

where expansion of the functions has been omitted for the sake of clarity. Integrating we get

$$
\begin{aligned}
f(\mu_1,\mu_2) &= \frac{1}{2\,\pi\,\sqrt{\sigma_1^2+\sigma_Z^2}\,\sqrt{\sigma_2^2+\sigma_Z^2}\,\sqrt{1-\rho^2}}\\
&\quad\times\exp\left\{-\frac{1}{2\,(1-\rho^2)}\left[\frac{(\mu_1-x_1)^2}{\sigma_1^2+\sigma_Z^2}\right.\right.\\
&\quad\left.\left.-2\,\rho\,\frac{(\mu_1-x_1)(\mu_2-x_2)}{\sqrt{\sigma_1^2+\sigma_Z^2}\,\sqrt{\sigma_2^2+\sigma_Z^2}}+\frac{(\mu_2-x_2)^2}{\sigma_2^2+\sigma_Z^2}\right]\right\}\,, \qquad (6.37)
\end{aligned}
$$

where

$$
\rho = \frac{\sigma_Z^2}{\sqrt{\sigma_1^2+\sigma_Z^2}\,\sqrt{\sigma_2^2+\sigma_Z^2}}\,. \qquad (6.38)
$$

If σ_Z vanishes then Eq. (6.37) has the simpler expression

$$
f(\mu_1,\mu_2)\xrightarrow[\sigma_Z\to 0]{}\frac{1}{\sqrt{2\,\pi}\,\sigma_1}\,\exp\left[-\frac{(\mu_1-x_1)^2}{2\,\sigma_1^2}\right]\frac{1}{\sqrt{2\,\pi}\,\sigma_2}\,\exp\left[-\frac{(\mu_2-x_2)^2}{2\,\sigma_2^2}\right], \qquad (6.39)
$$

i.e. if there is no uncertainty on the offset calibration then the joint density function $f(\mu_1,\mu_2)$ is equal to the product of two independent normal functions, i.e. μ_1 and μ_2 are independent. In the general case we have to conclude the following.

- The effect of the *common uncertainty* σ_Z makes the two values correlated, since they are affected by a common unknown systematic error.
- The joint density function is a bivariate Gaussian distribution of parameters x_1, $\sigma_{\mu_1} = \sqrt{\sigma_1^2 + \sigma_Z^2}$, x_2, $\sigma_{\mu_2} = \sqrt{\sigma_2^2 + \sigma_Z^2}$, and ρ (see example of Fig. 4.2).
- The marginal distributions are still normal:

$$\mu_1 \sim \mathcal{N}\left(x_1, \sqrt{\sigma_1^2 + \sigma_Z^2}\right), \tag{6.40}$$

$$\mu_2 \sim \mathcal{N}\left(x_2, \sqrt{\sigma_2^2 + \sigma_Z^2}\right). \tag{6.41}$$

- The covariance between μ_1 and μ_2 is

$$\begin{aligned} \mathrm{Cov}(\mu_1, \mu_2) &= \rho\, \sigma_{\mu_1} \sigma_{\mu_2} \\ &= \rho \sqrt{\sigma_1^2 + \sigma_Z^2}\sqrt{\sigma_2^2 + \sigma_Z^2} = \sigma_Z^2\,. \end{aligned} \tag{6.42}$$

- The correlation coefficient is always non-negative ($\rho \geq 0$), as intuitively expected from the definition of this kind of systematic error. The correlation coefficient vanishes when σ_Z is much smaller than σ_1 and σ_2, tends to 1 if σ_Z dominates (the uncertainties become 100% correlated).
- The distribution of any function $g(\mu_1, \mu_2)$ can be calculated using the standard methods of probability theory. For example, one can demonstrate that the sum $S = \mu_1 + \mu_2$ and the difference $D = \mu_1 - \mu_2$ are also normally distributed (see also the introductory discussion to the central limit theorem and Sec. 8.13 for the calculation of averages and standard deviations):

$$S \sim \mathcal{N}\left(x_1 + x_2, \sqrt{\sigma_1^2 + \sigma_2^2 + (2\,\sigma_Z)^2}\right), \tag{6.43}$$

$$D \sim \mathcal{N}\left(x_1 - x_2, \sqrt{\sigma_1^2 + \sigma_2^2}\right). \tag{6.44}$$

  The result can be interpreted in the following way.

  - The uncertainty on the difference does not depend on the common offset uncertainty: whatever the value of the true "zero" is, it cancels in differences.

– In the sum, instead, the effect of the common uncertainty is somewhat amplified since it enters "in phase" in the global uncertainty of each of the quantities.

6.11 Indirect calibration

Let us use the result of the previous section to solve another typical problem of measurements. Suppose that after (or before, it doesn't matter) we have done the measurements of x_1 and x_2 and we have the final result, summarized in Eq. (6.37), we know the "exact" value of μ_1 (for example we perform the measurement on a reference). Let us call it μ_1°. Will this information provide a better knowledge of μ_2? In principle yes: the difference between x_1 and μ_1° defines the systematic error (the true value of the "zero" Z). This error can then be subtracted from x_2 to get a corrected value. Also the overall uncertainty of μ_2 should change, intuitively it "should" decrease, since we are adding new information. But its value doesn't seem to be obvious, since the logical link between μ_1° and μ_2 is $\mu_1^\circ \to Z \to \mu_2$.

The problem can be solved exactly using the concept of conditional probability density function $f(\mu_2 \,|\, \mu_1^\circ)$ [see Eqs. (4.83)-(4.84)]. We get

$$\mu_{2\,|\,\mu_1^\circ} \sim \mathcal{N}\left(x_2 + \frac{\sigma_Z^2}{\sigma_1^2+\sigma_Z^2}\,(\mu_1^\circ - x_1),\; \sqrt{\sigma_2^2 + \left(\frac{1}{\sigma_1^2}+\frac{1}{\sigma_Z^2}\right)^{-1}}\right). \tag{6.45}$$

The best value of μ_2 is shifted by an amount Δ, with respect to the measured value x_2, which is not exactly $x_1 - \mu_1^\circ$, as was naïvely guessed, and the uncertainty depends on σ_2, σ_Z and σ_1. It is easy to be convinced that the exact result is more reasonable than the (suggested) first guess. Let us rewrite Δ in two different ways:

$$\Delta = \frac{\sigma_Z^2}{\sigma_1^2+\sigma_Z^2}\,(\mu_1^\circ - x_1) \tag{6.46}$$

$$= \frac{1}{\frac{1}{\sigma_1^2}+\frac{1}{\sigma_Z^2}}\left[\frac{1}{\sigma_1^2}\cdot(x_1-\mu_1^\circ) + \frac{1}{\sigma_Z^2}\cdot 0\right]. \tag{6.47}$$

- Equation (6.46) shows that one has to apply the correction $x_1 - \mu_1^\circ$ only if $\sigma_1 = 0$. If instead $\sigma_Z = 0$ there is no correction to be applied, since the instrument is perfectly calibrated. If $\sigma_1 \approx \sigma_Z$ the correction is half of the measured difference between x_1 and μ_1°.

- Equation (6.47) shows explicitly what is going on and why the result is consistent with the way we have modelled the uncertainties. In fact we have performed two independent calibrations: one of the offset and one of μ_1. The best estimate of the true value of the "zero" Z is the weighted average of the two measured offsets.
- The new uncertainty of μ_2 [see Eq. (6.45)] is a combination of σ_2 and the uncertainty of the weighted average of the two offsets. Its value is smaller than it would be with only one calibration and, obviously, larger than that due to the sampling fluctuations alone:

$$\sigma_2 \le \sqrt{\sigma_2^2 + \frac{\sigma_1^2\,\sigma_Z^2}{\sigma_1^2 + \sigma_Z^2}} \le \sqrt{\sigma_2^2 + \sigma_Z^2}\,. \tag{6.48}$$

6.12 The Gauss derivation of the Gaussian

It might be interesting to end this chapter in a historical vein, looking at how Gauss arrived at the distribution function which now carries his name. [68] Note that the Gaussian function was already known before Gauss, describing the asymptotical behavior of the binomial distribution, in a purely probabilistic context. The Gauss derivation arose in a more inferential framework and, indeed, Gauss used what we would nowadays call Bayesian reasoning.

Gauss's problem, expressed in modern terms, was: what is the more general form of the likelihood such that the maximum of the posterior of μ is equal to the arithmetic average of the observed values (and the function has some 'good' mathematical properties)?

In solving his problem, Gauss first derived a formula for calculating the probability of hypotheses given some observations had been made, under the assumption of equal prior probability of the hypotheses. In practice, he reobtained Bayes theorem (without citing Bayes) in the case of uniform prior. Note that the concept of prior (*"ante eventum cognitum"*) [4] was very clear and natural to him, opposed to the concept of posterior (*"post eventum cognitum"*). Then moving from discrete hypotheses to continuous observations x_i and true value μ (using our terminology), he looked for the functional form of φ, which describes the probability of obtaining x_i from μ (the likelihood, in our terms). Considering the observations to be

[4] All quotes in Latin are from Ref. [68].

independent, the joint distribution of the sample $\boldsymbol{x}$ is then given by

$$f(\boldsymbol{x}\,|\,\mu) = \varphi(x_1-\mu)\cdot\varphi(x_2-\mu)\cdot\;\cdots\;\cdot\varphi(x_n-\mu)\,. \qquad (6.49)$$

At this point, two hypotheses enter.

(1) All values of μ are considered *a priori* (*"ante illa observationes"*) equally likely (*"... aeque probabilia fuisse"*).
(2) The maximum *a posteriori* (*"post illas observationes"*) is given by $\mu = \overline{x}$, arithmetic average of the n observed values.

The first hypothesis gives

$$f(\mu\,|\,\boldsymbol{x}) \propto f(\boldsymbol{x}\,|\,\mu) = \varphi(x_1-\mu)\cdot\varphi(x_2-\mu)\cdot\;\cdots\;\cdot\varphi(x_n-\mu)\,. \qquad (6.50)$$

To use the second condition, he imposed that the first derivative of the posterior is null for $\mu = \overline{x}$:

$$\left.\frac{\mathrm{d}f(\mu\,|\,\boldsymbol{x})}{\mathrm{d}\mu}\right|_{\mu=\overline{x}} = 0 \;\Longrightarrow\; \left(\frac{\mathrm{d}}{\mathrm{d}\mu}\prod_i \varphi(x_i-\mu)\right)_{\mu=\overline{x}} = 0\,, \qquad (6.51)$$

i.e.

$$\sum_i \frac{\varphi'(x_i-\overline{x})}{\varphi(xi-\overline{x})} = 0\,, \qquad (6.52)$$

where φ' stands for the derivative of ϕ with respect to μ. Calling ψ the function φ'/φ and indicating with $z_i = x_i - \overline{x}$ the differences from the average, which have to follow the constraint $\sum_i z_i = 0$, we have

$$\begin{cases} \sum_i \psi(z_i) = 0 \\ \sum_i z_i = 0 \end{cases}. \qquad (6.53)$$

Since this relation must hold independently of n and the values of z_i, the functional form of $\psi(z)$ has to satisfy the following constraint:

$$\frac{1}{z}\,\psi(z) = k\,, \qquad (6.54)$$

where k is a constant (note that the limit $z \to 0$ is not a problem, for the derivative of φ at $z = 0$ vanishes and the condition $\psi(z)/z = k$ implies that numerator and denominator have to tend to zero with the same speed). It follows that

$$\frac{\mathrm{d}\varphi}{\varphi} = k\,z\,\mathrm{d}z\,,$$

i.e.

$$\varphi(z) \propto e^{\frac{k}{2}\,z^2} = e^{-h^2\,z^2}\,, \tag{6.55}$$

where Gauss replaced $k/2$ by $-h^2$ to make its negative sign evident, because φ is required to have a maximum in $z = 0$. Normalizing the function dividing by its integral from $-\infty$ to ∞, an integral acknowledged to be due to Laplace (*"ab ill. Laplace inventum"*), he finally gets the 'Gauss' error function (*"functio nostra fiet"*):

$$\varphi(z) = \frac{h}{\sqrt{\pi}}\,e^{-h^2\,z^2}\,. \tag{6.56}$$

Chapter 7

Counting experiments

"...have observed that, out of $p+q$ infants,
there is born p boys and q girls, and
...we seek the probability P that, out of $m+n$ infants
who must be born, there will be m boys and n girls.
...The probability that, in one year, the births of boys
will not be by a greater number in Paris than
those of girls, is therefore less than $\frac{1}{259}$;
...we can therefore wager with advantage
one against one that this will not happen in
the interval of one hundred seventy-five years"
(Pierre-Simone Laplace)

Measurement is not only reading a value on a scale, but also counting the occurrence of some events, for example when we are interested in measuring a cross section, an efficiency or a branching ratio in particle decay. The most important models for physics applications are the cases in which the number of counts are thought to be described by a binomial or a Poisson distribution. The purpose of the measurement consists of inferring the value of the parameter of these distributions.

7.1 Binomially distributed observables

Let us assume we have performed n trials and obtained x favorable events. What is the probability of the next event? This situation happens frequently when measuring efficiencies, branching ratios, etc. Stated more generally, one tries to infer the "constant and unknown probability"[1] of an event occurring.

Where we can assume that the probability is constant and the observed number of favorable events are binomially distributed, the unknown quantity to be measured is the parameter p of the binomial [see Eq. (4.18)]. Using Bayes' theorem we get

$$\begin{aligned} f(p \mid x, n, \mathcal{B}) &= \frac{f(x \mid \mathcal{B}_{n,p})\, f_\circ(p)}{\int_0^1 f(x \mid \mathcal{B}_{n,p})\, f_\circ(p)\, \mathrm{d}p} \\ &= \frac{\frac{n!}{(n-x)!\,x!}\, p^x\,(1-p)^{n-x}\, f_\circ(p)}{\int_0^1 \frac{n!}{(n-x)!\,x!}\, p^x\,(1-p)^{n-x}\, f_\circ(p)\, \mathrm{d}p} \\ &= \frac{p^x\,(1-p)^{n-x}}{\int_0^1 p^x\,(1-p)^{n-x}\, \mathrm{d}p}\,, \end{aligned} \qquad (7.1)$$

where an initial uniform distribution has been assumed. The final distribution is known to statisticians as beta distribution (see Sec. 4.2) since the integral at the denominator is the special function called β, defined also for real values of x and n (technically this is a beta with parameters $r = x + 1$

[1] This concept, which is very close to the physicist's mentality, is not correct from the probabilistic — *cognitive* — point of view. According to the Bayesian scheme, in fact, the probability changes with the new observations. The final inference of p, however, does not depend on the particular sequence yielding x successes over n trials. This can be seen in the next table where $f_n(p)$ is given as a function of the number of trials n, for the three sequences which give two successes (indicated by "1") in three trials [the use of Eq. (7.2) is anticipated]:

	Sequence		
n	011	101	110
0	1	1	1
1	$2\,(1-p)$	$2\,p$	$2\,p$
2	$6\,p\,(1-p)$	$6\,p\,(1-p)$	$3\,p^2$
3	$12\,p^2\,(1-p)$	$12\,p^2\,(1-p)$	$12\,p^2\,(1-p)$

This important result, related to the concept of ***exchangeability*** and to de Finetti's ***representation theorem***, [16, 27, 69, 70] "allows" a physicist who is reluctant to give up the concept "unknown constant probability" to see the problem from his point of view, ensuring that the same numerical result is obtained. Note that an approach which practically coincides with that based on exchangeability is used by Schrödinger in Ref. [41] in order to evaluate the probability of the $(n+1)$-th event, without having to speak of $f(p)$. Indeed, he obtains the Laplace's rule of succession (7.9), but, finally, he seems 'afraid' of the result, which *"can only be taken seriously for at least fairly large N, m and N − m"* [41] (these quantities correspond to our $n+1$, $x+1$ and $n-x$). The parametric-inferencial approach that we use solves easily this difficulty, by stating also how much we believe on the parameter p.

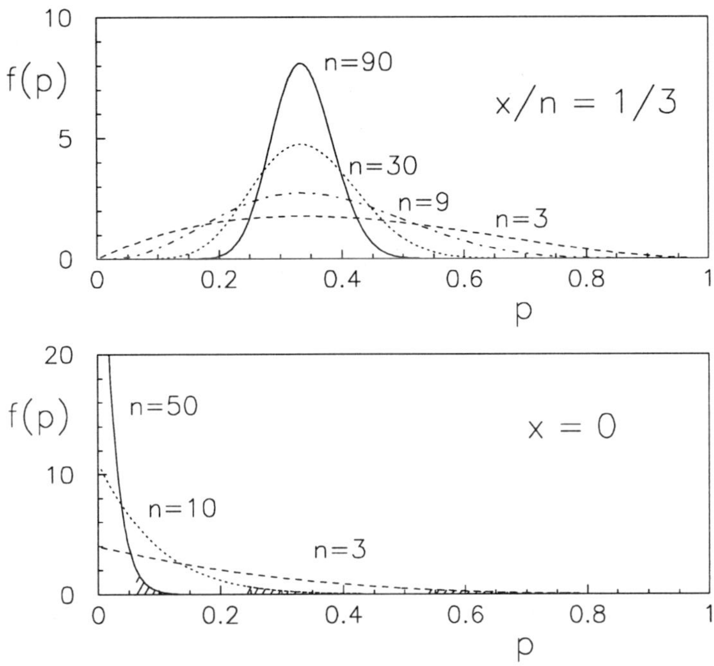

Fig. 7.1 Probability density function of the binomial parameter p, having observed x successes in n trials.

and $s = n - x + 1$). In our case these two numbers are integer and the integral becomes equal to $x!\,(n-x)!/(n+1)!$. We then get

$$f(p\,|\,x,n,\mathcal{B}) = \frac{(n+1)!}{x!\,(n-x)!}\,p^x\,(1-p)^{n-x}\,, \tag{7.2}$$

some examples of which are shown in Fig. 7.1. Expected value and variance of this distribution are:

$$\mathrm{E}(p) = \frac{x+1}{n+2} \tag{7.3}$$

$$\mathrm{Var}(p) = \frac{(x+1)(n-x+1)}{(n+3)(n+2)^2} \tag{7.4}$$

$$= \frac{x+1}{n+2}\left(\frac{n+2}{n+2} - \frac{x+1}{n+2}\right)\frac{1}{n+3}$$

$$= \mathrm{E}(p)\;(1-\mathrm{E}(p))\;\frac{1}{n+3}\,. \tag{7.5}$$

The value of p for which $f(p)$ has the maximum is instead $p_m = x/n$. The expected value E(p) has the meaning of probability of any of the next i-th events E_i $(i > n)$, given the present status of information. In fact, by definition, $P(E_i \,|\, p) = p$. But we are not sure about p. Hence, using probability rules we have:

$$\begin{aligned} P(E_{i>n} \,|\, x, n, \mathcal{B}) &= \int_0^1 P(E_i \,|\, p)\, f(p \,|\, x, n, \mathcal{B})\, \mathrm{d}p && (7.6)\\ &= \int_0^1 p\, f(p \,|\, x, n, \mathcal{B})\, \mathrm{d}p && (7.7)\\ &= \mathrm{E}(p) && (7.8)\\ &= \frac{x+1}{n+2} \quad \text{(for uniform prior)}\,. && (7.9) \end{aligned}$$

Equation (7.9) is known as "recursive Laplace formula", or "Laplace's rule of succession". Not that there is no magic if the formula gives a sensible result even for the extreme cases $x = 0$ and $x = n$ $\forall n$ (even if $n = 0\,!$): it is just a consequence of the prior.

When n, x and $n - x$ become "large" (in practice $\geq \mathcal{O}(10)$ is enough for many practical purposes) $f(p)$ has the following asymptotic properties:

$$\begin{aligned} \mathrm{E}(p) &\approx p_m = \frac{x}{n}\,, && (7.10)\\ \mathrm{Var}(p) &\approx \frac{x}{n}\left(1 - \frac{x}{n}\right)\frac{1}{n} = \frac{p_m\,(1-p_m)}{n}\,, && (7.11)\\ \sigma_p &\approx \sqrt{\frac{p_m\,(1-p_m)}{n}}\,, && (7.12)\\ p &\sim \mathcal{N}(p_m, \sigma_p)\,. && (7.13) \end{aligned}$$

Under these conditions the frequentistic "definition" (evaluation rule!) of probability (x/n) is recovered, but with several advantages: the concept of probability is kept well separated from the evaluation rule; the underlying hypotheses are well stated; we have a precise measure of how uncertain our evaluation is. We shall come back to this point in Sec. 7.3.

Note, finally, that Eq. (7.11) can also be obtained assuming an approximated normal distribution for $f(p)$ and using the general property (4.36), as

$$\mathrm{Var}^{-1}(p) \approx -\left.\frac{\partial^2 \ln f(p \,|\, x, n, \mathcal{B})}{\partial p^2}\right|_{p=x/n}\,. \qquad (7.14)$$

7.1.1 *Observing 0% or 100%*

Going back to practical applications, let us see two particular situations: when $x = 0$ and $x = n$. In these cases one usually gives the result as upper or lower limits, respectively. Let us sketch the solutions:

- x = n:

$$f(n\,|\,\mathcal{B}_{n,p}) = p^n, \tag{7.15}$$

$$f(p\,|\,x=n,\mathcal{B}) = \frac{p^n}{\int_0^1 p^n\,dp} = (n+1)\,p^n, \tag{7.16}$$

$$F(p\,|\,x=n,\mathcal{B}) = p^{n+1}\,. \tag{7.17}$$

To get the 95 % lower bound (*limit*):

$$F(p_\circ\,|\,x=n,\mathcal{B}) = 0.05\,,$$

$$p_\circ = \sqrt[n+1]{0.05}\,. \tag{7.18}$$

An increasing number of trials n constrain more and more p around the upper edge 1.

- x = 0:

$$f(0\,|\,\mathcal{B}_{n,p}) = (1-p)^n, \tag{7.19}$$

$$f(p\,|\,x=0,n,\mathcal{B}) = \frac{(1-p)^n}{\int_0^1 (1-p)^n\,dp} = (n+1)\,(1-p)^n, \tag{7.20}$$

$$F(p\,|\,x=0,n,\mathcal{B}) = 1-(1-p)^{n+1}\,. \tag{7.21}$$

To get the 95 % upper *bound (limit)*:

$$F(p_\circ\,|\,x=0,n,\mathcal{B}) = 0.95,$$

$$p_\circ = 1 - \sqrt[n+1]{0.05}\,. \tag{7.22}$$

The following table shows the 95 % probability limits as a function of n. The Poisson approximation, to be discussed in the next section, is also shown.

	Probability level = 95 %		
n	$x = n$	$x = 0$	
	binomial	binomial	Poisson approx. ($p_\circ = 3/n$)
3	$p \geq 0.47$	$p \leq 0.53$	$p \leq 1$
5	$p \geq 0.61$	$p \leq 0.39$	$p \leq 0.6$
10	$p \geq 0.76$	$p \leq 0.24$	$p \leq 0.3$
50	$p \geq 0.94$	$p \leq 0.057$	$p \leq 0.06$
100	$p \geq 0.97$	$p \leq 0.029$	$p \leq 0.03$
1000	$p \geq 0.997$	$p \leq 0.003$	$p \leq 0.003$

7.1.2 *Combination of independent measurements*

To show in this simple case how $f(p)$ is updated by the new information, let us imagine we have performed two experiments. The results are $x_1 = n_1$ and $x_2 = n_2$, respectively. Obviously the global information is equivalent to $x = x_1 + x_2$ and $n = n_1 + n_2$, with $x = n$. We then get

$$f(p \,|\, x = n, \mathcal{B}) = (n+1)\, p^n = (n_1 + n_2 + 1)\, p^{n_1+n_2}\,. \tag{7.23}$$

A different way of proceeding would have been to calculate the final distribution from the information $x_1 = n_1$,

$$f(p \,|\, x_1 = n_1, \mathcal{B}) = (n_1 + 1)\, p^{n_1}\,, \tag{7.24}$$

and feed it as initial distribution to the next inference:

$$f(p \,|\, x_1 = n_1, x_2 = n_2, \mathcal{B}) = \frac{p^{n_2} f(p \,|\, x_1 = n_1, \mathcal{B})}{\int_0^1 p^{n_2} f(p \,|\, x_1 = n_1, \mathcal{B})\, \mathrm{d}p} \tag{7.25}$$

$$= \frac{p^{n_2} (n_1 + 1)\, p^{n_1}}{\int_0^1 p^{n_2} (n_1 + 1)\, p^{n_1}\, \mathrm{d}p} \tag{7.26}$$

$$= (n_1 + n_2 + 1)\, p^{n_1+n_2}\,, \tag{7.27}$$

getting the same result.

7.1.3 *Conjugate prior and many data limit*

So far, we have used, for simplicity's sake, a uniform prior. If our beliefs differ substantially from those described by a uniform distribution (and we do not have many data, as will be clear in a while), we need to model our beliefs, insert then in the Bayes formula and do the calculations. As we have seen for the Gaussian case (Secs. 6.3 and 6.4) life gets easier if we choose a convenient mathematical form for the prior. The binomial case is particularly fortunate, in the sense that its conjugate prior is easy and flexible.

Apart from the binomial coefficient, $f(x\,|\,p)$ has the shape $p^x(1-p)^{n-x}$, having the same structure as the beta distribution (see Sec. 4.2) with parameters $r = x + 1$ and $s = n - x + 1$. Also the uniform prior is nothing but a beta function with parameters $r = s = 1$ (see Fig. 4.1). In general, if we choose an initial beta function with parameters r_i and s_i the inference will be

$$f(p\,|\,n, x, \text{Beta}(r, s)) \propto \left[p^x(1-p)^{n-x}\right] \times \left[p^{r_i-1}(1-p)^{s_i-1}\right] \quad (7.28)$$

$$\propto p^{x+r_i-1}(1-p)^{n-x+s_i-1}\,. \quad (7.29)$$

The final distribution is still a beta with $r_f = r_i + x$ and $s_f = s_i + (n - x)$, and expected value and standard deviation can be calculated easily from Eqs. (4.55) and (4.56).

Note that, contrary to the Gaussian case, the flexibility of the beta function (Fig. 4.1) allows several models of prior beliefs to be described, without the risk that mathematical convenience forces the solution (like the famous joke of the drunk man under a lamp, looking for his key lost in the darkness...). For example, a prior belief that p must be around 0.5, with 0.05 standard uncertainty corresponds to a beta function with $r_i = 49.5$ and $s_i = 49.5$. In other terms, our initial knowledge is equivalent to that which we would have reached starting from absolute indifference about p (uniform distribution) and having performed 97 trials, about 48-49 of which gave a success. If, given this condition, we perform $n = 10$ trials and register $x = 2$ successes, our knowledge will be updated into a beta of $r_f = 51.5$ and $s_f = 58.5$. The new expected value and uncertainty of p will be 0.472 and 0.048: we do not change our opinion much, although a relative frequency of 20% was observed.

The use of the conjugate prior in this problem demonstrates in a clear way how the inference becomes progressively independent of the prior information in the limit of a large amount of data: this happens when both $x \gg r_i$ and $n - x \gg s_i$. In this limit we get the same result we would get from a flat prior ($r_i = s_i = 1$).

7.2 The Bayes problem

The original "problem in the doctrine of chances", solved by Thomas Bayes [71] using for the first time the reasoning that presently carries his name, belongs to the kind of inferential problems with binomial likelihood. A billiard ball is rolled on a segment of unity length. The mechanism of the game is such that we consider all points p, where the ball might stop, equally likely (it would be more realistic to think of a pointer on a wheel). A second ball is then rolled n times under the same assumptions and somebody tells us the number of times x it stops in a position $p\prime \leq p$. The problem is to infer position p given n and x, for example to say what is the probability that p is between a and b, with $0 \leq a < b \leq 1$. Note that in this problem the uniform prior is specified in the assumptions of the problem.

7.3 Predicting relative frequencies – Terms and interpretation of Bernoulli's theorem

We have seen when and how it is possible to assess a probability using observed relative frequencies. A complementary problem is that of predicting relative frequency f_n in n "future" trials, under the hypothesis that $P(E_i) = p_0\ \forall\, i$. The uncertain number X of successes is described by the binomial distribution. The relative frequency of successes $f_n = X/n$ is an uncertain number too, with a probability function easily obtainable from the binomial one. Expected value and the standard deviation are

$$\mathrm{E}(f_n) \equiv \frac{1}{n}\,\mathrm{E}(X \,|\, \mathcal{B}_{n,p_0}) = \frac{n\,p_0}{n} = p_0 \tag{7.30}$$

$$\sigma(f_n) \equiv \frac{1}{n}\,\sigma((X \,|\, \mathcal{B}_{n,p_0}) = \frac{\sqrt{p_0\,(1-p_0)}}{\sqrt{n}}\,. \tag{7.31}$$

This result is at the basis of the well-known and often misunderstood *Bernoulli's theorem* (one of the "large number laws"): "as the number of trials is very large, we consider it highly improbable to observe values of f_n which differ much from p_0," or, if you like, "it is *practically certain* that the frequency becomes *practically equal* to the probability" [16]. The simplified expression "the relative frequency tends to the probability" might give the (wrong) idea that f_n tends to p_0 in a mathematical sense, like in the definition of a limit. Instead, for any n the range of f_n is always [0,1], though for large n we are practically sure that the extreme values will not be observed. Some comments are in order:

- The theorem does not imply any kind of "memory" of the trial mechanism that would influence future events to make the long term results "obey the large number law".[2] Let us take as an example an urn containing 70% white balls. We plan to make n extractions and have already made n_0 extractions, observing a relative frequency of white balls f_{W_0}. We are interested in the relative frequency we expect to observe when we reach n extractions. The crucial observation is that we can make probabilistic considerations only about the remaining $n - n_0$ extractions, the previous n belonging to the realm of certainty to which probability theory does not apply. Indicating by the subscript 1 the quantities referring to the remaining extractions, we have[3]

$$E[f_{W_1}] = p_0 \tag{7.32}$$

$$\sigma(f_{W_1}) = \frac{\sqrt{p_0\,(1-p_0)}}{\sqrt{n_1}}\,. \tag{7.33}$$

  Note, however, that the prevision of the relative frequency of the entire ensemble is in general different from that calculated a priori. Calling x_1 the uncertain number of favorable results in the next n_1 trials, we have the uncertain frequency $f_W = (f_{W_0}\, n_0 + x_1)/N$, and hence

$$E[f_W \,|\, n_0] = \frac{f_{W_0}\, n_0 + p_0\, n_1}{n} = \frac{f_{W_0}\, n_0 + p_0\,(n-n_0)}{n} \tag{7.34}$$

$$\sigma(f_W \,|\, n_0) = \sqrt{p_0\,(1-p_0)}\,\sqrt{\frac{n_1}{n}} = \sqrt{p_0\,(1-p_0)}\,\sqrt{\frac{n-n_0}{n}}\,. \tag{7.35}$$

  As n_0 approaches n, we are practically sure about the overall relative frequency, because it now belongs to the past.

---

[2] I think that the name 'law' itself should be avoided, because it gives the strong feeling of something to which nature has to obey, like in the case of Newton laws. Once I met a person who was disappointed about my interest in probability and statistics because *"the laws of statistics seems to me absolutely silly. Why, playing lotto, should one number have a higher chance of being extracted than the other numbers, if it has not shown up in the previous weeks?"*. Unfortunately, the number of reasonable persons of this kind is small, and most people believe in the mass-media interpretation of statistics 'laws'.

[3] Note that the reason why, contrary to what we have seen above, the observation of the relative frequency f_{W_0} does not change our belief in the future event is because in this example we are sure about the urn composition. This is equivalent to applying Bayes' theorem with a prior $f_\circ(p) = \delta(p - p_0)$, where $\delta()$ is the Dirac delta.

- Bernoulli's theorem cannot be used to justify the frequentistic definition of probability, since it is a theorem of probability theory, and hence cannot be used to define the basic concept of that theory, as sharply pointed out by de Finetti [16]:

> *"For those who seek to connect the notion of probability with that of frequency, results which relate probability and frequency in some way (and especially those results like the 'law of large numbers') play a pivotal rôle, providing support for the approach and for the identification of the concepts. Logically speaking, however, one cannot escape from the dilemma posed by the fact that the same thing cannot both be assumed first as a definition and then proved as a theorem; nor can one avoid the contradiction that arises from a definition which would assume as certain something that the theorem only states to be very probable."*

- Another law sometimes claimed to connect frequency with probability is the 'empirical law of chance'. In simple words, it says that "in all cases in which we are able to evaluate probability by symmetry and we can perform a large number of experiments, we note that the relative frequency approaches the probability." This 'law' is meaningless. It is just empirical evidence that Bernoulli's theorem 'works', since large deviations are very rare.

After having analyzed the special case of a precise value of $p = p_0$, let us see what happens if we include our uncertainty about it. Using the general rules of probability we get, for the number of successes X:

$$f(x) = \int_0^1 f(x \mid p)\, f(p)\, \mathrm{d}p\,. \tag{7.36}$$

Following what we did in Sec. 6.6, let us assume that our knowledge about p comes from a previous experiment of n_0 trials in which $X_0 = x_0$ successes were recorded (and before that experiment we considered all values of p equally likely). Thinking of n_1 future trials, our beliefs about the number

Table 7.1 Example of predictive distribution of the number of successes in n_1 trials, having observed x_0 successes in n_0 previous trials performed under the same conditions.

		$f(x_1 \mid n_0, x_0, n_1 = 10)$ in %			
X_1	$\frac{X_1}{n_1}$	$\begin{cases} x_0 = 1 \\ n_0 = 2 \end{cases}$	$\begin{cases} x_0 = 10 \\ n_0 = 20 \end{cases}$	$\begin{cases} x_0 = 100 \\ n_0 = 200 \end{cases}$	$\begin{cases} x_0 = 1000 \\ n_0 = 2000 \end{cases}$
0	0	3.85	0.42	0.12	0.10
1	0.1	6.99	2.29	1.11	0.99
2	0.2	9.44	6.51	4.67	4.42
3	0.3	11.19	12.54	11.88	11.74
4	0.4	12.24	18.07	20.21	20.48
5	0.5	12.59	20.33	24.02	24.55
6	0.6	12.24	18.07	20.21	20.48
7	0.7	11.19	12.54	11.88	11.74
8	0.8	9.44	6.51	4.67	4.42
9	0.9	6.99	2.29	1.11	0.99
10	1	3.84	0.42	0.12	0.10
$\mathrm{E}(X_1)$		5	5	5	5
$\sigma[X_1]$		2.64	1.87	1.62	1.58

of successes X_1 that we shall observe are given by

$$f(x_1 \mid n_0, x_0, n_1) = \int_0^1 \frac{n_1!}{x_1!\,(n_1 - x_1)!}\, p^{x_1} (1-p)^{n_1 - x_1} \times \frac{(n_0+1)!}{x_0!\,(n_0 - x_0)!}\, p^{x_0} (1-p)^{n_0 - x_0}\, \mathrm{d}p \tag{7.37}$$

$$= \frac{n_1!\,(n_0+1)!\,(x_0+x_1)!\,(n_0+n_1-x_0-x_1)!}{x_1!\,(n_1-x_1)!\,x_0!\,(n_0-x_0)!\,(n_0+n_1+1)!}\,. \tag{7.38}$$

This formula allows a straightforward calculation of the probability of the relative frequency X_1/n_1. Table 7.1 shows $f(x_1 \mid n_0, x_0, n_1)$ for $n_1 = 10$ and some values of n_0 and x_0 such that $x_0/n_0 = 1/2$. Expected value and standard deviation of X_1 are also shown (expected value and standard deviation of the relative frequencies are obtained dividing these numbers by $n_1 = 10$).

For large values of n_0 the distribution of X_1 tends to a binomial distribution of $p = 1/2$, consistent with the fact that we are becoming *practical certain* about the value of p. For small values of n_0 the distribution is broader than the binomial. Equation (7.38) can easily be extended to the more general case which takes into account a prior of p described by a beta function of parameters r_i and s_i (see Sec. 7.1.3), obtaining

$$f(x_1 \,|\, n_0, x_0, n_1, r_i, s_i) = \frac{n_1!\,(n_0+r_i+s_i-1)!\,(x_0+x_1+r_i-1)!}{x_1!\,(n_1-x_1)!\,(x_0+r_i-1)!\,(n_0-x_0+s_i-1)!} \times \frac{(n_0+n_1-x_0-x_1+s_i-1)!}{(n_0+n_1+r_i+s_i-1)!}\,. \tag{7.39}$$

Finally, when the Gaussian approximations hold [i.e. n_0, x_0, $n_0 - x_0$, n_1, $n_1\, x_0/n_0$ and $n_1\,(1 - x_0/n_0)$ are all larger than $\mathcal{O}(10)$] we recover Eq. (6.16), with x_f referring to the number of successes x_1, and where $x_p = x_0$, $\sigma_p = \sqrt{x_0/n_0\,(1-x_0/n_0)\,n_0}$ and $\sigma_f = \sqrt{x_0/n_0\,(1-x_0/n_0)\,n_1}$. Passing to the relative frequencies and making use of Eqs. (6.17)–(6.18), we get

$$\mathrm{E}(f_{n_1}) = \mathrm{E}\left(\frac{X_1}{n_1}\right) = \frac{x_0}{n_0} = f_{n_0} \tag{7.40}$$

$$\sigma(f_{n_1}) = \sigma\left(\frac{X_1}{n_1}\right) = \sqrt{f_{n_0}(1-f_{n_0})\left(\frac{1}{n_0}+\frac{1}{n_1}\right)}\,, \tag{7.41}$$

which tend to Eqs. (7.30)-(7.31) for $n_0 \to \infty$ (calling, in the latter equations, n the number of future trials, and identifying p_0 with f_{n_0}).

7.4 Poisson distributed observables

As is well known, the typical application of the Poisson distribution is in counting experiments such as source activity, cross-sections, etc. The unknown parameter to be inferred is λ [see Eq. (4.22)]. Applying the Bayes formula we get

$$f(\lambda \,|\, x, \mathcal{P}) = \frac{\frac{\lambda^x\, e^{-\lambda}}{x!}\, f_\circ(\lambda)}{\int_0^\infty \frac{\lambda^x\, e^{-\lambda}}{x!}\, f_\circ(\lambda)\, d\lambda}\,. \tag{7.42}$$

Assuming $f_\circ(\lambda)$ constant up to a certain $\lambda_{max} \gg x$ and making the integral by parts we obtain

$$f(\lambda \,|\, x, \mathcal{P}) = \frac{\lambda^x\, e^{-\lambda}}{x!} \tag{7.43}$$

$$F(\lambda \,|\, x, \mathcal{P}) = 1 - e^{-\lambda}\left(\sum_{n=0}^{x} \frac{\lambda^n}{n!}\right), \tag{7.44}$$

where the last result has been obtained by integrating Eq. (7.43) also by parts. Figure 7.2 shows some numerical examples. $f(\lambda)$ has the following

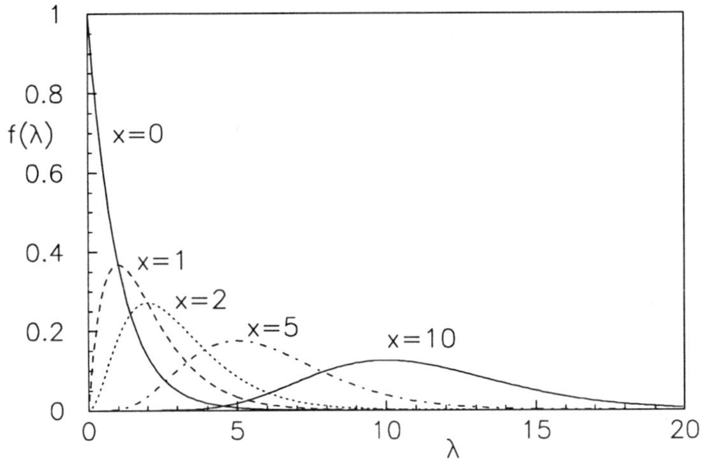

Fig. 7.2 Examples of $f(\lambda\,|\,x_i)$.

properties.

- Expected value, variance and mode of the probability distribution are

$$E(\lambda) = x + 1, \tag{7.45}$$

$$\mathrm{Var}(\lambda) = x + 1, \tag{7.46}$$

$$\lambda_m = x\,. \tag{7.47}$$

The fact that the best estimate of λ in the Bayesian sense is not the intuitive value x but $x+1$ should neither surprise, nor disappoint us. First, we should be used to distinguish of maximum belief (mode) from expected value (average) and "central value" (median). The reason why the expected value is shifted by one with respect to the mode is due to the uniform prior and the fact that λ (contrary to μ of the Gaussian) is limited on the left side to 0. In a certain sense, "there are always more possible values of λ on the right side than on the left side of x", and they pull the distribution to their side (the expression in quotation marks because we are dealing with infinites). Moreover, we should not forget that the full information is always given by $f(\lambda)$ and the use of the average is just a rough approximation. Finally, one as to notice that the difference between expected value and mode of λ, expressed in units of the standard deviation, is $1/\sqrt{x+1}$, and becomes immediately negligible with increasing x.

- When x becomes large we get

$$\begin{aligned} \mathrm{E}(\lambda) &\approx \lambda_m = x\,, &(7.48)\\ \mathrm{Var}(\lambda) &\approx \lambda_m = x\,, &(7.49)\\ \sigma_\lambda &\approx \sqrt{x}\,, &(7.50)\\ \lambda &\sim \mathcal{N}(x, \sqrt{x})\,. &(7.51) \end{aligned}$$

Equation (7.50) is one of the most familiar formulae used by physicists to assess the uncertainty of a measurement, although it is sometimes misused. As we have seen for the binomial case [Eq. (7.14)], Eq. (7.49) can be easily obtained assuming Gaussian approximation of $f(\lambda \,|\, x, \mathcal{P})$, i.e.

$$\mathrm{Var}^{-1}(\lambda) \approx -\left.\frac{\partial^2 \ln f(\lambda \,|\, x, \mathcal{P})}{\partial \lambda^2}\right|_{\lambda = x}. \qquad (7.52)$$

7.4.1 *Observation of zero counts*

Let us analyze the special case in which no event has been observed, i.e. $x = 0$. First, it should be clear that our state of information is not equivalent with not having performed the experiment. Very high values of λ are certainly ruled out. However, the precise inference on the low values of λ becomes highly sensitive to our priors. This is the typical problem in which there is no single objective solution, and we shall come back to it in Chapter 13. For the moment, instead of playing at random with mathematics, let us assume that *the experiment was planned with the hope of observing* something, i.e. that it could detect a handful of events within its lifetime. With this hypothesis one may use any vague prior function not strongly peaked at zero. We have already come across a similar case in Sec. 6.7, concerning the upper limit of the neutrino mass. There it was shown that reasonable hypotheses based on the *positive attitude* of the experimentalist are almost equivalent and that they give results consistent with detector performances. Let us use then the uniform distribution

$$\begin{aligned} f(\lambda \,|\, x = 0, \mathcal{P}) &= e^{-\lambda}, &(7.53)\\ F(\lambda \,|\, x = 0, \mathcal{P}) &= 1 - e^{-\lambda}, &(7.54)\\ &\lambda < 3 \text{ at } 95\,\% \text{ probability}\,. &(7.55) \end{aligned}$$

Note that many researchers are convinced (as I also was some years ago)

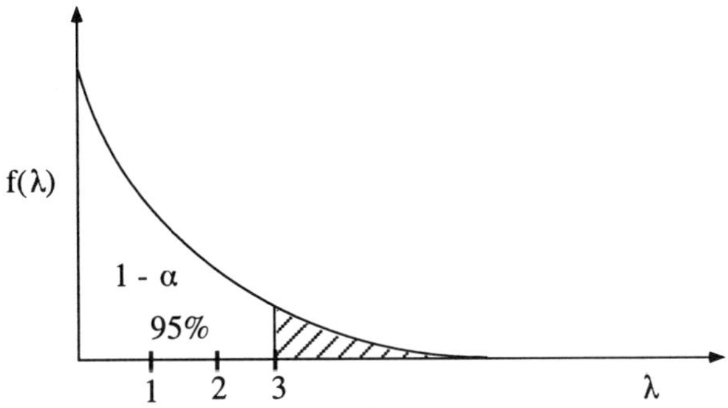

Fig. 7.3 Upper limit to λ having observed 0 events.

that this 95% probability limit is due to the fact that $f(x = 0 \,|\, \mathcal{P}_{\lambda=3}) = 0.05$ (the kind of arbitrary probability inversions criticized in Chapter 1). Instead, this is just a numerical coincidence, due to the known property of the exponential function under integration. What is bad is that the same reasoning is extended to cases in which this coincidence does not hold. [As a trivial example, think of a binomial likelihood with n and zero events observed. The value p_u, such that $\int_{p_u}^{1} f(p \,|\, \mathcal{B}_n, x = 0)\,\mathrm{d}p = 0.05$, does not imply, in general, that $f(x = 0 \,|\, \mathcal{B}_{np_u}) = 0.05$.]

7.5 Conjugate prior of the Poisson likelihood

Formally, the prior conjugate of the Poisson likelihood is given by the gamma distribution (Sec. 4.2), though this distribution is not as flexible as the beta met in the binomial case (Sec. 7.1.3). We have in fact,

$$\begin{aligned} f(\lambda \,|\, x, \mathrm{Gamma}(c_i, r_i)) &\propto \left[\lambda^x e^{-\lambda}\right] \times \left[\lambda^{c_i - 1} e^{-r_i\,\lambda}\right] \\ &\propto \lambda^{x + c_i - 1} e^{-(r_i + 1)\,\lambda}\,, \end{aligned} \tag{7.56}$$

where c_i and r_i are the initial parameters of the gamma distribution. The final distribution is therefore a gamma with $c_f = c_i + x$ and $r_f = r_i + 1$. The case of a flat prior is reobtained for $c = 1$ and $r \to 0$.

7.6 Predicting future counts

We have already seen predictive distributions for the Gaussian and binomial cases (see Secs. 6.6 and 7.3, respectively). Repeating the same reasoning for the Poisson case, and taking directly into account a conjugate prior modelled by a gamma distribution, we have

$$\begin{aligned} f(x_1 \mid x_0, c_i, r_i) &= \int_0^\infty f(x_1 \mid \lambda) f(\lambda \mid x_0, c_i, r_i) \, \mathrm{d}\lambda \\ &= \frac{\Gamma(x_0 + x_1 + c_i)}{x_1! \, \Gamma(x_0 + c_i)} \frac{(1 + r_i)^{x_0 + c_i}}{(2 + r_i)^{x_0 + x_1 + c_i}} \\ &\quad [\text{with } r_i > -2 \text{ and } c_i > -(x_0 + x_1)] \,, \end{aligned} \tag{7.57}$$

where x_0 refers to the past number of counts and x_1 to the future one, and the [always satisfied, see Eq. (4.45)] conditions on c_i and r_i are important in order to get the integral in such a closed form. In this case, expected value and variance also have a closed form:

$$\mathrm{E}(X_1) = \frac{x_0 + c_i}{1 + r_i} \xrightarrow[\textit{flat prior}]{} x_0 + 1 \tag{7.58}$$

$$\sigma(X_1) = \frac{\sqrt{2 + r_i}}{1 + r_i} \sqrt{c_i + x_0} \xrightarrow[\textit{flat prior}]{} \sqrt{2}\,\sqrt{x_0 + 1} \tag{7.59}$$

As usual, when x_0 is large, the Gaussian approximation holds, and we recover, once again, Eq. (6.16), with x_f referring to the number of successes x_1, and where $x_p \approx x_0$, $\sigma_p = \sigma_f \approx \sqrt{x_0}$. It follows that $\mathrm{E}(X_1) \approx x_0$ and $\sigma(X_1) \approx \sqrt{2}\,\sqrt{x_0}$. Note the $\sqrt{2}$ factor. It is not correct to state that, if we have observed 100 counts in an experiment, we are 68% confident of observing $\approx 100 \pm 10$ in another experiment performed under the same conditions: the range should be ± 14.

7.7 A deeper look to the Poissonian case

7.7.1 *Dependence on priors — practical examples*

One may worry how much the result changes if different priors are used in the analysis. Bearing in mind the rule of coherence, we are clearly interested only in reasonable[4] priors.

In frontier physics the choice of $f_\circ(\lambda) = k$ is often not reasonable. For example, searching for magnetic monopoles,[5] one does not believe that

[4] I insist on the fact that they must be reasonable, and not just any prior. The fact that absurd priors give absurd results does not invalidate the inferential framework based on subjective probability.

[5] Many formulae of this and the following sections have been derived to answer some question from colleagues of the MACRO underground experiment at Gran Sasso Laboratory, Italy.

$\lambda = 10^6$ and $\lambda = 1$ are equally possible. Realistically, one would expect to observe, with the planned experiment and running time, $\mathcal{O}(10)$ monopoles, if they exist at all. We follow the same arguments of Sec. 6.7 (negative neutrino mass), modelling the prior beliefs of a community of rational people who have planned and run the experiment. For reasons of mathematical convenience, we model $f_\circ(\lambda)$ with an exponential, but, extrapolating the results of Sec. 6.7, it is easy to understand that the exact function is not really crucial for the final result. The function

$$f_\circ(\lambda) = \frac{1}{10}\, e^{-\lambda/10}\,, \tag{7.60}$$

with $\mathrm{E}_\circ(\lambda) = 10$ and $\sigma_\circ(\lambda) = 10$ may be well suited to the case: the highest beliefs are for small values of λ, but also values up to 30 or 50 would not be really surprising. We obtain the following results:

$$f(\lambda\,|\,x=0) = \frac{e^{-\lambda}\frac{1}{10}\,e^{-\lambda/10}}{\int_0^\infty (\ldots)\,\mathrm{d}\lambda} \tag{7.61}$$

$$= \frac{11}{10}\,e^{-\frac{11}{10}\lambda} \tag{7.62}$$

$$\mathrm{E}(\lambda) = 0.91$$

$$P(\lambda \le 2.7) = 95\%$$

$$\lambda_u = 2.7 \text{ with } 95\% \text{ probability}\,. \tag{7.63}$$

The result is very stable. Changing $\mathrm{E}_\circ(\lambda)$ from '∞' to 10 has only a 10% effect on the upper limit. As far as the scientific conclusions are concerned, the two limit are "identical". For this reason one should not worry about using a uniform prior, instead of complicating one's life to model a more realistic prior.

As an exercise, we can extend this result to a generic expected value of events, still sticking to the exponential:

$$f_\circ(\lambda) = \frac{1}{\lambda_\circ} e^{-\lambda/\lambda_\circ}\,,$$

which has an expected value $\lambda_\circ$. The uniform distribution is recovered for

$\lambda_\circ \to \infty$. We get:

$$
\begin{aligned}
f(\lambda \,|\, x=0, \lambda_\circ) &\propto e^{-\lambda} \frac{1}{\lambda_\circ} e^{-\lambda/\lambda_\circ} \\
f(\lambda \,|\, x=0, \lambda_\circ) &= (1+\lambda_\circ)\, e^{-\lambda\,(1+\lambda_\circ)/\lambda_\circ} \\
&= \frac{1}{\lambda_1} e^{-\lambda/\lambda_1} \\
\text{with } \frac{1}{\lambda_1} &= \frac{1}{1} + \frac{1}{\lambda_\circ} \\
F(\lambda \,|\, x=0, \lambda_\circ) &= 1 - e^{-\lambda/\lambda_\circ}\,.
\end{aligned}
$$

The upper limit, at a probability level P_u, becomes:

$$
\lambda_u = -\lambda_1 \, \ln(1 - P_u)\,. \tag{7.64}
$$

7.7.2 *Combination of results from similar experiments*

As seen for the cases of Gaussian and binomial likelihoods, results may be combined in a natural way making an sequential use of Bayesian inference. As a first case we assume several experiments having the same efficiency and exposure time.

- Prior knowledge:

$$
f_\circ(\lambda \,|\, I_\circ)\,;
$$

- Experiment 1 provides Data$_1$:

$$
f_1(\lambda \,|\, I_\circ, \text{Data}_1) \propto f(\text{Data}_1 \,|\, \lambda, I_\circ)\, f_\circ(\lambda \,|\, I_\circ)\,;
$$

- Experiment 2 provides Data$_2$:

$$
f_2(\lambda \,|\, I_\circ, \text{Data}_1 \ldots) \propto f(\text{Data}_2 \,|\, \lambda, I_\circ)\, f_1(\lambda \,|\, \ldots)\,;
$$

$$
\Rightarrow f_2(\lambda \,|\, I_\circ, \text{Data}_1, \text{Data}_2)\,.
$$

- Combining n similar independent experiments we get

$$\begin{aligned} f(\lambda \mid \boldsymbol{x}) &\propto \prod_{i=1}^{n} f(x_i \mid \lambda)\, f_\circ(\lambda) \\ &\propto f(\underline{x} \mid \lambda)\, f_\circ(\lambda) \\ &\propto \prod_{i=1}^{n} \frac{e^{-\lambda}\lambda^{x_i}}{x_i!}\, f_\circ(\lambda) \\ &\propto e^{-n\,\lambda}\lambda^{\sum_{i=1}^{n} x_i} f_\circ(\lambda)\,. \end{aligned} \tag{7.65}$$

Then it is possible to evaluate expected value, standard deviation, and probability intervals.

As an exercise, let us analyze the two extreme cases, starting from a uniform prior:

Zero observations: if none of the n similar experiments has observed events we have

$$\begin{aligned} f(\lambda \mid n \text{ expts}, 0 \text{ evts}) &= n e^{-n\,\lambda} \\ F(\lambda \mid n \text{ expts}, 0 \text{ evts}) &= 1 - e^{-n\,\lambda} \\ \lambda_u &= -\frac{\ln(1-P_u)}{n} \quad \text{with probability } P_u\,. \end{aligned}$$

Large number of counts: If the number of observed events is large (and the prior flat), the result will be normally distributed:

$$f(\lambda) \sim \mathcal{N}(\mu_\lambda, \sigma_\lambda)\,.$$

Then, in this case it is convenient to evaluate expected value and standard deviation using general properties of the (multi-variate) Gaussian distribution shown in Sec. 4.3. (This is equivalent to recover a well-known maximum likelihood result, but under well-stated assumptions and with a more natural interpretation of the result, as discussed in Sec. 2.9.) From the maximum of $f(\lambda)$, in correspondence of $\lambda = \lambda_m$, we easily get:

$$\mu_\lambda = \mathrm{E}(\lambda) \approx \lambda_m = \frac{\sum_{i=1}^{n} x_i}{n}\,,$$

and from the second derivative of $\ln f(\lambda)$ around the maximum:

$$\left.\frac{\partial^2 \ln f(\lambda)}{\partial \lambda^2}\right|_{\lambda_m} = \frac{-n^2}{\sum_{i=1}^n x_i}$$

$$\sigma_\lambda^2 \approx -\left(\left.\frac{\partial^2 \ln f(\lambda)}{\partial \lambda}\right|_{\lambda_m}\right)^{-1} = \frac{1}{n}\frac{\sum_{i=1}^n x_i}{n}$$

$$\sigma_\lambda \approx \frac{\sqrt{\mu_\lambda}}{\sqrt{n}}\,.$$

7.7.3 *Combination of results: general case*

The previous case is rather artificial and can be used, at most, to combine several measurements of the same experiment repeated n times, each with the same running time. In general, experiments differ in size, efficiency, and running time. A result on λ is no longer meaningful. The quantity which is independent of these contingent factors is the rate, related to λ by

$$r = \frac{\lambda}{\epsilon\, S\, \Delta T} = \frac{\lambda}{\mathcal{L}}\,,$$

where ϵ indicates the efficiency, S the generic 'size' (either area or volume, depending on whatever is relevant for the kind of detection) and ΔT the running time: all the factors have been grouped into a generic 'integrated luminosity' $\mathcal{L}$ which quantify the effective exposure of the experiment.

As seen in the previous case, the combined result can be achieved using Bayes' theorem sequentially, but now one has to pay attention to the fact that:

- the observable is Poisson distributed, and each experiment can infer a λ parameter;
- the result on λ must be translated[6] into a result on r.

Starting from a prior on r (e.g. a monopole flux) and going from experiment 1 to n we have

[6]This two-step inference is not really needed, but it helps to follow the inferential flow. One could think more directly of

$$f(x\,|\,r, \mathcal{L}_i) = \frac{e^{-r\,\mathcal{L}_i}(r\,\mathcal{L}_i)^x}{x!}\,.$$

- from $f_\circ(r)$ and $\mathcal{L}_1$ we get $f_\circ(\lambda)$; then, from the data we perform the inference on λ and then on r:

$$\begin{aligned} f_\circ(r)\,\&\,\mathcal{L}_1 &\rightarrow f_{\circ_1}(\lambda) \\ \text{Data}_1 &\rightarrow f_1(\lambda\,|\,\text{Data}_1, f_{\circ_1}(\lambda)) \\ &\rightarrow f_1(r\,|\,\text{Data}_1, \mathcal{L}_1, f_\circ(r))\,. \end{aligned}$$

- The process is repeated for the second experiment:

$$\begin{aligned} f_1(r)\,\&\,\mathcal{L}_2 &\rightarrow f_{\circ_2}(\lambda) \\ \text{Data}_2 &\rightarrow f_2(\lambda\,|\,\text{Data}_2, f_{\circ_2}(\lambda)) \\ &\rightarrow f_2(r\,|\,\text{Data}_2, \mathcal{L}_2, f_1(r)) \\ &\rightarrow f_2(r\,|\,(\text{Data}_1, \mathcal{L}_1), (\text{Data}_2, \mathcal{L}_2), f_\circ(r))\,, \end{aligned}$$

- and so on for all the experiments.

Let us see in detail the case of null observation in all experiments ($\boldsymbol{x} = \boldsymbol{0} = \{0, 0, \ldots, 0\}$), starting from a uniform distribution.

Experiment 1:

$$\begin{aligned} f_1(\lambda\,|\,x_1 = 0) &= e^{-\lambda} \\ f_1(r\,|\,x_1 = 0) &= \mathcal{L}_1 e^{-\mathcal{L}_1 r} \qquad (7.66) \\ r_{u_1} &= \frac{-\ln 0.05}{\mathcal{L}_1} \text{ at 95\% probability}\,. \qquad (7.67) \end{aligned}$$

Experiment 2:

$$\begin{aligned} f_{\circ_2} &= \frac{\mathcal{L}_1}{\mathcal{L}_2} e^{-\frac{\mathcal{L}_1}{\mathcal{L}_2}\lambda} \\ f_2(\lambda\,|\,x_2 = 0) &\propto e^{-\lambda} \frac{\mathcal{L}_1}{\mathcal{L}_2} e^{-\frac{\mathcal{L}_1}{\mathcal{L}_2}\lambda} \propto e^{-\left(1+\frac{\mathcal{L}_1}{\mathcal{L}_2}\right)\lambda} \\ f_2(r\,|\,x_1 = x_2 = 0) &= (\mathcal{L}_1 + \mathcal{L}_2)\, e^{-(\mathcal{L}_1+\mathcal{L}_2)\, r}. \end{aligned}$$

Experiment n:

$$f_n(r\,|\,\boldsymbol{x} = \boldsymbol{0}, f_\circ(r) = k) = \sum_i \mathcal{L}_i e^{-\sum_i \mathcal{L}_i r}\,. \qquad (7.68)$$

The final result is insensitive to the data grouping. As the intuition suggests, many experiments give the same result of a single experiment with equivalent luminosity. To get the upper limit, we calculate, as usual, the cumulative distribution and require a certain probability P_u for r to be below r_u [i.e. $P_u = P(r \leq r_u)$]:

$$\begin{aligned} F_n(r \,|\, \boldsymbol{x} = \boldsymbol{0}, f_\circ(r) = k) &= 1 - e^{-\sum_i \mathcal{L}_i r} \\ r_u &= \frac{-\ln(1-P_u)}{\sum_i \mathcal{L}_i} \\ \frac{1}{r_u} &= \frac{-\sum_i \mathcal{L}_i}{\ln(1-P_u)} \\ &= \sum_i \frac{-\mathcal{L}_i}{\ln(1-P_u)} \\ &= \sum_i \frac{1}{r_{u_i}}, \end{aligned}$$

obtaining the following rule for the combination of upper limits on rates:

$$\frac{1}{r_u} = \sum_i \frac{1}{r_{u_i}}. \tag{7.69}$$

We have considered here only the case in which no background is expected, but it is not difficult to take background into account, following what has been said in Sec. 7.7.5.

7.7.4 *Including systematic effects*

A last interesting case is when there are systematic errors of uncertain size in the detector performance. Independently of where systematic errors may enter, the final result will be an uncertainty on $\mathcal{L}$. In the most general case, the uncertainty can be described by a probability density function:

$$f(\mathcal{L}) = f(\mathcal{L} \,|\, \text{best knowledge on experiment})\,.$$

For simplicity we analyze here only the case of a single experiment. In the case of many experiments, we only need to use the Bayesian inference several times, as has often been shown in the previous chapters.

Following the general lines given in Sec. 2.10.3, the problem can be solved by considering the conditional probability, obtaining :

$$f(r \,|\, \text{Data}) = \int f(r \,|\, \text{Data}, \mathcal{L})\, f(\mathcal{L})\, \mathrm{d}\mathcal{L}\,. \tag{7.70}$$

The case of absolutely precise knowledge of $\mathcal{L}$ is recovered when $f(\mathcal{L})$ is a Dirac delta.

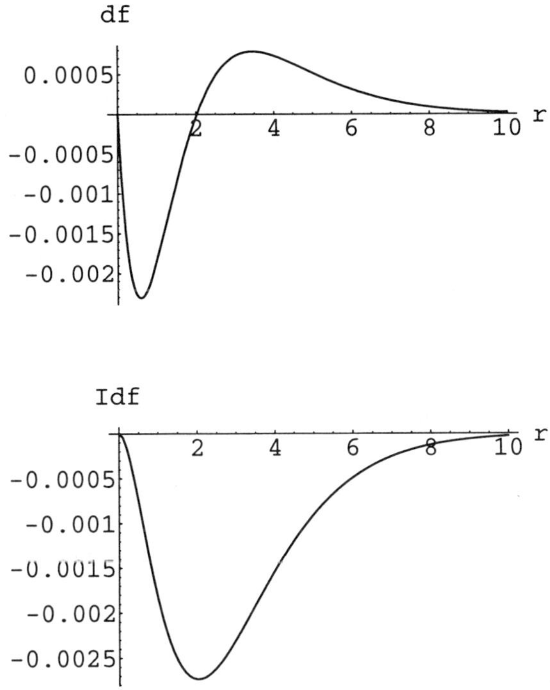

Fig. 7.4 Inference on the rate of a process, with and without taking into account systematic effects: upper plot: difference between $f(r\,|\,x=0,\mathcal{L}=1.0\pm 0.1)$ and $f(r\,|\,x=0,\mathcal{L}=1\pm 0)$, using a normal distribution of $\mathcal{L}$; lower plot: integral of the difference, to give a direct idea of the variation of the upper limit.

Let us treat in some more detail the case of null observation ($\boldsymbol{x}=\mathbf{0}$). For each possible value of $\mathcal{L}$ one has an exponential of expected value $1/\mathcal{L}$ [see Eq. (7.66)]. Each of the exponentials is weighted with $f(\mathcal{L})$. This means that, if $f(\mathcal{L})$ is rather symmetrical around its barycenter (expected value), in a first approximation the more or less steep exponentials will compensate, and the result of integral (7.70) will be close to $f(r)$ calculated in the barycenter of $\mathcal{L}$, i.e. in its nominal value $\mathcal{L}_\circ$:

$$f(r\,|\,\text{Data}) = \int f(r\,|\,\text{Data},\mathcal{L})\,f(\mathcal{L})\,\mathrm{d}\mathcal{L} \approx f(r\,|\,\text{Data},\mathcal{L}_\circ)$$

$$r_u\,|\,\text{Data} \approx r_u\,|\,\text{Data},\mathcal{L}_\circ\,.$$

To make a numerical example, let us consider $\mathcal{L}=1.0\pm 0.1$ (arbitrary units), with $f(\mathcal{L})$ following a normal distribution. The upper plot of Fig. 7.4 shows the difference between $f(r\,|\,\text{Data})$ calculated applying Eq. (7.70) and the result obtained with the nominal value $\mathcal{L}_\circ = 1$:

$$df = f(r\,|\,x=0, f(\mathcal{L})) - f(r\,|\,x=0, \mathcal{L}=1.0) \tag{7.71}$$

$$= \int f(r\,|\,x=0,\mathcal{L})\, f(\mathcal{L})\, \mathrm{d}\mathcal{L} - e^{-r}\,. \tag{7.72}$$

df is negative up to $r \approx 2$, indicating that systematic errors normally distributed tend to increase the upper limit. But the size of the effect is very tiny, and depends on the probability level chosen for the upper limit. This can be seen better in the lower plot of Fig. 7.4, which shows the integral of the difference of the two functions. The maximum difference is for $r \approx 2$. As far as the upper limits are concerned, we obtain (the large number of — non-significant — digits is only to observe the behavior in detail):

$$r_u(x=0, \mathcal{L}=1\pm 0, \text{ at } 95\%) = 2.996$$
$$r_u(x=0, \mathcal{L}=1.0\pm 0.1, \text{ at } 95\%) = 3.042\,.$$

An uncertainty of 10% due to systematics produces less than a 0.5% variation of the limits. For curiosity, here are also the 90% probability limits:

$$r_u(x=0, \mathcal{L}=1\pm 0, \text{ at } 90\%) = 2.304$$
$$r_u(x=0, \mathcal{L}=1.0\pm 0.1, \text{ at } 90\%) = 2.330\,.$$

To simplify the calculation (and also to get a feeling of what is going on) we can use some approximations.

(1) Since the dependence of the upper limit of r from $1/\mathcal{L}$ is given by

$$r_u = \frac{-\ln(1-P_u)}{\mathcal{L}}\,,$$

the upper limit averaged with the belief on $\mathcal{L}$ is given by

$$r_u = -\ln(1-P_u)\,\mathrm{E}\left(\frac{1}{\mathcal{L}}\right) = \int \frac{1}{\mathcal{L}} f(\mathcal{L})\, \mathrm{d}\mathcal{L}\,.$$

We need to solve an integral simpler than in the previous case. For the above example of $\mathcal{L} = 1.0 \pm 0.1$ we obtain $r_u = 2.326$ at 90% and $r_u = 3.026$ at 95%.

(2) Finally, as a real rough approximation, we can take into account the small asymmetry of r_u around the value obtained at the nominal value

of $\mathcal{L}$ averaging the two values of $\mathcal{L}$ at $\pm\sigma_\mathcal{L}$ from $\mathcal{L}_\circ$:

$$\begin{aligned} r_u &\approx \frac{-\ln(1-P_u)}{2}\left(\frac{1}{\mathcal{L}_\circ-\sigma_\mathcal{L}}+\frac{1}{\mathcal{L}_\circ+\sigma_\mathcal{L}}\right) \\ &\approx \frac{-\ln(1-P_u)}{\mathcal{L}_\circ}\left(1+\frac{\sigma^2_\mathcal{L}}{\mathcal{L}^2_\circ}\right) . \end{aligned}$$

We obtain numerically identical results to the previous approximation.

The main conclusion is that the uncertainty due to systematics plays only a second-order role, and it can be neglected for all practical purposes. A second observation is that this uncertainty increases slightly the limits if $f(\mathcal{L})$ is distributed normally, but the effect could also be negative if the $f(\mathcal{L})$ is asymmetric with positive skewness.

As a more general remark, one should not forget that the upper limit has the meaning of an uncertainty and not of a value of quantity. Therefore, as nobody really cares about an uncertainty of 10 or 20% on the uncertainty, the same is true for upper/lower limits. At the per cent level it is mere numerology (I have calculated it at the 10^{-4} level just to check the numerical sensitivity).

7.7.5 *Counting measurements in the presence of background*

As an example of a different kind of systematic effect, let us think of counting experiments in the presence of background. For example we are searching for a new particle, we make some selection cuts and count x events. But we also expect an average number of background events $\lambda_{B_\circ} \pm \sigma_B$, where σ_B is the standard uncertainty of $\lambda_{B_\circ}$, not to be confused with $\sqrt{\lambda_{B_\circ}}$. What can we say about λ_S, the true value of the average number associated with the signal? First we will treat the case in which the determination of the expected number of background events is well known ($\sigma_B/\lambda_{B_\circ} \ll 1$), and then the general case.

$\boxed{\sigma_B/\lambda_{B_\circ} \ll 1}$: Being the two processes incoherent, the true value of the sum of signal and background is given by their sum: $\lambda = \lambda_S + \lambda_{B_\circ}$. The likelihood is

$$P(x \mid \lambda) = \frac{e^{-\lambda}\lambda^x}{x!} . \tag{7.73}$$

Applying Bayes' theorem we have

$$f(\lambda_S \,|\, x, \lambda_{B_\circ}) = \frac{e^{-(\lambda_{B_\circ}+\lambda_S)}\,(\lambda_{B_\circ}+\lambda_S)^x\,f_\circ(\lambda_S)}{\int_0^\infty e^{-(\lambda_{B_\circ}+\lambda_S)}\,(\lambda_{B_\circ}+\lambda_S)^x\,f_\circ(\lambda_S)\,d\lambda_S}\,. \tag{7.74}$$

Choosing again $f_\circ(\lambda_S)$ uniform (in a reasonable interval) this gets simplified. The integral in the denominator can be calculated easily by parts and the final result is

$$f(\lambda_S \,|\, x, \lambda_{B_\circ}) = \frac{e^{-\lambda_S}\,(\lambda_{B_\circ}+\lambda_S)^x}{x!\,\sum_{n=0}^{x}\frac{\lambda_{B_\circ}^n}{n!}}\,, \tag{7.75}$$

$$F(\lambda_S \,|\, x, \lambda_{B_\circ}) = 1 - \frac{e^{-\lambda_S}\,\sum_{n=0}^{x}\frac{(\lambda_{B_\circ}+\lambda_S)^n}{n!}}{\sum_{n=0}^{x}\frac{\lambda_{B_\circ}^n}{n!}}\,. \tag{7.76}$$

From Eqs. (7.75)-(7.76) it is possible to calculate in the usual way the best estimate and the credibility intervals of λ_S. Two particular cases are of interest:

- If $\lambda_{B_\circ} = 0$ then formulae (7.43)-(7.44) are recovered. In such a case one measured count is enough to claim for a signal (if somebody is willing to believe that really $\lambda_{B_\circ} = 0$ without any uncertainty...).
- If $x = 0$ then

$$f(\lambda \,|\, x, \lambda_{B_\circ}) = e^{-\lambda_S}\,, \tag{7.77}$$

independently of $\lambda_{B_\circ}$. This behavior is not really obvious, and I must confess that it puzzled me for years, until Astone and Pizzella finally showed that the result is logically correct [72]. For further details about this result and for comparisons with what other methods produce, see Sec. 2.9.2 of Ref. [11]. It is interesting to note (Table 2 of Ref. [11]) that the PDG [51] blessed prescription [73] yields the manifestly absurd result that, given a null observation, the upper limit *decreases* with increasing background (a noisy measurement produces a tighter bound on a searched for rare phenomenon than a clean measurement!).

Any $g(\lambda_{B_\circ})$: In the general case, the true value of the average number of background events λ_B is unknown. We only known that it is distributed around $\lambda_{B_\circ}$ with standard deviation σ_B and probability density function $g(\lambda_B)$, not necessarily a Gaussian. What changes with respect to the previous case is the initial distribution, now a joint function of

λ_S and of λ_B. Assuming λ_B and λ_S independent, the prior density function is

$$f_\circ(\lambda_S, \lambda_B) = f_\circ(\lambda_S)\, g_\circ(\lambda_B)\,. \tag{7.78}$$

We leave $f_\circ$ in the form of a joint distribution to indicate that the result we shall get is the most general for this kind of problem. The likelihood, on the other hand, remains the same as in the previous example. The inference of λ_S is done in the usual way, applying Bayes' theorem and marginalizing with respect to λ_S:

$$f(\lambda_S \,|\, x, g_\circ(\lambda_B)) = \frac{\int e^{-(\lambda_B+\lambda_S)}\, (\lambda_B + \lambda_S)^x\, f_\circ(\lambda_S, \lambda_B)\, \mathrm{d}\lambda_B}{\iint e^{-(\lambda_B+\lambda_S)}\, (\lambda_B + \lambda_S)^x\, f_\circ(\lambda_S, \lambda_B)\, \mathrm{d}\lambda_S\, \mathrm{d}\lambda_B}\,. \tag{7.79}$$

The previous case [formula (7.75)] is recovered if the only value allowed for λ_B is $\lambda_{B_\circ}$ and $f_\circ(\lambda_S)$ is uniform:

$$f_\circ(\lambda_S, \lambda_B) = k\, \delta(\lambda_B - \lambda_{B_\circ})\,. \tag{7.80}$$

Chapter 8

Bypassing Bayes' theorem for routine applications

"Let us consider a dimensionless mass,
suspended from an inextensible massless wire,
free to oscillate without friction..."
(Any textbook)

In the previous chapters we have seen how to use the general formula (5.3) for practical applications. Unfortunately, when the problem becomes more complicated one starts facing integration problems. For this reason approximate and numerical methods are generally used. We shall concentrate our attention on approximations important for everyday use of probabilistic inference. Numerical methods, which in the most complicated problems mean *Monte Carlo* techniques, is a science in its own right and we shall not attempt to introduce it here, other than to give some hints and references in the appropriate places.

8.1 Maximum likelihood and least squares as particular cases of Bayesian inference

Let us continue with the case in which priors are so uninformative that a uniform distribution is a practical choice. Calling $\boldsymbol{\theta}$ the quantities to infer (i.e. the model parameters — the reason for this change of symbols is to use a notation which most readers are used to) and neglecting for a while systematic effects (i.e. we drop the influence quantities $\boldsymbol{h}$) Eq. (5.2) becomes:

$$f(\boldsymbol{\theta}\,|\,\boldsymbol{x},I) \propto f(\boldsymbol{x}\,|\,\boldsymbol{\theta},I)\,f_0(\boldsymbol{\theta}\,|\,I) \propto f(\boldsymbol{x}\,|\,\boldsymbol{\theta},I) = \mathcal{L}(\boldsymbol{\theta};\boldsymbol{x})\,, \tag{8.1}$$

where, we remember, the likelihood $\mathcal{L}(\boldsymbol{\theta};\boldsymbol{x})$ is a mathematical function of $\boldsymbol{\theta}$ (note that this function has no p.d.f. meaning and therefore normalization does not apply).

The set of $\boldsymbol{\theta}$ we believe most is that which maximizes $\mathcal{L}(\boldsymbol{\theta};\boldsymbol{x})$, a result known as the Maximum Likelihood principle. Here it has been reobtained as a special case of a more general framework, under clearly stated hypotheses, without the need of appealing principles. [1]

Also the usual Least Square formulae are easily derived if we take the well-known case of data points $\{x_i, y_i\}$, whose true values are related by a deterministic function $\mu_{y_i} = y(\mu_{x_i}, \boldsymbol{\theta})$ and with Gaussian errors only on the ordinates, i.e. we consider $x_i \approx \mu_{x_i}$. In the case of independence of the measurements, the likelihood dominated result becomes

$$f(\boldsymbol{\theta}\,|\,\boldsymbol{x},\boldsymbol{y},I) \propto \prod_i \exp\left[-\frac{(y_i - y(x_i,\boldsymbol{\theta}))^2}{2\sigma_{y_i}^2}\right], \tag{8.2}$$

or

$$f(\boldsymbol{\theta}\,|\,\boldsymbol{x},\boldsymbol{y},I) \propto \mathcal{L}(\boldsymbol{\theta};\boldsymbol{x},\boldsymbol{y}) = \exp\left[-\chi^2/2\right], \tag{8.3}$$

with

$$\chi^2 = \sum_i (y_i - y(x_i,\boldsymbol{\theta}))^2/\sigma_{y_i}^2\,, \tag{8.4}$$

being the well-known ‘chi-square’. Maximizing the likelihood is equivalent to minimizing the χ^2 and the most believed value of $\boldsymbol{\theta}$ is easily obtained, analytically in easy cases, or numerically for more complex ones. As far as the uncertainty on $\boldsymbol{\theta}$ is concerned, the widely used (and misused! — see Sec. 12.2) $\Delta\chi^2 = 1$ rule, or the

$$(V^{-1})_{ij}(\boldsymbol{\theta}) = \frac{1}{2}\frac{\partial^2\chi^2}{\partial\theta_i\partial\theta_j} \tag{8.5}$$

formula, with $\mathbf{V}(\boldsymbol{\theta})$ being the covariance matrix of $\boldsymbol{\theta}$, are just consequences of a multi-variate Gaussian distribution of $\boldsymbol{\theta}$ (and, hence a parabolic shape of χ^2). In fact, the generic multi-variate Gaussian p.d.f. of n variables $\boldsymbol{z}$

---

[1] There is another principle, which is considered to be a very good feature by frequentists, though not all frequentistic methods respect it [11]: the *Likelihood Principle*: In practice, it says that the result of an inference should not depend on multiplicative factors of the likelihood functions. This ‘principle’ too arises automatically in the Bayesian framework.

with expected values $\mathrm{E}(\boldsymbol{z}) = \boldsymbol{\mu}_z$ and covariance matrix $\mathbf{V}$ is

$$f(\boldsymbol{z}) = (2\pi)^{-n/2}|\mathbf{V}|^{-1/2} \exp\left[-\frac{1}{2}\boldsymbol{\Delta}^T\mathbf{V}^{-1}\boldsymbol{\Delta}\right], \tag{8.6}$$

where $\boldsymbol{\Delta}$ stands for the set of differences $z_i - \mu_{z_i}$ and $|\mathbf{V}|$ is the determinant of $\mathbf{V}$. Taking the logarithm of $f(\boldsymbol{z})$ and indicating $\boldsymbol{\Delta}^T\mathbf{V}^{-1}\boldsymbol{\Delta}$ by χ^2 (the usual chi-square) and all terms that do not depend on $\boldsymbol{z}$ by k, we have

$$\ln f(\boldsymbol{z}) = -\frac{1}{2}\chi^2 + k \tag{8.7}$$

$$= -\frac{1}{2}\sum_{ij}(z_i - \mu_{z_i})\,(V^{-1})_{ij}\,(z_j - \mu_{z_j}) + k\,, \tag{8.8}$$

from which Eq. (8.5) follows for the variables $\boldsymbol{z} = \boldsymbol{\theta}$.

In routine applications the hypotheses which lead to the maximum likelihood and least squares formulae often hold. But when these hypotheses are not justified we need to characterize the result by the multi-dimensional posterior distribution $f(\boldsymbol{\theta})$, going back to more general Eqs. (8.3), (8.1), or (5.2), depending on the hypotheses and approximations valid in each practical case, as sketched in Fig. 2.2.

The important conclusion from this section is that, as was the case for the 'definitions of probability', Bayesian methods often contain well-known conventional methods, but without introducing them as principles. The practitioner acquires then a superior degree of awareness about the range of validity of the methods and might as well use standard formulae with Bayesian spirit and with a more natural interpretation of the results, since we can speak about probability of model parameters, which is the usual way physicists think.

It is surprising that this rather natural thinking does not belong to the standard education of physicists, or at least this has been the case for most of the last century. But, as usual, there have been remarkable exceptions. Here is a recollection of a former Fermi student [74]:

> *"In my thesis I had to find the best 3-parameter fit to my data and the errors of those parameters in order to get the 3 phase shifts and their errors. Fermi showed me a simple analytic method. At the same time other physicists were using and publishing other cumbersome methods. Also Fermi taught me a general method, which he called Bayes Theorem, where one could easily derive the best-fit parameters and their errors as a special case of the maximum-likelihood method. I remember asking*

Fermi how and where he learned this. I expected him to answer R.A. Fisher or some other textbook on mathematical statistics. Instead he said 'perhaps it was Gauss'. I suspect he was embarrassed to admit that he had derived it all from his 'Bayes Theorem'."

8.2 Linear fit

Let us see, as a simple example, the case of a linear dependence between the true value of two quantities, i.e. $\mu_Y = m\,\mu_X + c$, which fits at best n data points $\{x_i, y_i, \sigma_i\}$, where the symbol σ_i stands, for σ_{y_i} of the previous section. Since there is no error on the X values, we can identify the observed x_i with μ_{X_i}, reducing the assumed law to $\mu_{Y_i} = m\,x_i + c$. We consider our knowledge about m and c sufficiently vague that a uniform prior can be used. Equation (8.2) becomes in this specific case

$$f(m, c\,|\,\boldsymbol{x}, \boldsymbol{y}, \boldsymbol{\sigma}, I) = K \prod_i \exp\left[-\frac{(y_i - m\,x_i - c)^2}{2\,\sigma_i^2}\right] \tag{8.9}$$

$$= K \exp\left[-\frac{1}{2}\sum_i \frac{(y_i - m\,x_i - c)^2}{\sigma_i^2}\right], \tag{8.10}$$

where K is a normalization constant. In principle, this is the end of the problem, at least conceptually: the constant K can be evaluated numerically; $f(m,c)$ can be plotted and inspected; expected value, standard deviation and all probability regions of interest can be calculated numerically. Nevertheless, formulae to calculate location and dispersion parameters of $f(m,c)$ can be useful for routine use. The mode, i.e. the set of parameter we believe in most, can be obtained analytically from the conditions

$$\frac{\partial f(m,c)}{\partial m} = 0 \tag{8.11}$$

$$\frac{\partial f(m,c)}{\partial c} = 0\,. \tag{8.12}$$

Alternatively, and with identical results, we can find the minimum of

$$\chi^2 = \frac{\sum_i (y_i - m\,x_i - c)^2}{\sigma_i^2} \tag{8.13}$$

with analogous conditions

$$\frac{\partial\chi^2}{\partial m} = \frac{\partial}{\partial m}\frac{\sum_i(y_i - m\,x_i - c)^2}{\sigma_i^2} = 0 \tag{8.14}$$

$$\frac{\partial\chi^2}{\partial c} = \frac{\partial}{\partial c}\frac{\sum_i(y_i - m\,x_i - c)^2}{\sigma_i^2} = 0\,. \tag{8.15}$$

Ignoring irrelevant factors, we get

$$\sum_i \frac{x_i y_i}{\sigma_i^2} - m\sum_i \frac{x_i^2}{\sigma_i^2} - c\sum_i \frac{x_i}{\sigma_i^2} = 0 \tag{8.16}$$

$$\sum_i \frac{y_i}{\sigma_i^2} - m\sum_i \frac{x_i}{\sigma_i^2} - c\sum_i \frac{1}{\sigma_i^2} = 0 \tag{8.17}$$

where all summations run from 1 to n. Dividing all terms by $\sum_i \frac{1}{\sigma_i^2}$, the equations to be solved can be rewritten as

$$\overline{x\,y} - m\,\overline{x^2} - c\overline{x} = 0 \tag{8.18}$$

$$\overline{y} - m\,\overline{x} - c = 0\,, \tag{8.19}$$

where $\overline{x}$, $\overline{x^2}$, $\overline{y}$ and $\overline{x\,y}$ are the averages weighted with $w_i = 1/\sigma_i^2$ [$\overline{x} = (\sum_i x_i/\sigma_i^2)/(\sum_i 1/\sigma_i^2)$, and so on]. Solving the two equations we get:

$$\text{mode}(m) = \frac{\overline{x\,y} - \overline{x}\,\overline{y}}{\overline{x^2} - \overline{x}} \tag{8.20}$$

$$\text{mode}(c) = \overline{y} - \overline{x}\,\text{mode}(m)\,. \tag{8.21}$$

The direct calculation of the expected value is usually much more complicated, because one has to perform an integral. The same is true for the standard deviation. At this point we can simply *assume* that $f(m,c)$ is approximately a bivariate Gaussian p.d.f. to obtain:

$$\text{E}(m) = \text{mode}(m) \tag{8.22}$$

$$\text{E}(c) = \text{mode}(c). \tag{8.23}$$

Indeed, in the case of a linear fit, the Gaussian solution is exact, because the term at the exponent of $f(m,c)$ can be reduced to a negative quadratic form and, finally, to the canonical bivariate Gaussian form (4.80).[2] To

[2]This property holds not only in linear fits, but also for all models in which the parameters appear linearly in $y(m_x, \boldsymbol{\theta})$. In other words, linear dependence on $\boldsymbol{\theta}$, not on x, is required.

calculate variances and covariance we can make use of Eq. (8.5), obtaining

$$\mathbf{V}^{-1}(m,c) = \begin{pmatrix} \sum_i \frac{x_i^2}{\sigma_i^2} & \sum_i \frac{x_i}{\sigma_i^2} \\ \sum_i \frac{x_i}{\sigma_i^2} & \sum_i \frac{1}{\sigma_i^2} \end{pmatrix} = \left(\sum_i \frac{1}{\sigma_i^2}\right) \begin{pmatrix} \overline{x^2} & \overline{x} \\ \overline{x} & 1 \end{pmatrix} \tag{8.24}$$

and, hence,

$$\mathbf{V}(m,c) = \frac{1}{\overline{x^2} - \overline{x}^2} \frac{1}{\sum_i \frac{1}{\sigma_i^2}} \begin{pmatrix} 1 & -\overline{x} \\ -\overline{x} & \overline{x^2} \end{pmatrix}, \tag{8.25}$$

i.e.

$$\sigma(m) = \frac{1}{\sqrt{\overline{x^2} - \overline{x}^2}} \sqrt{\frac{1}{\sum_i \frac{1}{\sigma_i^2}}} = \frac{1}{\sqrt{\mathrm{Var}(\boldsymbol{x})}} \sqrt{\frac{1}{\sum_i \frac{1}{\sigma_i^2}}} \tag{8.26}$$

$$\sigma(c) = \sqrt{\overline{x^2}}\,\sigma(m) = \frac{1}{\sqrt{1 - \overline{x}^2 \,/\, \overline{x^2}}} \sqrt{\frac{1}{\sum_i \frac{1}{\sigma_i^2}}} \tag{8.27}$$

$$\rho(m,c) = -\frac{\overline{x}}{\sqrt{\overline{x^2}}} = -\frac{\overline{x}}{\sqrt{\mathrm{Var}(\boldsymbol{x}) + \overline{x}^2}}\,. \tag{8.28}$$

If the standard deviations which model the y_i are all equal, $1/\sum_i 1/\sigma_i^2$ becomes equal to σ^2/n, showing clearly that the uncertainty on the parameters depend on σ and $\sqrt{n}$. Note also the dependence of $\sigma(m)$ and $\sigma(c)$ on $\sqrt{\mathrm{Var}(\boldsymbol{x})} = \sqrt{\overline{x^2} - \overline{x}^2}$, the standard deviation of the statistical distribution of the data points on the x-axis.[3] $\sqrt{\mathrm{Var}(\boldsymbol{x})}$ can be associated to the intuitive concept of 'lever arm' of the data points: the parameter of the straight line are better determined if the measurements are performed over a wide range. If the lever arm vanishes, then $\sigma(m)$ and $\sigma(c)$ diverge, unless all data points are concentrated at $X = 0$. In this latter case ($\overline{x} = \overline{x^2} = \mathrm{Var}(\boldsymbol{x}) = 0$, but $\overline{x}\,/\,\overline{x^2} \to 1$) Eq. (8.27) shows that c is indeed well measured, as it is easy to understand.

A few other remarks are important for practical applications: The correlation coefficient $\rho(m,c)$ vanishes if the fit is performed in the barycenter of the data points. In the case when the σ_i are all believed to be equal (though unknown) the formulae for calculating the 'best values' of the parameters do not depend on σ, and 'un-weighted' least square formulae are recovered. An initially unknown common σ can be inferred from the same data set

[3]This standard deviation should not be confused with the standard deviation describing the error on X, which has been assumed to be negligible in this model. For this reason it has been indicated by $\sqrt{\mathrm{Var}(\boldsymbol{x})}$.

using Bayesian reasoning, as will be shown in a while, or estimated from the residuals with respect to the best fit ($\sigma^2 \approx \sum_i [y_i - \mathrm{E}(m) y_i - \mathrm{E}(c)]^2/n$) — the two results coincide for a large data set, where 'large' could mean just a few dozen data points).

8.3 Linear fit with errors on both axes

If the x-values are also affected by independent Gaussian errors, the likelihood of observing the data points becomes

$$\begin{aligned} f(x_i, y_i \,|\, \mu_{X_i}, m, c) &= f(x_i \,|\, \mu_{X_i}) \cdot f(y_i \,|\, \mu_{X_i}, m, c) \\ &= \frac{1}{\sqrt{2\,\pi}\sigma_{X_i}} \exp\left[-\frac{(x_i - \mu_{X_i})^2}{2\,\sigma^2_{X_i}}\right] \\ &\quad \times \frac{1}{\sqrt{2\,\pi}\sigma_{Y_i}} \exp\left[-\frac{(y_i - m\,\mu_{X_i} - c)^2}{2\,\sigma^2_{Y_i}}\right], \end{aligned} \tag{8.29}$$

which depends on the unknown μ_{X_i}. These values can be inferred, together with m and c, from the data:

$$f(\boldsymbol{\mu}_X, m, c \,|\, \boldsymbol{x}, \boldsymbol{y}) \propto \left[\prod_i f(x_i, y_i \,|\, \mu_{X_i}, m, c)\right] \cdot f_\circ(\boldsymbol{\mu}_X, m, c)\,. \tag{8.30}$$

Marginalization over the μ_{X_i} yields the result searched for:

$$f(m, c \,|\, \boldsymbol{x}, \boldsymbol{y}) \propto \int f(\boldsymbol{\mu}_X, m, c \,|\, \boldsymbol{x}, \boldsymbol{y})\, \mathrm{d}\boldsymbol{\mu}_X\,. \tag{8.31}$$

Assuming a uniform $f_\circ(\boldsymbol{\mu}_X, m, c)$, we get the following result:

(1) if $\sigma_{X_i} \to 0$ the Gaussian distribution describing the probability of the observed X_i around μ_{X_i} tends to Dirac delta's $\delta(x_i - \mu_{X_i})$ and the integral gives

$$f(m, c \,|\, \boldsymbol{x}, \boldsymbol{y}) \propto \prod_i \frac{1}{\sqrt{2\,\pi}\,\sigma_{Y_i}} \exp\left[-\frac{(y_i - m\,x_i - c)^2}{2\,\sigma^2_{Y_i}}\right], \tag{8.32}$$

thus recovering Eq. (8.10).

(2) In the general case we have

$$f(m, c \,|\, \boldsymbol{x}, \boldsymbol{y}) \propto \prod_i \frac{1}{\sqrt{2\,\pi}\sqrt{\sigma^2_{Y_i} + m^2\sigma^2_{X_i}}} \exp\left[-\frac{(y_i - m\,x_i - c)^2}{2\,(\sigma^2_{Y_i} + m^2\sigma^2_{X_i})}\right]. \tag{8.33}$$

Essentially, σ_{Y_i} is replaced by an effective standard deviation which is the quadratic sum of σ_{Y_i} with σ_{X_i} rescaled ('propagated') with the derivative $\partial Y/\partial X$ calculated in x_i (equal to m in the linear case). This observation allows this result to be used in nonlinear cases, too, at least as an approximate method.
Calculating expected values and covariance matrix of m and c can be more complicated in this case, but this is just a technical question. As stressed several times, the full solution is given by Eq. (8.33), after normalization. As an approximation, expected values and covariance matrix can be determined iteratively, evaluating $\mathrm{E}(m)$ neglecting σ_{X_i} and using this value in Eqs. (8.20–8.20). Usually, the convergence is so fast that one can estimate m graphically ('by eye'), and the first iteration is accurate enough.

8.4 More complex cases

The aim of Secs. 8.1 and 8.2 was to show how to reproduce well-known formulae starting from general Bayesian ideas, under a certain number of well-defined conditions, including the uniform prior about the fit parameters. If some of these hypotheses do not hold because, for example, there are constraints on the value of the parameter, or the error function is not Gaussian, the reader now knows what to do, at least in principle. As an example of a bit more complicated situation, the case of non-negligible error on the x-values was analyzed in detail in Sec. 8.3. As has been mentioned previously several times (I really want to stress this) nowadays, given the power of numerical and computational methods, it is not essential to arrive at nice closed formulae for the expected values and covariance matrix of the parameters.

Let us see, for example, what happens if we do not know the values of σ_i which enter in the fit, but we have good reason to think ('we believe') that they are the same for all y_i. We use the data to infer σ too, and Eq. (8.10) becomes

$$f(m, c, \sigma \,|\, \boldsymbol{x}, \boldsymbol{y}, I) = \frac{K}{\sigma^n} \exp\left[-\frac{1}{2}\sum_i \frac{(y_i - m\,x_i - c)^2}{\sigma^2}\right] f_0(\sigma)\,, \tag{8.34}$$

where $f_0(\sigma)$ has been written explicitly, to remind us that we should at least constrain σ to be positive, and the $1/\sigma^n$ factor has been made explicit since it can no longer be absorbed by the normalization constant. The inference on the fit parameters and on σ is achieved with marginalizations:

$$f(m, c \,|\, \boldsymbol{x}, \boldsymbol{y}, I) = \int f(m, c, \sigma \,|\, \boldsymbol{x}, \boldsymbol{y}, I)\, \mathrm{d}\sigma \qquad (8.35)$$

$$f(\sigma \,|\, \boldsymbol{x}, \boldsymbol{y}, I) = \int f(m, c, \sigma \,|\, \boldsymbol{x}, \boldsymbol{y}, I)\, \mathrm{d}m\, \mathrm{d}c\,. \qquad (8.36)$$

If we think that σ is not constant, but, for example, depends linearly on x, it is enough to replace σ with $\sigma_0 + \alpha\, x_i$ in Eq. (8.34), infer $f(m, c, \sigma_0, \alpha \,|\, \boldsymbol{x}, \boldsymbol{y}, I)$ and perform the marginalizations of interest.

Finally, I would like to point out that Bayesian methods are particularly suited to solving more complex cases of 'regression'. We cannot enter here into advanced applications, and I recommend Refs. [75, 76] as a starting points.

8.5 Systematic errors and 'integrated likelihood'

Systematic effects are easily included. Calling, as usual, $\boldsymbol{h}$ the influence quantities, we have

$$f(\boldsymbol{\theta} \,|\, \boldsymbol{x}, \boldsymbol{h}, I) \propto f(\boldsymbol{x} \,|\, \boldsymbol{\theta}, \boldsymbol{h}, I)\, f_\circ(\boldsymbol{\theta} \,|\, I) \propto f(\boldsymbol{x} \,|\, \boldsymbol{\theta}, \boldsymbol{h}, I) = \mathcal{L}(\boldsymbol{\theta}; \boldsymbol{x}, \boldsymbol{h})\,, \qquad (8.37)$$

from which

$$\begin{aligned} f(\boldsymbol{\theta} \,|\, \boldsymbol{x}, I) &\propto \int f(\boldsymbol{x} \,|\, \boldsymbol{\theta}, \boldsymbol{h}, I)\, f_\circ(\boldsymbol{\theta} \,|\, I)\, f_\circ(\boldsymbol{h} \,|\, I)\, \mathrm{d}\boldsymbol{h} \\ &\propto \int \mathcal{L}(\boldsymbol{\theta}; \boldsymbol{x}, \boldsymbol{h})\, f_\circ(\boldsymbol{\theta})\, \mathrm{d}\boldsymbol{h} \equiv \mathcal{L}_I(\boldsymbol{\theta}; \boldsymbol{x})\ . \end{aligned} \qquad (8.38)$$

All approximate results of the previous section are recovered, just replacing the likelihood with what is sometimes called, incorrectly, *integrated likelihood*, in practice an average likelihood weighted with the p.d.f. of $\boldsymbol{h}$. In real cases the integral must often be performed by Monte Carlo and all expected values (together with mode, variance, covariances and probability intervals) can only be performed numerically, but the simplicity of the basic reasoning still holds.

8.6 Linearization of the effects of influence quantities and approximate formulae

It is important to derive the approximation rules consistent with the Bayesian approach to handle uncertainties due to systematic errors for everyday use. The resulting formulae will be compared with the ISO recommendations [5] and suggestions about the modelling of uncertainty influence quantities ('systematic effects') will be given.

Let us ignore for a while all quantities of influence which could produce unknown systematic errors. In this case Eq. (5.3) can be replaced by Eq. (5.4), which can be further simplified if we remember that correlations between the results are originated by unknown systematic errors. In the absence of these, the joint distribution of all quantities $\boldsymbol{\mu}$ is simply the product of marginal ones:

$$f_R(\boldsymbol{\mu}) = \prod_i f_{R_i}(\mu_i)\,, \tag{8.39}$$

with

$$f_{R_i}(\mu_i) = f_{R_i}(\mu_i \,|\, x_i, \boldsymbol{h}_\circ)\,. \tag{8.40}$$

The symbol $f_{R_i}(\mu_i)$ indicates that we are dealing with *raw values* evaluated at $\boldsymbol{h} = \boldsymbol{h}_\circ$ (the choice of the adjective 'raw' will become clearer in a while). Since for any variation of $\boldsymbol{h}$ the inferred values of μ_i will change, it is convenient to name with the same subscript R the quantities obtained for $\boldsymbol{h}_\circ$:

$$f_{R_i}(\mu_i) \longrightarrow f_{R_i}(\mu_{R_i})\,. \tag{8.41}$$

Let us indicate with $\widehat{\mu}_{R_i}$ and σ_{R_i} the best estimates and the standard uncertainty of the raw values:

$$\widehat{\mu}_{R_i} = \mathrm{E}(\mu_{R_i}) \tag{8.42}$$

$$\sigma^2_{R_i} = \mathrm{Var}(\mu_{R_i})\,. \tag{8.43}$$

For any possible configuration of conditioning hypotheses $\boldsymbol{h}$, *corrected* values μ_i are obtained:

$$\mu_i = \mu_{R_i} + g_i(\boldsymbol{h})\,. \tag{8.44}$$

The function which relates the corrected value to the raw value and to the systematic effects has been denoted by g_i so as not to be confused with a probability density function. Expanding Eq. (8.44) in series around $\boldsymbol{h}_\circ$ we finally arrive at the expression which will allow us to make the approximate evaluations of uncertainties:

$$\mu_i = \mu_{R_i} + \sum_l \frac{\partial g_i}{\partial h_l}\,(h_l - h_{\circ_l}) + \dots \,. \tag{8.45}$$

(All derivatives are evaluated at $\{\widehat{\mu}_{R_i}, \boldsymbol{h}_\circ\}$. To simplify the notation a similar convention will be used in the following formulae.)

Neglecting the terms of the expansion above the first order, and taking the expected values, we get

$$\begin{aligned} \widehat{\mu}_i &= \mathrm{E}(\mu_i) \\ &\approx \widehat{\mu}_{R_i}\,, \end{aligned} \tag{8.46}$$

$$\begin{aligned} \sigma^2_{\mu_i} &= \mathrm{E}\left[(\mu_i - \mathrm{E}[\mu_i])^2\right] \\ &\approx \sigma^2_{R_i} + \sum_l \left(\frac{\partial g_i}{\partial h_l}\right)^2 \sigma^2_{h_l} \\ &\quad \left\{ + 2\sum_{l<m} \left(\frac{\partial g_i}{\partial h_l}\right)\left(\frac{\partial g_i}{\partial h_m}\right) \rho_{lm}\,\sigma_{h_l}\,\sigma_{h_m} \right\}; \end{aligned} \tag{8.47}$$

$$\begin{aligned} \mathrm{Cov}(\mu_i,\mu_j) &= \mathrm{E}\left[(\mu_i - \mathrm{E}[\mu_i])(\mu_j - \mathrm{E}[\mu_j])\right] \\ &\approx \sum_l \left(\frac{\partial g_i}{\partial h_l}\right)\left(\frac{\partial g_j}{\partial h_l}\right) \sigma^2_{h_l} \\ &\quad \left\{ + 2\sum_{l<m} \left(\frac{\partial g_i}{\partial h_l}\right)\left(\frac{\partial g_j}{\partial h_m}\right) \rho_{lm}\,\sigma_{h_l}\,\sigma_{h_m} \right\}. \end{aligned} \tag{8.48}$$

The terms included within $\{\cdot\}$ vanish if the unknown systematic errors are uncorrelated, and the formulae become simpler. Unfortunately, very often this is not the case, as when several calibration constants are simultaneously obtained from a fit (for example, in most linear fits slope and intercept have a correlation coefficient close to -0.9).

Sometimes the expansion (8.45) is not performed around the best values of $\boldsymbol{h}$ but around their nominal values, in the sense that the correction for the known value of the systematic errors has not yet been applied (see

Sec. 6.9). In this case Eq. (8.45) should be replaced by

$$\mu_i = \mu_{R_i} + \sum_l \frac{\partial g_i}{\partial h_l}\,(h_l - h_{N_l}) + \ldots\,, \tag{8.49}$$

where the subscript N stands for *nominal.* The best value of μ_i is then

$$\begin{aligned} \widehat{\mu}_i &= \mathrm{E}(\mu_i) \\ &\approx \widehat{\mu}_{R_i} + \mathrm{E}\left(\sum_l \frac{\partial g_i}{\partial h_l}\,(h_l - h_{N_l})\right) \\ &= \widehat{\mu}_{R_i} + \sum_l \delta\mu_{i_l}\,. \end{aligned} \tag{8.50}$$

Instead, Eqs. (8.47) and (8.48) remain valid, with the condition that the derivative is calculated at $\boldsymbol{h}_N$. If $\rho_{lm} = 0$, it is possible to rewrite Eqs. (8.47) and (8.48) in the following way, which is very convenient for practical applications:

$$\sigma^2_{\mu_i} \approx \sigma^2_{R_i} + \sum_l \left(\frac{\partial g_i}{\partial h_l}\right)^2 \sigma^2_{h_l} \tag{8.51}$$

$$= \sigma^2_{R_i} + \sum_l u^2_{i_l}\,; \tag{8.52}$$

$$\mathrm{Cov}(\mu_i, \mu_j) \approx \sum_l \left(\frac{\partial g_i}{\partial h_l}\right)\left(\frac{\partial g_j}{\partial h_l}\right) \sigma^2_{h_l} \tag{8.53}$$

$$= \sum_l s_{ij_l} \left|\frac{\partial g_i}{\partial h_l}\right| \sigma_{h_l} \left|\frac{\partial g_j}{\partial h_l}\right| \sigma_{h_l} \tag{8.54}$$

$$= \sum_l s_{ij_l}\, u_{i_l}\, u_{j_l} \tag{8.55}$$

$$= \sum_l \mathrm{Cov}_l(\mu_i, \mu_j)\,. \tag{8.56}$$

u_{i_l} is the component of the standard uncertainty due to effect h_l. s_{ij_l} is equal to the product of signs of the derivatives, which takes into account whether the uncertainties are positively or negatively correlated.

To summarize, when systematic effects are not correlated with each other, the following quantities are needed to evaluate the corrected result, the combined uncertainties and the correlations:

- the raw $\widehat{\mu}_{R_i}$ and σ_{R_i};
- the best estimates of the corrections $\delta\mu_{i_l}$ for each systematic effect h_l;

- the best estimate of the standard deviation u_{i_l} due to the imperfect knowledge of the systematic effect;
- for any pair $\{\mu_i, \mu_j\}$ the sign of the correlation s_{ij_l} due to the effect h_l.

In physics applications it is frequently the case that the derivatives appearing in Eqs. (8.50)–(8.54) cannot be calculated directly, as for example when h_l are parameters of a simulation program, or acceptance cuts. Then variations of μ_i are usually studied by varying a particular h_l within a *reasonable* interval, holding the other influence quantities at the nominal value. $\delta\mu_{i_l}$ and u_{i_l} are calculated from the interval $\pm\Delta_i^{\pm}$ of variation of the true value for a given variation $\pm\Delta_{h_l}^{\pm}$ of h_l and from the probabilistic meaning of the intervals (i.e. from the assumed distribution of the true value). This empirical procedure for determining $\delta\mu_{i_l}$ and u_{i_l} has the advantage that it can take into account nonlinear effects, since it directly measures the difference $\widehat{\mu}_i - \widehat{\mu}_{R_i}$ for a given difference $h_l - h_{N_l}$.

Some simple examples are given in Sec. 8.9, and two typical experimental applications will be discussed in more detail in Sec. 8.13. More details on the subject, including the approximate treatment of nonlinear effects, will be shown in Chapter 12.

8.7 BIPM and ISO recommendations

In this section we compare the results obtained in the previous section with the recommendations [4] of the Bureau International des Poids et Mesures (BIPM) and the International Organization for Standardization (ISO) on *"the expression of experimental uncertainty"*. [5]

"(1) *The uncertainty in the result of a measurement generally consists of several components which may be grouped into two categories according to the way in which their numerical value is estimated:*

A: *those which are evaluated by statistical methods;*
B: *those which are evaluated by other means.*

There is not always a simple correspondence between the classification into categories A or B and the previously used classification into 'random' and 'systematic' uncertainties. The term 'systematic uncertainty' can be misleading and should be avoided.
The detailed report of the uncertainty should consist of a complete list of the components, specifying for each the method used to obtain its numerical result. "

Essentially the first recommendation states that all uncertainties can be treated probabilistically. The distinction between types A and B is subtle and can be misleading if one thinks of 'statistical methods' as synonymous with 'probabilistic methods', as is currently the case in Physics. Here 'statistical' has the classical meaning of repeated measurements. The names 'A' and 'B' do not reveal much fantasy, but according to Klaus Weise it was the only agreement to which the ISO committee could come [35].

"(2) *The components in category A are characterized by the estimated variances s_i^2 (or the estimated 'standard deviations' s_i) and the number of degrees of freedom ν_i. Where appropriate, the covariances should be given."*

The estimated variances correspond to $\sigma_{R_i}^2$ of the previous section. The degrees of freedom are related to small samples and to the *Student t* distribution. The problem of small samples is not discussed in these notes, but clearly this recommendation is a relic of frequentistic methods.[4] With the approach followed in this primer there is no need to talk about degrees of freedom, since the Bayesian inference defines the final probability function $f(\mu)$ completely.

"(3) *The components in category B should be characterized by quantities u_j^2, which may be considered as approximations to the corresponding variances, the existence of which is assumed. The quantities u_j^2 may be treated like variances and the quantities u_j like standard deviations. Where appropriate, the covariances should be treated in a similar way."*

Clearly, this recommendation is meaningful only in a Bayesian framework.

"(4) *The combined uncertainty should be characterized by the numerical value obtained by applying the usual method for the combination of variances. The combined uncertainty and its components should be expressed in the form of 'standard deviations'. "*

This is what we have found in Eqs. (8.47) and (8.48).

"(5) *If, for particular applications, it is necessary to multiply the combined uncertainty by a factor to obtain an overall uncertainty, the multiplying factor used must always be stated. "*

This last recommendation states once more that the uncertainty is 'by default' the standard deviation of the true value distribution. Any other quantity calculated to obtain a credibility interval with a certain probability level should be clearly stated.

---

[4]For criticisms about the standard treatment of the small-sample problem see Ref. [33].

To summarize, these are the basic ingredients of the BIPM/ISO recommendations.

subjective definition of probability: it allows variances to be assigned conceptually to any physical quantity which has an uncertain value;

uncertainty as standard deviation

- it is 'standard';
- the rule of combination (4.99)-(4.99) applies to standard deviations and not to confidence intervals;

combined standard uncertainty: it is obtained by the usual formula of 'error propagation' and it makes use of variances, covariances and first derivatives;

central limit theorem: it makes, under proper conditions, the true value normally distributed if one has several sources of uncertainty.

Consultation of the *Guide* [5] is recommended for further explanations about the justification of the standards, for the description of evaluation procedures, and for examples. I would just like to end this section with some examples of the evaluation of type B uncertainties and with some words of caution concerning the use of approximations and of linearization.

8.8 Evaluation of type B uncertainties

The ISO *Guide* states that

> *"For estimate x_i of an input quantity[5] X_i that has not been obtained from repeated observations, the ... standard uncertainty u_i is evaluated by scientific judgment based on all the available information on the possible variability of X_i. The pool of information may include*
>
> - *previous measurement data;*
> - *experience with or general knowledge of the behaviour and properties of relevant materials and instruments;*
> - *manufacturer's specifications;*
> - *data provided in calibration and other certificates;*
> - *uncertainties assigned to reference data taken from handbooks".*

[5] By 'input quantity' the ISO *Guide* means any of the contributions h_l or μ_{R_i} which enter into Eqs. (8.47) and (8.48).

8.9 Examples of type B uncertainties

In practice, we need to model our uncertainty on each influence quantity which acts as systematic effects. Figure 8.1 shows some simple models. Let us make some examples.

(1) Previous measurements of other particular quantities, performed in similar conditions, have provided a repeatability standard deviation of σ_r:

$$u = \sigma_r \, .$$

This example shows a type B uncertainty originated by random errors.

(2) We have measured n counts, with n large, and evaluate the uncertainty on the 'theoretical average number of events' to be $u = \sqrt{n}$ (see Sec. 7.4). This is another example of type B uncertainty that is caused by random errors. In fact, $u = \sqrt{n}$ has not been evaluated as a standard deviation from a sample of data (we might have read in the detector display just the number n), but results from believing a probabilistic model of detector response. The same is true when we infer an efficiency using a binomial model (see Sec. 7.1).

(3) A manufacturer's calibration certificate states that the uncertainty, defined as k standard deviations, is "$\pm\Delta$":

$$u = \frac{\Delta}{k} \, .$$

(4) A result is reported in a publication as $\overline{x} \pm \Delta$, stating that the average has been performed on four measurements and the uncertainty is a 95 % confidence interval. One has to conclude that the confidence interval has been calculated using the *Student t*:

$$u = \frac{\Delta}{3.18} \, .$$

(5) A manufacturer's specification states that the error on a quantity should not exceed Δ. With this limited information one has to assume a uniform distribution:

$$u = \frac{2\,\Delta}{\sqrt{12}} = \frac{\Delta}{\sqrt{3}} \, .$$

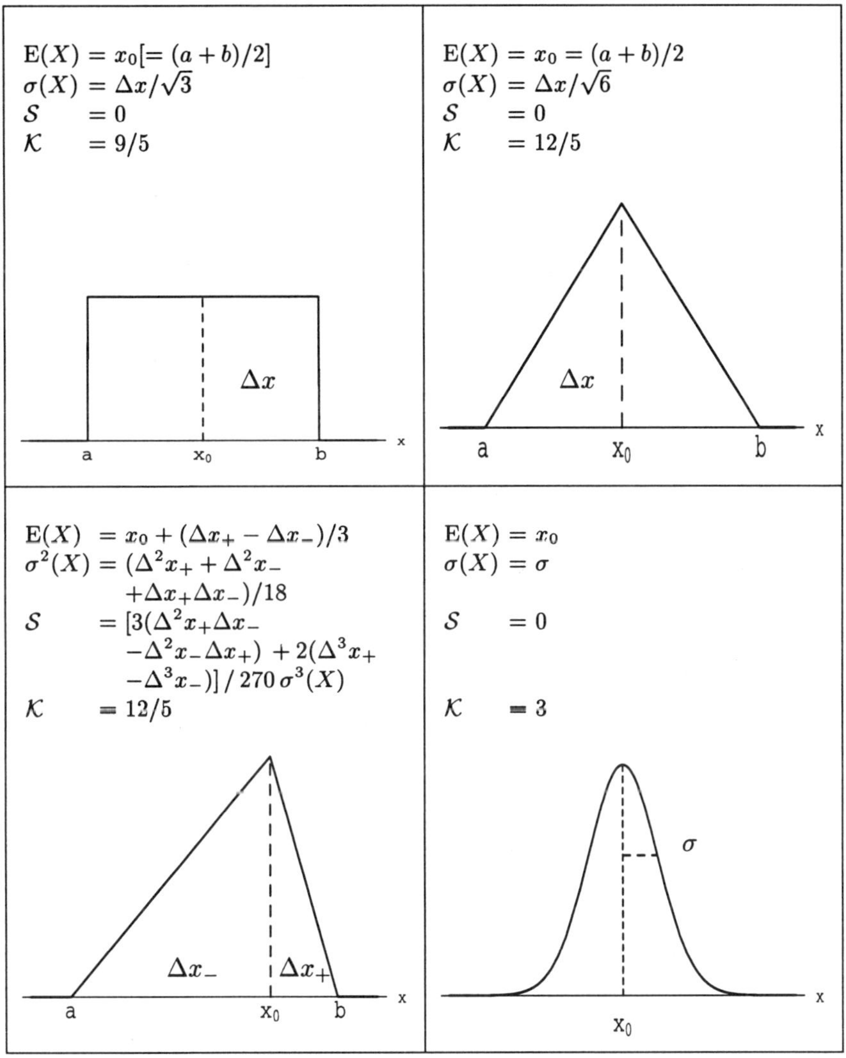

Fig. 8.1 Typical models to assess type-B uncertainties: uniform distribution, symmetric triangular distribution, asymmetric triangular distribution, and Gaussian distribution. The expressions of the most relevant statistical parameters are reported ($\mathcal{S}$ stands for skewness, $\mathcal{K}$ for kurtosis).

(6) A physical parameter of a Monte Carlo is believed to lie in the interval of $\pm\Delta$ around its best value, but not with uniform distribution: the degree of belief that the parameter is at center is higher than the degree of belief that it is at the edges of the interval. With this information a triangular distribution can be reasonably assumed:

$$u = \frac{\Delta}{\sqrt{6}}\,.$$

Note that the coefficient in front of Δ changes from the 0.58 of the previous example to the 0.41 of this. If the interval $\pm\Delta$ were a 3σ interval then the coefficient would have been equal to 0.33. These variations — to be considered extreme — are smaller than the statistical fluctuations of empirical standard deviations estimated from ≈ 10 measurements. This shows that one should not be worried that the type B uncertainties are less accurate than type A, especially if one tries to model the distribution of the physical quantity *honestly.*

8.10 Comments on the use of type B uncertainties

I know by personal experience that physicists, including myself at the very beginning, are reluctant to evaluate type B uncertainty, especially when they are not easily obtainable by a model (e.g. the '$\pm\sqrt{n}$' case encountered in point (1) of the previous Section) but required much *scientific judgment.* I find the situation paradoxical, because this is the only way most of 'systematic errors' can be — and actually are — evaluated. But, unfortunately without the guidance of subjective probability these 'errors' are often *arbitrary* numbers of obscure meaning: they are not 'intervals of certainty' (who would never state it?); they are not interval expressing some degree of belief. Then, which rules should be used to combine them? Not those of probability, if these 'objects' do not belong to the probability theory. Perhaps those of certainty (assuming such interval really indicate some certainty), but then *"the error bars become too large!"*... (and 'prescriptions' are preferred).

Probability intervals determined by coherence One of the reasons of hesitation is the model to choose: should it be a Gaussian or a uniform distribution? Coherence suggests to try to think of possible bets to determine *roughly* the interval of 'practical certainty', the 50% interval, and so on. In this way we get some ideas of how the probability distribution looks like.

Role of central limit theorem At this point, we have to turn our vague ideas into numbers. Fortunately, if we have many contributions to the uncertainty, as is often the case, the central limit theorem makes the result only depending on expected value and standard deviation, and not on the details of the distribution. Therefore, the psychological resistance to choose a model should become weaker. As a numerical example, let us consider the standard deviations of input quantities believed to be, with certainty or with high probability, in the interval between -1 and $+1$.

$$\begin{aligned}
\text{Uniform:}\ \sigma(X) &= \frac{1}{\sqrt{3}} \approx 0.58\,. \\
\text{Symmetric triangular:}\ \sigma(X) &= \frac{1}{\sqrt{6}} \approx 0.41\,. \\
\text{Asymmetric triangular peaked at } 1/2\text{:}\ \sigma(X) &= \sqrt{\frac{13}{72}} \approx 0.42\,. \\
\text{Gaussian, 90\% probability interval:}\ \sigma(X) &= \frac{1}{1.64} \approx 0.61\,. \\
\text{Gaussian, 95\% probability interval:}\ \sigma(X) &= \frac{1}{1.96} \approx 0.51\,.
\end{aligned}$$

We see that, for practical purposes, the differences between the σ's are irrelevant. Nevertheless, in order to avoid a bias of the overall uncertainty one should try to model each component according to the best knowledge of the physics case, rather than by choosing systematically the model which gives the most conservative uncertainty.[6] Note that in the case of asymmetric triangular distribution, the expected value of X is neither the center of the interval, nor the peak of the distribution. In this case we have $\mathrm{E}(X) = 1/6 \approx 0.17$. If one uses, incorrectly, the peak value, one introduces a bias which is $\approx 80\%$ of a standard deviation.

[6]In case of doubt between some models, the probability theory teaches that one should use $f(x) = \sum_i f_i(x) p_i$ where p_i is our confidence in the different models. It follows:

$$\begin{aligned}
\mathrm{E}(X) &= \sum_i p_i \int x f_i(x)\,\mathrm{d}x = \sum_i \mathrm{E}_i(X)\,p_i \\
\mathrm{E}(X^2) &= \sum_i p_i \int x^2 f_i(x)\,\mathrm{d}x = \sum_i \mathrm{E}_i(X^2)\,p_i \\
\sigma^2(X) &= \sum_i \sigma_i^2(X)\,p_i\,.
\end{aligned}$$

As an example, Fig. 4.3 shows the resulting uncertainty on the quantity $Y = X_1 + X_2$, where the X_i are independent and their uncertainty is described by identical asymmetrical triangular distributions. The combined result is obtained analytically using Eq. (4.95). One can see how good the Gaussian approximation already is and how biased a result could be, if the best estimate of the sum is performed using mode or median, and if the final uncertainty is evaluated with ad hoc rules of the kind shown in the introduction.
Obviously, one has to be careful about the condition of validity of the central limit theorem. In the most general case the final distribution will not be (multivariate) Gaussian and the combination of uncertainty must be done evaluating the integral (4.95) by Monte Carlo methods. Nevertheless, the compensations upon which the central limit theorem relies make the result highly model independent in this case too. An analysis of important frontier physics quantities in which these ideas (though in my opinion in a rather conservative way — see Ref. [77]) are applied throughout, can be found in Ref. [78].

Sensitivity analysis Finally, in case of doubts, a *sensitivity analysis* is recommended, i.e. changing models and model parameters in a reasonable way. In the case where there is large sensitivity to a parameter, a detailed account should be given.

My frank recommendation is the following. Try to model at best, honestly, the uncertainties in your field of expertise, use probability rules and you will get sensible results. If you think you 'know nothing' and you do not want to be committed, you should not publish any result. Perhaps you would feel less anxious working in a mathematics department.

Let us conclude with a practical example based on realistic numbers in which the methods described here are compared with naïve considerations.

Example The absolute energy calibration of an electromagnetic calorimeter module is not known exactly and is estimated to be between the nominal one and +10 %. The 'statistical' error is known by test beam measurements to be $18\%/\sqrt{E/\text{GeV}}$. What is the uncertainty on the energy measurement of an electron which has apparently released 30 GeV?

- There is no type A uncertainty, since only one measurement has been performed.
- The energy has to be corrected for the best estimate of the cali-

bration constant: +5 %, with an uncertainty of $10\,\%/\sqrt{12}$ due to sampling (the 'statistical' error):

$$E = 31.5 \pm 1.0\,\mathrm{GeV}\,.$$

- Then one has to take into account the uncertainty due to absolute energy scale calibration:
  - assuming a uniform distribution of the true calibration constant, $u = 31.5 \times 0.1/\sqrt{12} = 0.9\,\mathrm{GeV}$:

$$E = 31.5 \pm 1.3\,\mathrm{GeV}\,,$$

  - assuming, more reasonably, a triangular distribution, $u = 31.5 \times 0.05/\sqrt{6} = 0.6\,\mathrm{GeV}$,

$$E = 31.5 \pm 1.2\,\mathrm{GeV}\,,$$

- Interpreting the maximum deviation from the nominal calibration as uncertainty (see comment at the end of Sec. 6.9),

$$E = 30.0 \pm 1.0 \pm 3.0\,\mathrm{GeV} \rightarrow E = 30.0 \pm 3.2\,\mathrm{GeV}\,;$$

  As already mentioned earlier in these notes, while reasonable assumptions (in this case the first two) give consistent results, this is not true if one makes inconsistent use of the information just for the sake of giving 'safe' uncertainties.

8.11 Caveat concerning the blind use of approximate methods

The mathematical apparatus of variances and covariances of Eqs. (8.47)-(8.48) is often seen as the most complete description of uncertainty and in most cases used blindly in further uncertainty calculations. It must be clear, however, that this is just an approximation based on linearization. If the function which relates the corrected value to the raw value and the systematic effects is not linear then the linearization may cause trouble. An interesting case is discussed in Sec. 8.13.

There is another problem which may arise from the simultaneous use of Bayesian estimators <u>and</u> approximate methods[7]. Let us introduce the

[7]This is exactly the presumed paradox reported by the PDG [79] as an argument against Bayesian statistics (Sec. 29.6.2, p. 175: *"If Bayesian estimates are averaged, they do not converge to the true value, since they have all been forced to be positive"*).

problem with an example.

Example 1: 1000 independent measurements of the efficiency of a detector have been performed (or 1000 measurements of branching ratio, if you prefer). Each measurement was carried out on a base of 100 events and each time 10 favorable events were observed (this is obviously strange — though not impossible — but it simplifies the calculations). The result of each measurement will be (see Eqs. (7.3)–(7.5)):

$$\widehat{\epsilon}_i = \frac{10+1}{100+2} = 0.1078, \tag{8.57}$$

$$\sigma(\epsilon_i) = \sqrt{\frac{11 \times 91}{103 \times 102^2}} = 0.031\,. \tag{8.58}$$

Combining the 1000 results using the standard weighted average procedure gives

$$\epsilon = 0.1078 \pm 0.0010\,. \tag{8.59}$$

Alternatively, taking the complete set of results to be equivalent to 100 000 trials with 10 000 favorable events, the combined result is

$$\epsilon' = 0.10001 \pm 0.0009 \tag{8.60}$$

(the same as if one had used Bayes' theorem sequentially to infer $f(\epsilon)$ from the partial 1000 results). The conclusions are in disagreement and the first result is clearly mistaken (the solution will be given after the following example).

The same problem arises in the case of inference of the Poisson distribution parameter λ and, in general, whenever $f(\mu)$ is not symmetrical around $\mathrm{E}(\mu)$.

Example 2: Imagine an experiment running continuously for one year, searching for magnetic monopoles and identifying none. The consistency with zero can be stated either quoting $\mathrm{E}(\lambda) = 1$ and $\sigma_\lambda = 1$, or a 95 % upper limit $\lambda < 3$. In terms of rate (number of monopoles per day) the result would be either $\mathrm{E}(r) = 2.7 \cdot 10^{-3}$, $\sigma(r) = 2.7 \cdot 10^{-3}$, or an upper limit $r < 8.2 \cdot 10^{-3}$. It is easy to show that, if we take the 365 results for each of the running days and combine them using the standard weighted average, we get $r = 1.00 \pm 0.05$ monopoles per day! This absurdity is not caused by the Bayesian method, but by the standard rules for combining the results (the weighted average formulae (6.9) and (6.10) are derived from the normal distribution hypothesis). Using Bayesian inference would have led to a consistent and reasonable result no matter how the 365 days of running had been subdivided for partial analysis.

This suggests that in some cases it could be preferable to give the result in terms of the value of μ which maximizes $f(\mu)$ (p_m and λ_m of Secs. 7.1 and 7.4). This way of presenting the results is similar to that suggested by the maximum likelihood approach, with the difference that for $f(\mu)$ one should take the final probability density function and not simply the likelihood. Since it is practically impossible to summarize the outcome of an inference in only two numbers (best value and uncertainty), a description of the method used to evaluate them should be provided, except when $f(\mu)$ is approximately normally distributed (fortunately this happens most of the time).

8.12 Propagation of uncertainty

We have seen how to infer the value of the generic quantity μ in several cases, using also approximations. Conceptually this is a very simple task in the Bayesian framework, whereas the frequentistic one requires a lot of gymnastics, going back and forth from the logical level of true values to the logical level of estimators. If one accepts that the true values are just uncertain numbers[8], then, calling Y a function of other quantities X, each having a probability density function $f(x)$, the probability density function of Y $f(y)$ can be calculated with the standard formulae which follow from the rules probability (see Sec. 4.4).

In particular, it is very important for practical application the case in which the linearization approximation holds (Sec. 4.4) and well-known 'error propagation formulae' are recovered. But one has to be very careful in checking the validity of the approximation, as the following (counter-) example shows.

Example: The speed of a proton is measured with a time-of-flight system. Find the 68, 95 and 99 % probability intervals for the energy, knowing that $\beta = v/c = 0.9971$, and that distance and time have been measured

[8]To make the formalism lighter, let us call both the uncertain numbers, or 'random variable', associated with the quantity and the quantity itself by the same name X_i (instead of μ_{x_i}).

with a 0.2 % accuracy.
The relation

$$E = \frac{mc^2}{\sqrt{1-\beta^2}}$$

is strongly nonlinear. The results given by the approximate method and the correct one are shown in the table below.

Probability (%)	Linearization E (GeV)	Correct result E (GeV)
68	$6.4 \le E \le 18$	$8.8 \le E \le 64$
95	$0.7 \le E \le 24$	$7.2 \le E < \infty$
99	$0. \le E \le 28$	$6.6 \le E < \infty$

This argument will be treated in more detail in Chapter 12, where also practical formulae for second order expansion will be given.

8.13 Covariance matrix of experimental results – more details

This section, based on Ref. [80], shows once more practical rules to build the covariance matrix associated with experimental data with correlated uncertainty (see also Secs. 6.10 and 8.6), treating explicitly also the case of normalization uncertainty. Then it will be shown that, in this case, the covariance matrix evaluated in this way produces biased χ^2 fits.

8.13.1 *Building the covariance matrix of experimental data*

In physics applications, it is rarely the case that the covariance between the best estimates of two physical quantities[9], each given by the arithmetic average of direct measurements ($x_i = \overline{X_i} = \frac{1}{n}\sum_{k=1}^{n} X_{ik}$), can be evaluated

[9]In this section the symbol X_i will indicate the variable associated to the i-th physical quantity and X_{ik} its k-th direct measurement; x_i the best estimate of its value, obtained by an average over many direct measurements or indirect measurements, σ_i the standard deviation, and y_i the value corrected for the calibration constants. The weighted average of several x_i will be denoted by $\overline{x}$.

from the sample covariance[10] of the two averages:

$$\mathrm{Cov}(x_i, x_j) = \frac{1}{n\,(n-1)} \sum_{k=1}^{n} (X_{ik} - \overline{X}_i)(X_{jk} - \overline{X}_j) \,. \qquad (8.61)$$

More frequent is the well-understood case in which the physical quantities are obtained as a result of a χ^2 minimization, and the terms of the inverse of the covariance matrix are related to the curvature of χ^2 at its minimum:

$$\left(V^{-1}\right)_{ij} = \frac{1}{2} \left. \frac{\partial^2 \chi^2}{\partial X_i \partial X_j} \right|_{x_i, x_j} \,. \qquad (8.62)$$

In most cases one determines independent values of physical quantities with the same detector, and the correlation between them originates from the detector calibration uncertainties. Frequentistically, the use of Eq. (8.61) in this case would correspond to having a 'sample of detectors', each of which is used to perform a measurement of all the physical quantities.

A way of building the covariance matrix from the direct measurements is to consider the original measurements and the calibration constants as a common set of independent and uncorrelated measurements, and then to calculate corrected values that take into account the calibration constants. The variance/covariance propagation will automatically provide the full covariance matrix of the set of results. Let us derive it for two cases that occur frequently, and then proceed to the general case.

8.13.1.1 *Offset uncertainty*

Let $x_i \pm \sigma_i$ be the $i = 1, \ldots, n$ results of independent measurements and $\mathbf{VF}_X$ the (diagonal) covariance matrix. Let us assume that they are all affected by the same calibration constant c, having a standard uncertainty σ_c. The corrected results are then $y_i = x_i + c$. We can assume, for simplicity, that the most probable value of c is 0, i.e. the detector is well calibrated. One has to consider the calibration constant as the physical quantity X_{n+1}, whose best estimate is $x_{n+1} = 0$. A term $V_{X_{n+1,n+1}} = \sigma_c^2$ must be added to the covariance matrix.

[10]The '$n-1$' at the denominator of Eq. (8.61) is for the same reason as the '$n-1$' of the sample standard deviation. Although I do not agree with the rationale behind it, this formula can be considered a kind of standard and, anyhow, replacing '$n-1$' by 'n' has no effect in normal applications. As already said, I will not discuss the small-sample problem; anyone is interested in my worries concerning default formulae for small samples, as well as Student t distribution may have a look at Ref. [33].

The covariance matrix of the corrected results is given by the transformation

$$\mathbf{V}_Y = \mathbf{M}\mathbf{V}_X\mathbf{M}^T\,, \tag{8.63}$$

where $M_{ij} = \left.\frac{\partial Y_i}{\partial X_j}\right|_{x_j}$. The elements of $\mathbf{V}_Y$ are given by

$$V_{Y_{kl}} = \sum_{ij} \left.\frac{\partial Y_k}{\partial X_i}\right|_{x_i} \left.\frac{\partial Y_l}{\partial X_j}\right|_{x_j} V_{X_{ij}}\,. \tag{8.64}$$

In this case we get

$$\sigma^2(Y_i) = \sigma_i^2 + \sigma_c^2, \tag{8.65}$$

$$\mathrm{Cov}(Y_i, Y_j) = \sigma_c^2 \qquad (i \neq j), \tag{8.66}$$

$$\rho_{ij} = \frac{\sigma_c^2}{\sqrt{\sigma_i^2+\sigma_c^2}\sqrt{\sigma_j^2+\sigma_c^2}} \tag{8.67}$$

$$= \frac{1}{\sqrt{1+\left(\frac{\sigma_i}{\sigma_c}\right)^2}\sqrt{1+\left(\frac{\sigma_j}{\sigma_c}\right)^2}}\,, \tag{8.68}$$

reobtaining the results of Sec. 6.10. The total uncertainty on the single measurement is given by the combination in quadrature of the individual and the common standard uncertainties, and all the covariances are equal to σ_c^2. To verify, in a simple case, that the result is reasonable, let us consider only two independent quantities X_1 and X_2, and a calibration constant $X_3 = c$, having an expected value equal to zero. From these we can calculate the correlated quantities Y_1 and Y_2 and finally their sum ($S \equiv Z_1$) and difference ($D \equiv Z_2$). The results are

$$\mathbf{V}_Y = \begin{pmatrix} \sigma_1^2+\sigma_c^2 & \sigma_c^2 \\ \sigma_c^2 & \sigma_2^2+\sigma_c^2 \end{pmatrix}, \tag{8.69}$$

$$\mathbf{V}_Z = \begin{pmatrix} \sigma_1^2+\sigma_2^2+4\,\sigma_c^2 & \sigma_1^2-\sigma_2^2 \\ \sigma_1^2-\sigma_2^2 & \sigma_1^2+\sigma_2^2 \end{pmatrix}. \tag{8.70}$$

It follows that

$$\sigma^2(S) = \sigma_1^2+\sigma_2^2+(2\,\sigma_c)^2, \tag{8.71}$$

$$\sigma^2(D) = \sigma_1^2+\sigma_2^2\,, \tag{8.72}$$

as intuitively expected.

8.13.1.2 *Normalization uncertainty*

Let us consider now the case where the calibration constant is the scale factor f, known with a standard uncertainty σ_f. Also in this case, for simplicity and without losing generality, let us suppose that the most probable value of f is 1. Then $X_{n+1} = f$, i.e. $x_{n+1} = 1$, and $V_{X_{n+1,n+1}} = \sigma_f^2$. Then

$$\sigma^2(Y_i) = \sigma_i^2 + \sigma_f^2\, x_i^2\,, \tag{8.73}$$

$$\mathrm{Cov}(Y_i, Y_j) = \sigma_f^2\, x_i\, x_j \qquad (i \neq j)\,, \tag{8.74}$$

$$\rho_{ij} = \frac{x_i\, x_j}{\sqrt{x_i^2 + \frac{\sigma_i^2}{\sigma_f^2}}\,\sqrt{x_j^2 + \frac{\sigma_j^2}{\sigma_f^2}}}\,, \tag{8.75}$$

$$|\rho_{ij}| = \frac{1}{\sqrt{1 + \left(\frac{\sigma_i}{\sigma_f\, x_i}\right)^2}\,\sqrt{1 + \left(\frac{\sigma_j}{\sigma_f\, x_j}\right)^2}}\,. \tag{8.76}$$

To verify the results let us consider two independent measurements X_1 and X_2; let us calculate the correlated quantities Y_1 and Y_2, and finally their product ($P \equiv Z_1$) and their ratio ($R \equiv Z_2$):

$$\mathbf{V}_Y = \begin{pmatrix} \sigma_1^2 + \sigma_f^2\, x_1^2 & \sigma_f^2\, x_1\, x_2 \\ & \\ \sigma_f^2\, x_1\, x_2 & \sigma_2^2 + \sigma_f^2\, x_2 \end{pmatrix}, \tag{8.77}$$

$$\mathbf{V}_Z = \begin{pmatrix} \sigma_1^2\, x_2^2 + \sigma_2^2\, x_1^2 + 4\,\sigma_f^2\, x_1^2\, x_2^2 & \sigma_1^2 - \sigma_2^2\, \frac{x_1^2}{x_2^2} \\ & \\ \sigma_1^2 - \sigma_2^2\, \frac{x_1^2}{x_2^2} & \frac{\sigma_1^2}{x_2^2} + \sigma_2^2\, \frac{x_1^2}{x_2^4} \end{pmatrix}. \tag{8.78}$$

It follows that

$$\sigma^2(P) = \sigma_1^2\, x_2^2 + \sigma_2^2\, x_1^2 + (2\,\sigma_f\, x_1\, x_2)^2\,, \tag{8.79}$$

$$\sigma^2(R) = \frac{\sigma_1^2}{x_2^2} + \sigma_2^2\, \frac{x_1^2}{x_2^4}\,. \tag{8.80}$$

Just as an unknown common offset error cancels in differences and is enhanced in sums, an unknown normalization error has a similar effect on the ratio and the product. It is also interesting to calculate the standard uncertainty of a difference in the case of a normalization error:

$$\sigma^2(D) = \sigma_1^2 + \sigma_2^2 + \sigma_f^2\,(x_1 - x_2)^2\,. \tag{8.81}$$

The contribution from an unknown normalization error vanishes if the two values are equal.

8.13.1.3 *General case*

Let us assume that there are n independently measured values x_i and m *calibration constants* c_j with their covariance matrix $\mathbf{V}_c$. The latter can also be theoretical parameters influencing the data, and moreover they may be correlated, as usually happens if, for example, they are parameters of a calibration fit. We can then include the c_j in the vector that contains the measurements and $\mathbf{V}_c$ in the covariance matrix $\mathbf{V}_X$:

$$\boldsymbol{x} = \begin{pmatrix} x_1 \\ \vdots \\ x_n \\ c_1 \\ \vdots \\ c_m \end{pmatrix}, \qquad \mathbf{V}_X = \left(\begin{array}{cccc|c} \sigma_1^2 & 0 & \cdots & 0 & \\ 0 & \sigma_2^2 & \cdots & 0 & \\ \cdots & \cdots & \cdots & \cdots & \mathbf{0} \\ 0 & 0 & \cdots & \sigma_n^2 & \\ \hline & & \mathbf{0} & & \mathbf{V}_c \end{array}\right). \tag{8.82}$$

The corrected quantities are obtained from the most general function

$$Y_i = Y_i(X_i, \boldsymbol{c}) \qquad (i = 1, 2, \ldots, n)\,, \tag{8.83}$$

and the covariance matrix $\mathbf{V}_Y$ from the covariance propagation $\mathbf{V}_Y = \mathbf{M}\mathbf{V}_X\mathbf{M}^T$.

As a frequently encountered example, we can think of several normalization constants, each affecting a subsample of the data — as is the case where each of several detectors measures a set of physical quantities. Let us consider just three quantities (X_i) and three uncorrelated normalization standard uncertainties (σ_{f_j}), the first common to X_1 and X_2, the second to X_2 and X_3 and the third to all three. We get the following covariance matrix:

$$\begin{pmatrix} \sigma_1^2 + \left(\sigma_{f_1}^2 + \sigma_{f_3}^2\right) x_1^2 & \left(\sigma_{f_1}^2 + \sigma_{f_3}^2\right) x_1\, x_2 & \sigma_{f_3}^2\, x_1\, x_3 \\ \left(\sigma_{f_1}^2 + \sigma_{f_3}^2\right) x_1\, x_2 & \sigma_2^2 + \left(\sigma_{f_1}^2 + \sigma_{f_2}^2 + \sigma_{f_3}^2\right) x_2^2 & \left(\sigma_{f_2}^2 + \sigma_{f_3}^2\right) x_2\, x_3 \\ \sigma_{f_3}^2\, x_1\, x_3 & \left(\sigma_{f_2}^2 + \sigma_{f_3}^2\right) x_2\, x_3 & \sigma_3^2 + \left(\sigma_{f_2}^2 + \sigma_{f_3}^2\right) x_3^2 \end{pmatrix}.$$

8.14 Use and misuse of the covariance matrix to fit correlated data

We have already seen in Sec. 8.11 paradoxical results obtained using uncritically approximate formulae. Some of those affects are well known. Less known is a curious effect which might arise in minimum χ^2 fits, which we have seen in Sec. 8.1 that can be considered for many practical purposes good approximations of Bayesian analysis.

8.14.1 *Best estimate of the true value from two correlated values*

Once the covariance matrix is built one uses it in a χ^2 fit to get the parameters of a function. The quantity to be minimized is χ^2, defined as

$$\chi^2 = \mathbf{\Delta}^T\, \mathbf{V}^{-1}\, \mathbf{\Delta}\,, \tag{8.84}$$

where $\mathbf{\Delta}$ is the vector of the differences between the experimental and the theoretical values. Let us consider the simple case in which two results of the same physical quantity are available, and the individual and the common standard uncertainty are known. The best estimate of the true value of the physical quantity is then obtained by fitting the constant $Y = k$ through the data points. In this simple case the χ^2 minimization can be performed easily. We will consider the two cases of offset and normalization uncertainty. As before, we assume that the detector is well calibrated, i.e. the most probable value of the calibration constant is, respectively for the two cases, 0 and 1, and hence $y_i = x_i$.

8.14.2 *Offset uncertainty*

Let $x_1 \pm \sigma_1$ and $x_2 \pm \sigma_2$ be the two measured values, and σ_c the common standard uncertainty:

$$\chi^2 = \frac{1}{D}\left[(x_1-k)^2(\sigma_2^2+\sigma_c^2)+(x_2-k)^2(\sigma_1^2+\sigma_c^2)\right.$$
$$\left.-2(x_1-k)(x_2-k)\,\sigma_c^2\right], \tag{8.85}$$

where $D = \sigma_1^2\,\sigma_2^2 + (\sigma_1^2+\sigma_2^2)\,\sigma_c^2$ is the determinant of the covariance matrix.

Minimizing χ^2 and using the second derivative calculated at the minimum we obtain the best value of k and its standard deviation:

$$\widehat{k} = \frac{x_1\,\sigma_2^2 + x_2\,\sigma_1^2}{\sigma_1^2+\sigma_2^2} \quad (=\overline{x}), \tag{8.86}$$

$$\sigma^2(\widehat{k}) = \frac{\sigma_1^2\,\sigma_2^2}{\sigma_1^2+\sigma_2^2} + \sigma_c^2\,. \tag{8.87}$$

The most probable value of the physical quantity is exactly that which one obtains from the average $\overline{x}$ weighted with the inverse of the individual variances. Its overall uncertainty is the quadratic sum of the standard deviation of the weighted average and the common one. The result coincides with the simple expectation.

8.14.3 *Normalization uncertainty*

Let $x_1 \pm \sigma_1$ and $x_2 \pm \sigma_2$ be the two measured values, and σ_f the common standard uncertainty on the scale:

$$\chi^2 = \frac{1}{D}\left[(x_1-k)^2(\sigma_2^2+x_2^2\,\sigma_f^2)+(x_2-k)^2(\sigma_1^2+x_1^2\,\sigma_f^2)\right.$$
$$\left.-2\cdot(x_1-k)\cdot(x_2-k)\cdot x_1\cdot x_2\cdot\sigma_f^2\right], \tag{8.88}$$

where $D = \sigma_1^2\,\sigma_2^2 + (x_1^2\,\sigma_2^2 + x_2^2\,\sigma_1^2)\,\sigma_f^2$. We obtain in this case the following result:

$$\widehat{k} = \frac{x_1\,\sigma_2^2 + x_2\,\sigma_1^2}{\sigma_1^2+\sigma_2^2+(x_1-x_2)^2\,\sigma_f^2}, \tag{8.89}$$

$$\sigma^2(\widehat{k}) = \frac{\sigma_1^2\,\sigma_2^2 + (x_1^2\,\sigma_2^2+x_2^2\,\sigma_1^2)\,\sigma_f^2}{\sigma_1^2+\sigma_2^2+(x_1-x_2)^2\,\sigma_f^2}\,. \tag{8.90}$$

With respect to the previous case, $\widehat{k}$ has a new term $(x_1-x_2)^2\,\sigma_f^2$ in the denominator. As long as this is negligible with respect to the individual variances we still get the weighted average $\overline{x}$, otherwise a smaller value is obtained. Calling r the ratio between $\widehat{k}$ and $\overline{x}$, we obtain

$$r = \frac{\widehat{k}}{\overline{x}} = \frac{1}{1 + \frac{(x_1 - x_2)^2}{\sigma_1^2 + \sigma_2^2}\,\sigma_f^2}\,. \qquad (8.91)$$

Written in this way, one can see that the deviation from the simple average value depends on the compatibility of the two values and on the normalization uncertainty. This can be understood in the following way: as soon as the two values are in some disagreement, the fit starts to vary the normalization factor (in a hidden way) and to squeeze the scale by an amount allowed by σ_f, in order to minimize the χ^2. The reason the fit prefers normalization factors smaller than 1 under these conditions lies in the standard formalism of the covariance propagation, where only first derivatives are considered. This implies that the individual standard deviations are not rescaled by lowering the normalization factor, but the points get closer.

Example 1. Consider the results of two measurements, $8.0 \cdot (1 \pm 2\,\%)$ and $8.5 \cdot (1 \pm 2\,\%)$, having a $10\,\%$ common normalization error. Assuming that the two measurements refer to the same physical quantity, the best estimate of its true value can be obtained by fitting the points to a constant. Minimizing χ^2 with $\mathbf{V}$ estimated empirically by the data, as explained in the previous section, one obtains a value of 7.87 ± 0.81, which is surprising to say the least, since the most probable result is outside the interval determined by the two measured values.

Example 2. A real life case of this strange effect which occurred during the global analysis of the 'R ratio' in e^+e^- annihilation performed by the CELLO Collaboration [81], is shown in Fig. 8.2. The data points represent the averages in energy bins of the results of the PETRA and PEP experiments. They are all correlated and the bars show the total uncertainty (see Ref. [82] for details). In particular, at the intermediate stage of the analysis shown in the figure, an overall $1\,\%$ systematic error due to theoretical uncertainties was included in the covariance matrix. The R values above 36 GeV show the first hint of the rise of the e^+e^- cross-section due to the Z° pole. At that time it was very interesting to prove that the observation was not just a statistical fluctuation. In order to test this, the R measurements were fitted with a theoretical function having <u>no</u> Z° contributions, using only data below a certain energy. It was expected that a fast increase of χ^2 per number of degrees of freedom ν would be observed above 36 GeV, indicating that a the-

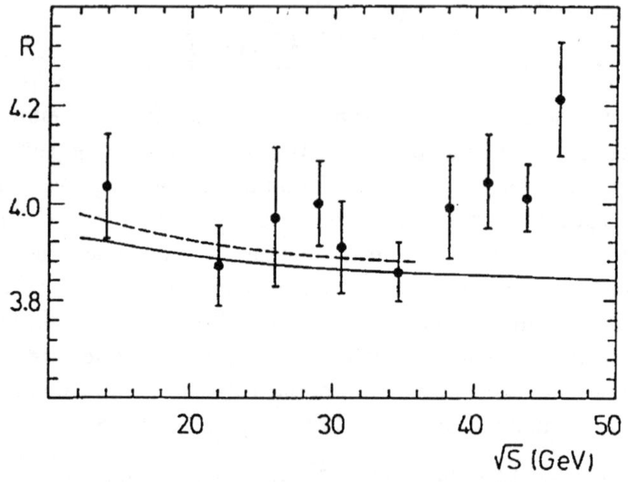

Fig. 8.2 R measurements from PETRA and PEP experiments with the best fits of QED+QCD to all the data (full line) and only below 36 GeV (dashed line). All data points are correlated (see text).

oretical prediction without Z° would be inadequate for describing the high-energy data. The surprising result was a 'repulsion' (see Fig. 8.2) between the experimental data and the fit: Including the high-energy points with larger R a lower curve was obtained, while χ^2/ν remained almost constant.

To see the source of this effect more explicitly let us consider an alternative way often used to take the normalization uncertainty into account. A scale factor f, by which all data points are multiplied, is introduced to the expression of the χ^2:

$$\chi^2_A = \frac{(f\,x_1 - k)^2}{(f\,\sigma_1)^2} + \frac{(f\,x_2 - k)^2}{(f\,\sigma_2)^2} + \frac{(f-1)^2}{\sigma_f^2}\,. \tag{8.92}$$

Let us also consider the same expression when the individual standard deviations are not rescaled:

$$\chi^2_B = \frac{(f\,x_1 - k)^2}{\sigma_1^2} + \frac{(f\,x_2 - k)^2}{\sigma_2^2} + \frac{(f-1)^2}{\sigma_f^2}\,. \tag{8.93}$$

The use of χ_A^2 always gives the result $\widehat{k} = \overline{x}$, because the term $(f-1)^2/\sigma_f^2$ is harmless[11] as far as the value of the minimum χ^2 and the determination on $\widehat{k}$ are concerned. Its only influence is on $\sigma(\widehat{k})$, which turns out to be equal to quadratic combination of the weighted average standard deviation with $\sigma_f\,\overline{x}$, the normalization uncertainty on the average. This result corresponds to the usual one when the normalization factor in the definition of χ^2 is not included, and the overall uncertainty is added at the end.

Instead, the use of χ_B^2 is equivalent to the covariance matrix: The same values of the minimum χ^2, of $\widehat{k}$ and of $\sigma(\widehat{k})$ are obtained, and $\widehat{f}$ at the minimum turns out to be exactly the r ratio defined above. This demonstrates that the effect happens when the data values are rescaled independently of their standard uncertainties. The effect can become huge if the data show mutual disagreement. The equality of the results obtained with χ_B^2 with those obtained with the covariance matrix allows us to study, in a simpler way, the behavior of r $(=\widehat{f})$ when an arbitrary number of data points are analyzed. The fitted value of the normalization factor is

$$\widehat{f} = \frac{1}{1+\sum_{i=1}^{n}\frac{(x_i-\overline{x})^2}{\sigma_i^2}\,\sigma_f^2}\,. \tag{8.94}$$

If the values of x_i are consistent with a common true value it can be shown that the expected value of $\widehat{f}$ is

$$<\widehat{f}> = \frac{1}{1+(n-1)\,\sigma_f^2}\,. \tag{8.95}$$

Hence, there is a bias on the result when for a non-vanishing σ_f a large number of data points are fitted. In particular, the fit on average produces a bias larger than the normalization uncertainty itself if $\sigma_f > 1/(n-1)$. One can also see that $\sigma^2(\widehat{k})$ and the minimum of χ^2 obtained with the covariance matrix or with χ_B^2 are smaller by the same factor r than those obtained with χ_A^2.

---

[11]This can be seen by rewriting Eq. (8.92) as

$$\frac{(x_1-k/f)^2}{\sigma_1^2}+\frac{(x_2-k/f)^2}{\sigma_2^2}+\frac{(f-1)^2}{\sigma_f^2}\,.$$

For any f, the first two terms determine the value of k, and the third one binds f to 1.

8.14.4 *Peelle's Pertinent Puzzle*

To summarize, when there is an overall uncertainty due to an unknown systematic error and the covariance matrix is used to define χ^2, the behavior of the fit depends on whether the uncertainty is on the offset or on the scale. In the first case the best estimates of the function parameters are exactly those obtained without overall uncertainty, and only the parameters' standard deviations are affected. In the case of unknown normalization errors, biased results can be obtained. The size of the bias depends on the fitted function, on the magnitude of the overall uncertainty and on the number of data points.

It has also been shown that this bias comes from the linearization performed in the usual covariance propagation. This means that, even though the use of the covariance matrix can be very useful in analyzing the data in a compact way using available computer algorithms, care is required if there is one large normalization uncertainty which affects all the data.

The effect discussed above has also been observed independently by R.W. Peelle and reported the year after the analysis of the CELLO data [81]. The problem has been extensively discussed among the community of nuclear physicists, where it is currently known as "Peelle's Pertinent Puzzle" [83]. Cases in Physics in which this effect has been found to have biased the result are discussed in Refs. [84, 85]. A recent report of a similar *"pathological best fit"* can be found in Ref. [86].

Chapter 9

Bayesian unfolding

"Now we see but a poor reflection as in a mirror..."
"Now I know in part..."
(1 Cor 13,12)

9.1 Problem and typical solutions

In any experiment the distribution of the measured observables differs from that of the corresponding *true* physical quantities due to physics and detector effects. For example, one may be interested in measuring the variables x and Q^2 in deep-inelastic scattering events. In such a case one is able to build statistical estimators which in principle have a physical meaning similar to the true quantities, but which have a non-vanishing variance and are also distorted due to QED and QCD radiative corrections, parton fragmentation, particle decay and limited detector performances. The aim of the experimentalist is to *unfold* the observed distribution from all these distortions so as to extract the true distribution (see also Refs. [87, 88]). This requires a satisfactory knowledge of the overall effect of the distortions on the true physical quantity.

When dealing with only one physical variable the usual method for handling this problem is the so-called *bin-to-bin* correction: one evaluates a *generalized efficiency* (it may even be larger than unity) by calculating the ratio between the number of events falling in a certain bin of the reconstructed variable and the number of events in the same bin of the true variable with a Monte Carlo simulation. This efficiency is then used to estimate the number of true events from the number of events observed in that bin. Clearly this method requires the same subdivision in bins of the true and the experimental variable and hence it cannot take into account large migrations of events from one bin to the others. Moreover it neglects the unavoidable correlations between adjacent bins. This approximation is valid only if the amount of migration is negligible and if the standard deviation of the smearing is smaller than the bin size.

An attempt to solve the problem of migrations is sometimes made by building a matrix which connects the number of events generated in one bin to the number of events observed in the other bins. This matrix is then inverted and applied to the measured distribution. This immediately produces inversion problems if the matrix is singular. On the other hand, there is no reason from a probabilistic point of view why the inverse matrix should exist. This can easily be seen by taking the example of two bins of the true quantity both of which have the same probability of being observed in each of the bins of the measured quantity. It follows that treating probability distributions as vectors in space is not correct, even in principle. Moreover the method is not able to handle large statistical fluctuations even if the matrix can be inverted (if we have, for example, a very large number of events with which to estimate its elements and we choose the binning in such a way as to make the matrix not singular). The easiest way to see this is to think of the unavoidable negative terms of the inverse of the matrix which in some extreme cases may yield negative numbers of unfolded events. Quite apart from these theoretical reservations, the actual experience of those who have used this method is rather discouraging, the results being highly unstable.

9.2 Bayes' theorem stated in terms of causes and effects

Let us state Bayes' theorem in terms of several independent *causes* (C_i, $i = 1, 2, \ldots, n_C$) which can produce one *effect* (E). For example, if we consider deep-inelastic scattering events, the effect E can be the observation of an event in a cell of the measured quantities $\{\Delta Q^2_{meas}, \Delta x_{meas}\}$. The causes C_i are then all the possible cells of the true values $\{\Delta Q^2_{true}, \Delta x_{true}\}_i$. Let us assume we know the *initial probability* of the causes $P(C_i)$ and the conditional probability that the i-th cause will produce the effect $P(E\,|\,C_i)$. The Bayes formula is then

$$P(C_i\,|\,E) = \frac{P(E\,|\,C_i)\,P(C_i)}{\sum_{l=1}^{n_C} P(E\,|\,C_l)\,P(C_l)}\,. \tag{9.1}$$

$P(C_i \mid E)$ depends on the initial probability of the causes. If one has no better prejudice concerning $P(C_i)$ the process of inference can be started from a uniform distribution.

The final distribution depends also on $P(E \mid C_i)$. These probabilities must be calculated or estimated with Monte Carlo methods. One has to keep in mind that, in contrast to $P(C_i)$, these probabilities are not updated by the observations. So if there are ambiguities concerning the choice of $P(E \mid C_i)$ one has to try them all in order to evaluate their *systematic effects* on the results.

9.3 Unfolding an experimental distribution

If one observes $n(E)$ events with effect E, the expected number of events assignable to each of the causes is

$$\widehat{n}(C_i) = n(E)\, P(C_i \mid E)\,. \tag{9.2}$$

As the outcome of a measurement one has several possible effects E_j ($j = 1, 2, \ldots, n_E$) for a given cause C_i. For each of them the Bayes formula (9.1) holds, and $P(C_i \mid E_j)$ can be evaluated. Let us write Eq. (9.1) again in the case of n_E possible effects[1], indicating the initial probability of the causes with $P_\circ(C_i)$:

$$P(C_i \mid E_j) = \frac{P(E_j \mid C_i)\, P_\circ(C_i)}{\sum_{l=1}^{n_C} P(E_j \mid C_l)\, P_\circ(C_l)}\,. \tag{9.3}$$

One should note the following.

- $\sum_{i=1}^{n_C} P_\circ(C_i) = 1$, as usual. Note that if the probability of a cause is initially set to zero it can never change, i.e. if a cause does not exist it cannot be invented.
- $\sum_{i=1}^{n_C} P(C_i \mid E_j) = 1$. This normalization condition, mathematically trivial since it comes directly from Eq. (9.3), indicates that each effect must come from one or more of the causes under examination. This means that if the observables also contain a non-negligible amount of background, this needs to be included among the causes.

[1]The broadening of the distribution due to the smearing suggests a choice of n_E larger than n_C. It is worth mentioning that there is no need to reject events where a measured quantity has a value outside the range allowed for the physical quantity. For example, in the case of deep-inelastic scattering events, cells with $x_{meas} > 1$ or $Q^2_{meas} < 0$ give information about the true distribution too.

- $0 \leq \epsilon_i \equiv \sum_{j=1}^{n_E} P(E_j \,|\, C_i) \leq 1$. There is no need for each cause to produce at least one of the effects. ϵ_i gives the *efficiency* of finding the cause C_i in any of the possible effects.

After N_{obs} experimental observations one obtains a distribution of frequencies $\boldsymbol{n}(E) \equiv \{n(E_1), n(E_2), \ldots, n(E_{n_E})\}$. The expected number of events to be assigned to each of the causes (taking into account only the observed events) can be calculated by applying Eq. (9.2) to each effect:

$$\widehat{n}(C_i)|_{obs} = \sum_{j=1}^{n_E} n(E_j)\, P(C_i \,|\, E_j)\,. \tag{9.4}$$

When inefficiency[2] is also brought into the picture, the best estimate of the true number of events becomes

$$\widehat{n}(C_i) = \frac{1}{\epsilon_i} \sum_{j=1}^{n_E} n(E_j)\, P(C_i \,|\, E_j) \qquad \epsilon_i \neq 0\,. \tag{9.5}$$

From these unfolded events we can estimate the true total number of events, the final probabilities of the causes and the overall efficiency:

$$\begin{aligned} \widehat{N}_{true} &= \sum_{i=1}^{n_C} \widehat{n}(C_i)\,, \\ \widehat{P}(C_i) \equiv P(C_i \,|\, \boldsymbol{n}(E)) &= \frac{\widehat{n}(C_i)}{\widehat{N}_{true}}\,, \\ \widehat{\epsilon} &= \frac{N_{obs}}{\widehat{N}_{true}}\,. \end{aligned}$$

If the initial distribution $\boldsymbol{P}_\circ(C)$ is not consistent with the data, it will not agree with the final distribution $\widehat{\boldsymbol{P}}(C)$. The closer the initial distribution is to the true distribution, the better the agreement is. For simulated data one can easily verify that the distribution $\widehat{\boldsymbol{P}}(C)$ lies between $\boldsymbol{P}_\circ(C)$ and the true one. This suggests proceeding iteratively. Figure 9.1 shows an example of a bidimensional distribution unfolding.

More details about iteration strategy, evaluation of uncertainty, etc. can be found in Ref. [89]. I would just like to comment on an obvious criticism that may be made: "*the iterative procedure is against the Bayesian spirit*, since the same data are used many times for the same inference". In principle the objection is valid, but in practice this technique is a "trick"

[2] If $\epsilon_i = 0$ then $\widehat{n}(C_i)$ will be set to zero, since the experiment is not sensitive to the cause C_i.

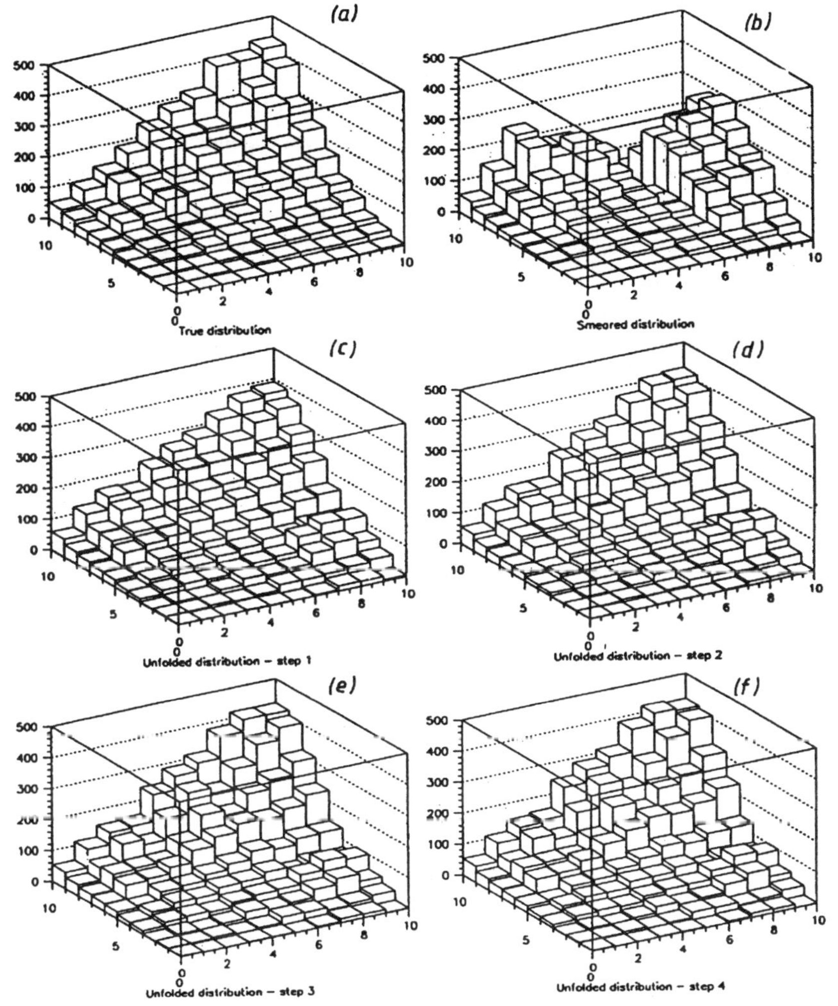

Fig. 9.1 Example of a two-dimensional unfolding: true distribution (a), smeared distribution (b) and results after the first four steps [(c) to (f)].

to give to the experimental data a weight (an importance) larger than that of the priors. A more rigorous procedure which took into account uncertainties and correlations of the initial distribution would have been much more complicated. An attempt of this kind can be found in Ref. [90]. Examples of unfolding procedures performed with non-Bayesian methods are described in Refs. [87, 88, 15].

What is usually known in (especially particle) physics as unfolding belongs, more in general, to the class of problems elsewhere known as image reconstruction, or image restoration. When we think of images, we think immediately of millions of pixels. Thus, the very general procedure discussed here becomes infeasible (image working with a most general $10^7 \times 10^7$ *smearing matrix*!). Simplifications and modelizations are then mandatory, for example assuming that the smearing only affects close cells (pixels), taking a Gaussian for the smearing function, assuming a uniform noise (or at most described by some simple function), and so on. The subject becomes complicated and goes beyond the purpose of this text and we refer to more specialized literature. Within the Bayesian framework, an extra ingredient which is often used is Maximum Entropy. Therefore, a starting points to search for material on the subject are Maximum Entropy books and on line resources, like Refs. [91, 92, 93]. A nice introduction to the subject is given by Ken Hanson [94].

Part 3

Further comments, examples and applications

Chapter 10

Miscellanea on general issues in probability and inference

"You see, a question has arisen,
about which we cannot come to an agreement,
probably because we have read too many books"
(Brecht's Galileo)

10.1 Unifying role of subjective approach

I would like to give some examples to clarify what I mean by 'linguistic schizophrenia' (see Sec. 3.2). Let us consider the following:

(1) probability of a '6' when tossing a die;
(2) probability that the 100 001st event will be accepted in the acceptance cuts of the analysis of simulated events, if I know that 91 245 out of 100 000 generated events[1] have already been accepted;
(3) probability that a real event will be accepted in the analysis, given the knowledge of point 2, and assuming that exactly the same analysis program is used, and that the Monte Carlo describes best the physics and the detector;
(4) probability that an observed track is π^+, if I have learned from the Monte Carlo that ...;
(5) probability that the Higgs mass is greater than 400 GeV;
(6) probability that the 1000th decimal digit of π is 5;
(7) probability of rain tomorrow;
(8) probability that the US dollar will be exchanged at $\geq 2\,\mathrm{DM}$ before the end of 1999 (statement made in spring 1998).

[1]Please note that 'event' is also used here according to HEP jargon (this is quite a case of homonymy to which one has to pay attention, but it has nothing to do with the linguistic schizophrenia I am talking about).

Let us analyze in detail the statements.

- The evaluation of point 1 is based on considerations of physical symmetry, using the combinatorial evaluation rule. The first remark is that a convinced frequentist should abstain from assessing such a probability until he has collected statistical data on that die. Otherwise he is implicitly assuming that the frequency-based definition is not a definition, but one of the possible evaluation rules (and then the concept can only be that related to the degree of belief...).
  For those who, instead, believe that probability is only related to symmetry the answer appears to be absolutely objective: 1/6. But it is clear that one is in fact giving a very precise and objective answer to something that is not real ('the idealized die'). Instead, we should only talk about reality. This example should help to clarify the de Finetti sentence quoted in Sec. 2.2 (*"The classical view ..."*, in particular, *"The original sentence becomes meaningful if reversed..."*).
- Point 2 leads to a consistent answer within the frequentistic approach, which is numerically equal to the subjective one [see, for example, Eqs. (7.3) and (7.10)], whilst it has no solution in a combinatorial definition.
- Points 3 and 4 are different from point 2. The frequentistic definition is not applicable. The translation from simulated events to real events is based on beliefs, which may be as firmly based as you like, but they remain beliefs. So, although this operation is routinely carried out by every experimentalist, it is meaningful only if the probability is meant as a degree of belief and not a limit of relative frequency.
- Points 3–7 are only meaningful if probability is interpreted as a degree of belief.[2]

The unifying role of subjective probability should be clear from these examples. All those who find statements 1–7 meaningful, are implicitly using subjective probability. If not, there is nothing wrong with them, on condition that they make probabilistic statements only in those cases where their definition of probability is applicable (essentially never in real life and in research). If, however, they still insist on speaking about probability outside the condition of validity of their definition, refusing the point of view of subjective probability, they fall into the self-declared linguistic

[2]In fact, one could use the combinatorial evaluation in point 6 as well, because of the discussed cultural reasons, but not everybody is willing to speak about the probability of something which has a very precise value, although unknown.

schizophrenia of which I am talking, and they generate confusion.[3]

Another very important point is the crucial role of coherence (see Sec. 3.2), which allows the exchange of the value of the probability between rational individuals: if someone tells me that he judges the probability of a given event to be 68%, then I imagine that he is as confident about it as he would be about extracting a white ball from a box which contains 100 balls, 68 of which are white. This event could be related, for example, to the result of a measurement:

$$\mu = \mu_\circ \pm \sigma(\mu)\,,$$

assuming a Gaussian model. If an experimentalist feels ready to place a 2:1 bet[4] in favor of the statement, but not a 1:2 bet against it, it means that his assessment of probability is not coherent. In other words, he is cheating, for he knows that his result will be interpreted differently from what he really believes (he has consciously overestimated the 'error bar', because he is afraid of being contradicted). If you want to know whether a result is coherent, take an interval given by 70% of the quoted uncertainty and ask the experimentalist if he is ready to place a 1:1 bet in either direction.

10.2 Frequentists and combinatorial evaluation of probability

In the previous section it was said that frequentists should abstain from assessing probabilities if a long-run experiment has not been carried out. But frequentists do, using a sophisticated reasoning, of which perhaps not everyone is aware. I think that the best way to illustrate this reasoning is with an example of an authoritative exponent, Polya [95], who adheres to von Mises' views [54].

> *"A bag contains p balls of various colors among which there are exactly f white balls. We use this simple apparatus to produce a random mass phenomenon. We draw a ball, we look at its color and we write W if the ball is white, but we write D if it is of a different color. We put back the ball just drawn into the bag, we shuffle the balls in the bag, then we draw again one and note the color of this second ball, W or D. In*

[3]See for example Refs. [79] and [73], where it is admitted that the Bayesian approach is good for decision problems, although they stick to the frequentistic approach.

[4]This corresponds to a probability of $2/3 \approx 68\%$.

proceeding so, we obtain a random sequence (...):

$$W\ D\ D\ D\ W\ D\ D\ W\ W\ D\ D\ D\ W\ W\ D\,.$$

*What is the long range relative frequency of the white balls?
Let us assume that the balls are homogeneous and exactly spherical, made of the same material and having the same radius. Their surfaces are equally smooth, and their different coloration influences only negligibly their mechanical behavior, if it has any influence at all. The person who draws the balls is blindfolded or prevented in some other manner from seeing the balls. The position of the balls in the bag varies from one drawing to the other, is unpredictable, beyond our control. Yet the permanent circumstances are well under control: the balls are all the same shape, size, and weight; they are* indistinguishable *by the person who draws them.
Under such circumstances we see no reason why one ball should be preferred to another and we naturally expect that, in the long run, each ball will be drawn approximately* equally often. *Let us say that we have the patience to make 10 000 drawings. Then we should expect that each of the p balls will appear about*

$$\frac{10\,000}{p}\ \textit{times}\,.$$

There are f white balls. Therefore, in 10 000 drawings, we expect to get white

$$f\,\frac{10\,000}{p} = 10\,000\,\frac{f}{p}\ \textit{times};$$

this is the expected frequency of the white balls. To obtain the relative frequency, we have to divide by the number of observations, or drawings, that is, 10 000. And so we are led to the statement: the long range relative frequency, or probability, *of the white balls is f/p.
The letters f and p are chosen to conform to the traditional mode of expression. As we have to draw one of the p balls, we have to choose one of p* possible *cases. We have good reasons (equal condition of the p balls) not to prefer any of these p possible cases to any other. If we wish that a white ball should be drawn (for example, if we are betting on white), the f white balls appear to us as* favorable *cases. Hence we can describe the probability f/p as the* ratio of the number of favorable cases to the number of possible cases."

The approach sketched in the above example is based on the refusal of calling probability (the intuitive concept of it) by its name. The term 'probability' is used instead for 'long-range relative frequency'. Nevertheless, the value of probability is not evaluated from the information about past frequency, but from the hypothetical long-range relative frequency, based on: a) plausible (and subjective!) reasoning on equiprobability (although not stated with this term) of the possible outcomes; b) the expectation ($\equiv$ belief) that the relative frequency will be equal to the fraction of white balls in the bag.[5] The overall effect is to confuse the matter, without any philosophical or practical advantages (compare the twisted reasoning of the above example with Hume's lucid exposure of the concept of probability and its evaluation by symmetry arguments, reported in Sec. 2.2).

10.3 Interpretation of conditional probability

As repeated throughout these notes, and illustrated with many examples, probability is always conditioned probability. Absolute probability makes no sense. Nevertheless, there is still something in the 'primer' which can be misleading and that needs to be clarified, namely the so-called 'formula of conditional probability' (Sec. 3.5.2):

$$P(E \mid H) = \frac{P(E \cap H)}{P(H)} \qquad (P(H) \neq 0)\,. \tag{10.1}$$

What does it mean? Textbooks present it as a definition (a kind of 4th axiom), although very often, a few lines later in the same book, the formula $P(E \cap H) = P(E \mid H) \cdot P(H)$ is presented as a theorem (!).

In the subjective approach, one is allowed to talk about $P(E \mid H)$ independently of $P(E \cap H)$ and $P(H)$. In fact, $P(E \mid H)$ is just the assessment of the probability of E, under the condition that H is true. Then it cannot depend on the probability of H. It is easy to show with an example that this point of view is rather natural, whilst that of considering Eq. (10.1) as a definition is artificial. Let us take

- H = the mass of the Higgs particle is 250 GeV;
- E = the Higgs decay products which detected in a LHC detector (the Large Hadron Collider is the proton-proton collider under construction at CERN laboratory in Geneva);
- the evaluation of $P(E \mid H)$ is a standard PhD student task. He chooses $m_H = 250\,\text{GeV}$ in the Monte Carlo and counts how many events pass the cuts (for the interpretation of this operation, see the previous section). No one would think that $P(E \mid H)$ must be evaluated only from

[5]Sometimes this expectation is justified advocating the law of large numbers, expressed by the Bernoulli theorem. This is unacceptable, as discussed in Sec. 7.3.

$P(E \cap H)$ and $P(H)$, as the definition (10.1) would imply. Moreover, the procedure is legitimate even if we knew with certainty that the Higgs mass was below 200 GeV and, therefore, $P(H) = 0$.

In the subjective approach, Eq. (10.1) is a true theorem required by coherence. It means that although one can speak of each of the three probabilities independently of the others, once two of them have been elicited, the third is constrained. It is interesting to demonstrate the theorem to show that it has nothing to do with the kind of heuristic derivation of Sec. 3.5.2:

- Let us imagine a coherent bet on the conditional event $E \,|\, H$ to win a unitary amount of money ($B = 1$, as the scale factor is inessential). Remembering the meaning of conditional probability in terms of bets (see Sec. 3.5.2), this means that
  - we pay (with certainty) $A = P(E \,|\, H)$;
  - we win 1 if E and H are both verified (with probability $P(E \cap H)$);
  - we get our money back (i.e. A) if H does not happen (with probability $P(\overline{H})$).
- The expected value of the 'gain' G is given by the probability of each event multiplied by the gain associated with each event:

$$\mathrm{E}(G) = 1 \cdot (-P(E \,|\, H)) + P(E \cap H) \cdot 1 + P(\overline{H}) \cdot P(E \,|\, H)\,,$$

  where the first factors of the products on the right-hand side of the formula stand for probability, the second for the amount of money. It follows that

$$\begin{aligned} \mathrm{E}(G) &= -P(E \,|\, H) + P(E \cap H) + (1 - P(H)) \cdot P(E \,|\, H) \\ &= P(E \cap H) - P(E \,|\, H) \cdot P(H)\,. \end{aligned} \tag{10.2}$$

- Coherence requires the rational better to be indifferent to the direction of the bet, i.e. $\mathrm{E}(G) = 0$. Applying this condition to Eq. (10.2) we obtain Eq. (10.1).

10.4 Are the beliefs in contradiction to the perceived objectivity of physics?

This is one of the most important points to be clarified since it is felt by many to be the biggest obstacle, preventing them from understanding the Bayesian approach: is there a place for beliefs in science? The usual criticism is that science must be objective and, hence, that there should be no room for subjectivity. A colleague once told me: *"I do not believe something. I assess it. This is not a matter for religion!"*

As I understand it, there are two possible ways to surmount the obstacle. The first is to try to give a more noble status of objectivity to the Bayesian approach, for example by formulating objective priors. In my opinion the main result of this attempt is to spoil the original nature of the theory, by adding dogmatic ingredients [33]. The second way consists, more simply, in recognizing that beliefs are a natural part of doing science.[6] Admitting that they exist does not spoil the perceived objectivity of well-established science. In other words, one needs only to look closely at how frontier science makes progress, instead of seeking refuge in an idealized concept of objectivity.[7]

Clearly this discussion would require another book, and not just some side remarks, but I am confident that the reader for whom this report is intended, and who is supposed to have working experience in frontier research, is already prepared for what I am going to say. I find it hard to discuss these matters with people who presume to teach us about the way physics, and science in general, proceeds, without having the slightest direct experience of what they are talking about.

First of all, I would like to invite you to pay attention to the expressions we use in private and public discussions, and in written matter too. Here are some examples:

- "I believe that ...";
- "We have to get experience with ...";
- "I don't trust that guy (or that collaboration, or that procedure)";
- "Oh yes, if this has been told you by ..., then you can rely on it";
- "We have only used the calorimeter for this analysis, because we are not yet confident with the central detector";
- The evening before I had to talk about this subject, I overheard the following conversation in the CERN cafeteria:

[6]The Franklin's book *"Experiment, right or wrong"* [96] presents an interesting historical analysis of several classical XX century particle physics experiments. The physicist activity is seen as a *"set of strategies of reasonable beliefs in experimental results"*, which can be *"explained in terms of Bayesian confirmation theory"*.

[7]My preferred motto on this matter is *"no one should be allowed to speak about objectivity unless he has had 10–20 years working experience in frontier science, economics, or any other applied field"*.

– Young fellow: *"I have measured the resistivity, and it turns out to be* $10^{11}\,\Omega$*"*;
– Senior: *"No, it cannot be. Tomorrow I will make the measurement and I am sure to get the right value. ...By the way, have you considered that ...?"*

The statistician Don Berry [97] amused himself by counting how many times Stephen Hawking uses 'belief', 'to believe', or synonyms, in his *'A brief history of time'*. The book could have been entitled *'A brief history of beliefs'*, Berry pointed out in his talk. By the way, as other famous physicists cited in this book, Hawking too likes to express his beliefs in terms of bets. Here is a bet between Kip Thorne and Hawking as to whether Cygnus X-1 is a black hole:

> *"Whereas Stephen Hawking has a large investment in General Relativity and Black Holes and desires an insurance policy, and whereas Kip Thorne likes to live dangerously without an insurance policy,*
> *Therefore be it resolved that Stephen Hawking bets 1 year's subscription to 'Penthouse' as against Kip Thorne's wager of a 4-year subscription to 'Private Eye', that Cignus X-1 does not contain a black hole of mass above the Chandrasekhar limit."* [98]

The role of beliefs in physics has been highlighted in a particularly efficient way by the science historian Peter Galison [61]:

> *"Experiments begin and end in a matrix of beliefs. ...beliefs in instrument type, in programs of experiment enquiry, in the trained, individual judgments about every local behavior of pieces of apparatus."*

Then, taking as an example the discovery of the positron, he remarks:

> *"Taken out of time there is no sense to the judgment that Anderson's track 75* [see Fig. 10.1] *is a positive electron; its textbook reproduction has been denuded of the prior experience that made Anderson confident in the cloud chamber, the magnet, the optics, and the photography."*[8]

This means that pure observation does not create, or increase, knowledge without personal inputs which are needed to elaborate the information.[9]

[8] For an accurate historical account of the positron discovery Ref. [99] is recommended.

[9] A few years ago, I met an elderly physicist at a meeting of the Italian Physical Society, who was nostalgic about the good old times when *"we could see* $\pi \to \mu \to e$ *decay in emulsions"*, and complained that at present the sophisticated electronic experiments are based on models. It took me a while to convince him that in emulsions too he had a

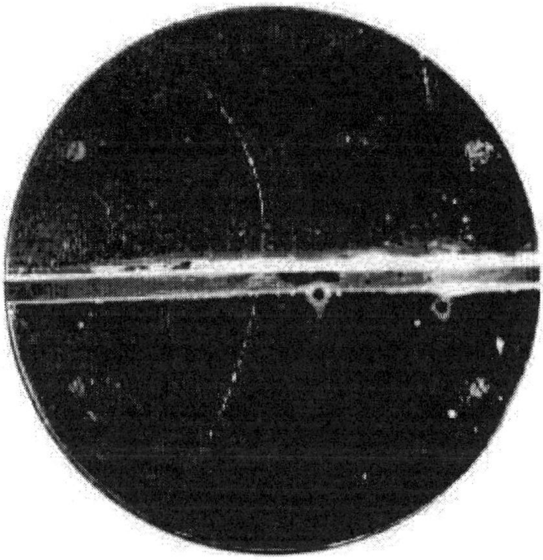

Fig. 10.1 Anderson's first picture of a positron track.

In fact, there is nothing really objective in physics, if by objective we mean that something follows necessarily from observation, like the proof of a theorem. There are, instead, beliefs everywhere. Nevertheless, physics is objective, or at least that part of it that is at present well established, if we mean by 'objective', that a rational individual cannot avoid believing it. This is the reason why we can talk in a relaxed way about beliefs in physics without even remotely thinking that it is at the same level as the stock exchange, betting on football scores, or ...New Age. The reason is that, after centuries of experimentation, theoretical work and successful predictions, there is such a consistent network of beliefs, it has acquired the status of an objective construction: one cannot mistrust one of the elements of the network without contradicting many others. Around this solid core of objective knowledge there are fuzzy borders which correspond to areas of present investigations, where the level of intersubjectivity is still very low. Nevertheless, when one proposes a new theory or model, one has to check immediately whether it contradicts some well-established beliefs. An interesting example comes from the 1997 HERA high Q^2 events, already discussed in Sec. 1.9. A positive consequence of this claim was to trigger a kind of mega-exercise undertaken by many theorists, consisting

model and that he was not seeing these particles either.

of systematic cross-checks of HERA data, candidate theories, and previous experimental data. The conclusion is that the most influential physicists[10] tend not to believe a possible explanation in terms of new physics [100, 101]. But this has little to do with the 'statistical significance' of the events. It is more a question of the difficulty of inserting this evidence into what is considered to be the most likely network of beliefs.

I would like to conclude this section with a Feynman quotation [102].

> *"Some years ago I had a conversation with a layman about flying saucers - because I am scientific I know all about flying saucers! I said 'I don't think there are flying saucers'. So my antagonist said, 'Is it impossible that there are flying saucers? Can you prove that it's impossible?' 'No', I said, 'I can't prove it's impossible. It's just very unlikely'. At that he said, 'You are very unscientific. If you can't prove it impossible then how can you say that it's unlikely?' But that is the way that is scientific. It is scientific only to say what is more likely and what less likely, and not to be proving all the time the possible and impossible. To define what I mean, I might have said to him, 'Listen, I mean that from my knowledge of the world that I see around me, I think that it is much more likely that the reports of flying saucers are the results of the known irrational characteristics of terrestrial intelligence than of the unknown rational efforts of extra-terrestrial intelligence'. It is just more likely. That is all."*

10.5 Frequentists and Bayesian 'sects'

Many readers may be interested in how the problem 'to Bayes or not to Bayes' is viewed by statisticians. In order to thoroughly analyze the situation, one should make a detailed study not only of the probability theory, but also of the history and sociology of statistical science. The most I can do here is to give personal impressions, certainly biased, and some references. I invite the reader to visit the statistics department in his University, browse their journals and books, and talk to people (and to judge the different theses by the logical strength of their arguments, not weighing them just by numbers...).

[10] Outstanding physicists have no reluctance in talking explicitly about beliefs. Then, paradoxically, objective science is for those who avoid the word 'belief' nothing but the set of beliefs of the influential scientists to which they believe...

10.5.1 *Bayesian versus frequentistic methods*

An often cited paper for a reasonably balanced discussion [79] on the subject is the article *"Why isn't everyone a Bayesian?"*, by B. Efron [103]. Key words of the paper are: *Fisherian inference*; *Frequentistic theory*; *Neyman–Pearson–Wald*; *Objectivity*. For this reason, pointing out this paper as 'balanced' is not really fair. Nevertheless, I recommend reading the article, together with the accompanying comments and the reply by the author published in the same issue of the journal (a typical practice amongst statisticians).

So, it is true that *"Fisherian and Neyman–Pearson–Wald ideas have shouldered Bayesian theory aside in statistical practice"* [103], but *"The answer is simply that statisticians do not know what the statistical paradigm says. Why should they? There are very few universities in the world with statistics departments that provides a good course on the subject."* [104] Essentially, the main point of the Efron paper is to maintain traditional methods, despite the *"disturbing catalog of inconsistencies"* [103], and the *"powerful theoretical reasons for preferring Bayesian inference"* [103]. Moreover, perhaps not everybody who cites the Efron paper is aware of further discussions about it, like the letter in which Zellner [105] points out that one of the problems posed by Efron already had a Bayesian solution (in the Jeffreys' book [49]), that Efron admitted to knowing and even to having used [106]. As a kind of final comment on this debated paper, I would like to cite Efron's last published reply I am aware of [106]:

> *"First of all let me thank the writers for taking my article in its intended spirit: not as an attack on the Bayesian enterprise, but rather as a critique of its preoccupation with philosophical questions, to the detriment of statistical practice. Meanwhile I have received some papers, in particular one from A.F.M. Smith, which show a healthy Bayesian interest in applications, so my worries were overstated if not completely groundless."*

There are some other references which I would like to suggest if you are interested in forming your own opinion on the subject. They have also appeared in The American Statistician, where in 1997 an entire Teaching Corner section of the journal [97] was devoted to three papers presented in a round table on 'Bayesian possibilities for introductory statistics' at the 156th Annual Meeting of the American Statistical Association, held in Chicago, in August 1996. For me these articles are particularly important because I was by chance in the audience of the round table (really 'by chance'!). At the end of the presentations I was finally convinced that frequentism was dead, at least as a philosophical idea. I must say, I was persuaded by the non-arguments of the defender of frequentism even more than by the arguments of the defenders of the Bayesian approach. I report here the abstract[11] of Moore, who presented the 'reason to hesitate' to teach Bayesian statistics:

> *"The thesis of this paper is that Bayesian inference, important though it is for statisticians, is among the mainly important statistical topics that it is wise to avoid in most introductory instruction. The first reason is pragmatic (and empirical): Bayesian methods are as yet relatively little used in practice. We have an obligation to prepare students to understand the statistics they will meet in their further studies and work, not the statistics we may hope will someday replace now-standard methods. A second argument also reflects current conditions: Bayesians do not agree on standard approaches to standard problem settings. Finally, the reasoning of Bayesian inference, depending as it does on ideas of conditional probability, is quite difficult for beginners to appreciate. There is of course no easy path to a conceptual grasp of inference, but standard inference at least rests on repetition of one straightforward question, What would happen if I did this many times? "*

Even if some arguments might be valid, thinking about statisticians who make surveys in a standardized form (in fields that they rarely understand, such as medicine and agriculture), surely they do not hold in physics, even less in frontier physics. As I commented to Moore after his talk, what is important for a physicist is not "what would happen if I did this many times?", but "what am I learning by the experiment?".[12]

10.5.2 *Subjective or objective Bayesian theory?*

Once you have understood that probability and frequencies are different concepts, that probability of hypothesis is a useful and natural concept for

[11] I quote here the original abstract, which appears on page 18 of the conference abstract book.

[12] I also made other comments on the general illogicality of his arguments, which you may easily imagine by reading the abstract. For these comments I even received applause from the audience, which really surprised me, until I learned that David Moore is one of the most authoritative American statisticians: only a outsider like me would have said what I said...

reporting results, that Bayes' theorem is a powerful tool for updating probability and learning from data, that priors are important and pretending that they do not exist is equivalent to assuming them flat, and so on, it is difficult to then take a step back. However, it is true that there is no single shared point of view among those who, generally speaking, support the Bayesian approach. I don't pretend that I can provide an exhaustive analyze of the situation here, or to be unbiased about this matter either.

The main schools of thought are the 'subjectivists' and the 'objectivists'. The dispute may look strange to an outsider, if one thinks that both schools use probability to represent degrees of belief. Nevertheless, objectivists want to minimize the person's contribution to the inference, by introducing reference priors (for example Jeffreys' priors [49]) or other constraints, such as maximum entropy (for an overview see Refs. [27] and [107]). The motto is *"let the data speak for themselves"*. I find this subject highly confusing, and even Bernardo and Smith (Bernardo is one of the key persons behind reference priors) give the impression of contradicting themselves often on this point as, for example, when the subject of reference analysis is introduced:

> *"to many attracted to the formalism of the Bayesian inferential paradigm, the idea of a non-informative prior distribution, representing 'ignorance' and 'letting the data speak for themselves' has proved extremely seductive, often being regarded as synonymous with providing objective inferences. It will be clear from the general subjective perspective we have maintained throughout this volume, that we regard this search for 'objectivity' to be misguided. However, it will also be clear from our detailed development in Section 5.4 that we recognize the rather special nature and role of the concept of a 'minimal informative' prior specification - appropriately defined! In any case, the considerable body of conceptual and theoretical literature devoted to identifying 'appropriate' procedures for formulating prior representations of 'ignorance' constitutes a fascinating chapter in the history of Bayesian Statistics. In this section we shall provide an overview of some of the main directions followed in this search for a Bayesian 'Holy Grail'.* [27]

In my point of view, the extreme idea along this line is represented by the Jaynes' 'robot' (*"In order to direct attention to constructive things and away from controversial irrelevance, we shall invent an imaginary being. Its brain is to be designed by us, so that it reasons according to certain defined rules. These rules will be deduced from simple desiderata which, it appears to us, would be desirable in human brains"* [108]).

As far as I understand it, I see only problems with objectivism, although I do agree on the notion of a commonly perceived objectivity, in the sense of intersubjectivity (see Sec. 10.4). Frankly, I find probabilistic evaluations made by a coherent subjectivist, assessed under personal responsibility, to be more trustworthy and more objective than values obtained in a mechanical way using objective prescriptions [33].

Moving to a philosophical level deeper than this kind of angels' sex debate (see Sec. 3.11), there is the important issue of what an event is. All events listed in Sec. 10.1 (apart from that of point 4) are somehow verifiable. Perhaps one will have to wait until tomorrow, the end of 1999, or 2010, but at a certain point the event may become certain, either true or false. However, one can think about other events, examples of which have been shown in these notes, that are not verifiable, either for a question of principle, or by accident.

- The old friend could die, carrying with him the secret of whether he had been cheating, or simply lucky (Sec. 3.6).
- The particle interacts with the detector (Sec. 3.5.4) and continues its flight: was it really a π or a μ?
- Using our best knowledge about temperature measurement we can state that the temperature of a room at a certain instant is $21.7 \pm 0.3\,^\circ\mathrm{C}$ with 95% probability (Sec. 10.1); after the measurement the window is opened, the weather changes, the thermometer is lost: how is it possible to verify the event '$21.4 \le T/^\circ\mathrm{C} \le 22.0$'?

This problem is present every time we make a probabilistic statement about physics quantities. It is present not only when a measurand is critically time dependent (the position of a plane above the Atlantic), but also in the case of fundamental constants. In this latter case we usually believe in the progress of science and thus we hope that the quantity will be measured so well in the future that it will one day become a kind of exact value, in comparison to today's uncertainty. But it is absurd to think that one day we will be able to 'open an electron' and read on a label all its properties with an infinite number of digits. This means that for scientific applications it is convenient to enlarge the concept of an event (see Sec. 3.2), releasing the condition of verifiability.[13] At this point the normative role of the

[13]It is interesting to realize, in the light of this reflection, that the ISO definition of true value (*"a value compatible with the definition of a given particular quantity"*, see Secs. 1.2 and 1.3) can accommodate this point of view.

hypothetical coherent bet becomes crucial. A probability evaluation, made by an honest person well-trained in applying coherence on verifiable events, becomes, in my opinion, the only means by which degrees of belief can be exchanged among rational people.[14] We have certainly reached a point in which the domain of physics, metaphysics and moral overlap, but it looks to me that this is exactly the way in which science advances.

It seems to me that almost all Bayesian schools support this idea of the extended meaning of an event, explicitly or tacitly (anyone who speaks about $f(\theta)$, with θ a parameter of a distribution, does it). A more radical point of view, which is very appealing from the philosophical perspective, but more difficult to apply (especially as long as the quantification of the uncertainty in measurements is concerned) is the predictive approach (or operational subjectivism), along the lines of de Finetti's thinking. The concept of probability is strictly applied only to real observables, very precisely ('operationally') defined.[15] The events are all associated with discrete uncertain numbers (integer or rational), in the simplest case 1 or 0 if there are only two possibilities (true or false). Having excluded non-observables, it makes no sense to speak of $f(\mu \,|\, \text{data})$, but only of $f(x \,|\, \text{data})$, where X stands for a future (or, in general, not yet known) observation (see Sec. 6.6). For the moment I prefer to stick to our 'metaphysical' true values, but I encourage anyone who is interested in this subject to read Lad's book [69], which also contains a very interesting philosophical and historical introduction to the subject. For a recent presentation of this approach, see also Coletti-Scozzafava treatise [70].

---

[14]Take, for example, the bet used by Laplace to report his conclusion about boy/girl birth chance in Paris (see quote at the beginning of Chapter 7). A bet which needs hundred seventy-five years to be settled is meaningless. Moreover, nobody would consider Laplace so naïve to believe that the population in Paris would remain stationary hundreds of years. Therefore, that bet can only be understood as a virtual bet to express his degree of belief.

[15]This point of view is followed not only by mathematicians or philosophers. Here is how Schrödinger defines the *event* in his "Foundation of the theory of probability" [41]. "*By event we understand for the present purpose a simple or arbitrary complicated individual state of affairs (or fact or occurrence or happening) which either does or conceivably might obtain in the real world around us and of which we are given a description in words, clear and accurate enough to leave us no doubt, that by taking (or having taken at the time or times in question) sufficient cognizance of the relevant part of the world it would be possible to decide unambiguously, whether this particular fact (or state of affairs, etc.) actually obtains or not, any third possibility being excluded. ... As verbal descriptions not fulfilling the requirement and thus, in my opinion, not specifying an event, let me mention 'The distance between the towns D. and G. is between 157.357124 and 157.357125'.*"

10.5.3 *Bayes' theorem is not everything*

Finally, I would like to recall that Bayes' theorem is a very important tool, but it can be used only when the scheme of prior, likelihood, and final is set up, and the distributions are properly normalized. This happens very often in measurement uncertainty problems, but less frequently in other applications, such as assessing the probabilities of hypotheses. When Bayes' theorem is not applicable, conclusions may become strongly dependent on individuals and the only guidance remains the normative rule of the hypothetical coherent bet.

10.6 Biased Bayesian estimators and Monte Carlo checks of Bayesian procedures

This problem has already been raised in Secs. 5.2 and 5.3. We have seen there that the expected value of a parameter can be considered, somehow, to be analogous to the estimators[16] of the frequentistic approach. It is well known, from courses on conventional statistics, that one of the nice properties an estimator should have is that of being free of bias.

Let us consider the case of Poisson and binomial distributed observations, exactly as they have been treated in Secs. 7.1 and 7.4, i.e. assuming a uniform prior. Using the typical notation of frequentistic analysis, let us indicate with θ the parameter to be inferred, with $\hat{\theta}$ its estimator.

Poisson: $\theta = \lambda$; X indicates the possible observation and $\hat{\theta}$ is the estimator in the light of X:

$$\begin{aligned} \hat{\theta} &= \mathrm{E}(\lambda \,|\, X) = X + 1\,, \\ \mathrm{E}(\hat{\theta}) &= \mathrm{E}(X+1) = \lambda + 1 \neq \lambda\,. \end{aligned} \tag{10.3}$$

The estimator is biased, but consistent (the bias become negligible when

[16]It is worth remembering that, in the Bayesian approach, the complete answer is given by the final distribution. The prevision ('expected value') is just a way of summarizing the result, together with the standard uncertainty. Besides motivations based on penalty rules, which we cannot discuss, a practical justification is that what matters for any further approximate analysis, are expected values and standard deviation, whose properties are used in uncertainty propagation. There is nothing wrong in providing the mode(s) of the distribution or any other quantity one finds it sensible to summarize $f(\mu)$ as well. What I dislike is the reduction of one of these summaries of the final probability density function to a 'Bayesian estimator' in the frequentistic sense. And, unfortunately, when many practitioners say to use Bayesian statistics, they refer to orrible things of this kind.

X is large).

Binomial: $\theta = p$; after n trials one may observe X favorable results, and the estimator of p is then

$$\hat{\theta} = \mathrm{E}(p \,|\, X) = \frac{X+1}{n+2}\,,$$
$$\mathrm{E}(\hat{\theta}) = \mathrm{E}\left(\frac{X+1}{n+2}\right) = \frac{n\,p+1}{n+2} \neq p\,. \tag{10.4}$$

In this case as well the estimator is biased, but consistent.

What does it mean? The result looks worrying at first sight, but, in reality, it is the analysis of bias that is misleading. In fact:

- the initial intent is to reconstruct at best the parameter, i.e. the true value of the physical quantity identified with it;
- the freedom from bias requires only that the expected value of the estimator should equal the value of the parameter, for a given value of the parameter,

$$\mathrm{E}(\hat{\theta} \,|\, \theta) = \theta \qquad (\text{e.g. } \mathrm{E}(\hat{\lambda} \,|\, \lambda) = \lambda)\,,$$
$$(\text{i.e. } \int \hat{\theta}\, f(\hat{\theta} \,|\, \theta)\, \mathrm{d}\hat{\theta} = \theta\ )\,. \tag{10.5}$$

But what is the true value of θ? We don't know, otherwise we would not be wasting our time trying to estimate it (always keep real situations in mind!). For this reason, our considerations cannot depend only on the fluctuations of $\hat{\theta}$ around θ, but also on the different degrees of belief of the possible values of θ. Therefore they must depend also on $f_\circ(\theta)$. For this reason, the Bayesian result is that which makes the best use[17] of the state of knowledge about θ and of the distribution of $\hat{\theta}$ for each possible value θ. This can be easily understood by going back to the examples of Sec. 1.7. It is also easy to see that the freedom from bias of the frequentistic approach requires $f_\circ(\theta)$ to be uniformly distributed from $-\infty$ to $+\infty$ (implicitly, as frequentists refuse the very concept of probability of θ). Essentially,

[17] I refer to the steps followed in the proof of Bayes' theorem given in Sec. 2.7. They should convince the reader that $f(\theta \,|\, \hat{\theta})$ calculated in this way is the best we can say about θ. Some say that *"in the Bayesian inference the answer is the answer"* (I have heard this sentence from Adrian Smith at the Valencia-6 conference), in the sense that one can use all his best knowledge to evaluate the probability of an event, but then, whatever happens, cannot change the assessed probability, but, at most, it can — and must — be taken into account for the next assessment of a different, although analogous event.

whenever a parameter has a limited range, the frequentistic analysis decrees that Bayesian estimators are biased.

There is another important and subtle point related to this problem, namely that of the Monte Carlo check of Bayesian methods. Let us consider the case depicted in Fig. 1.3 and imagine making a simulation, choosing the value $\mu_\circ = 1.1$, generating many (e.g. 10000) events, and considering three different analyses:

(1) a maximum likelihood analysis;
(2) a Bayesian analysis, using a flat distribution for μ;
(3) a Bayesian analysis, using a distribution of μ 'of the kind' $f_\circ(\mu)$ of Fig. 1.3, assuming that we have a good idea of the kind of physics we are doing.

Which analysis will reconstruct a value closest to $\mu_\circ$? You don't really need to run the Monte Carlo to realize that the first two procedures will perform equally well, while the third one, advertised as the best in these notes, will systematically underestimate $\mu_\circ$!

Now, let us assume we have observed a value of x, for example $x = 1.1$. Which analysis would you use to infer the value of μ? Considering only the results of the Monte Carlo simulation it seems obvious that one should choose one of the first two, but certainly not the third!

This way of thinking is wrong, but unfortunately it is often used by practitioners who have no time to understand what is behind Bayesian reasoning, who perform some Monte Carlo tests, and decide that the Bayesian statistics "does not work" (this is an actual statement I have heard by some particle physics colleague). The solution to this apparent paradox is simple. If you believe that μ is distributed like $f_\circ(\mu)$ of Fig. 1.3, then you should use this distribution in the analysis and also in the generator. Making a simulation based only on a single true value, or on a set of points with equal weight, is equivalent to assuming a flat distribution for μ and, therefore, it is not surprising that the most grounded Bayesian analysis is that which performs worst in the simple-minded frequentistic checks. It is also worth remembering that priors are not just mathematical objects to be plugged into Bayes' theorem, but must reflect prior knowledge. Any inconsistent use of them leads to paradoxical results.

10.7 Frequentistic coverage

Another prejudice toward Bayesian inference shared by practitioners who have grown up with conventional statistics is related to the so-called 'frequentistic coverage'. Since, in my opinion, this is a kind of condensate of frequentistic nonsense,[18] I avoid summarizing it in my own words, as the risk of distorting something in which I cannot see any meaning is too high. A quotation[19] taken from Ref. [110] should 'clarify' the issue:

> *"Although particle physicists may use the words 'confidence interval' loosely, the most common meaning is still in terms of original classical concept of "coverage" which follows from the method of construction suggested in Fig. ... This concept is usually stated (too narrowly, as noted below) in terms of a hypothetical ensemble of similar experiments, each of which measures m and computes a confidence interval for m_t with say, 68% C.L. Then the classical construction guarantees that in the limit of a large ensemble, 68% of the confidence intervals contain the unknown true value m_t, i.e., they 'cover' m_t. This property, called coverage in the frequentistic sense, is the defining property of classical confidence intervals. It is important to see this property as what it is: it reflects the relative frequency with which the statement, 'm_t is in the interval (m_1, m_2)', is a true statement. The probabilistic variables in this statement are m_1 and m_2; m_t is fixed and unknown. It is equally important to see what frequentistic coverage is not: it is not a statement about the degree of belief that m_t lies within the confidence interval of a particular experiment. The whole concept of 'degree of belief' does not exist with respect to classical confidence intervals, which are cleverly (some would say devilishly) defined by a construction which keeps strictly*

---

[18]Günter Zech says, more optimistically: *"Coverage is the magic objective of classical confidence bounds. It is an attractive property from a purely aesthetic point of view but it is not obvious how to make use of this concept."* [109] But I think that Aristoteles would have gotten mad if somebody had tried to convince him of the proposition "the interval contains $\theta_\circ$ with probability β" does not imply "$\theta_\circ$ is in that interval with probability β". Zech seems to me to be overly patient in comparing frequentist and Bayesian methods [11] from a pragmatic physicist's point of view (in the sense Ref. [30]), reaching at the conclusion that *"classical methods are not recommended because they violate the Likehood Principle, they can produce inconsistent results, suffer from lack of precision and generality."* Instead, I usually refuse to make systematic comparisons of frequentist versus Bayesian methods for solving the same problem, simply because I was taught in elementary school – and I still believe it is correct – not to compare or add non-homogeneous objects, like apples and potatoes, meters and liters ... and then frequentistic CL's and probability intervals. For a clear and concise introduction to what frequentistic confidences mean and of what they do not mean (together with some historical remarks) see Ref. [12].

[19]The translation of the symbols is as follows: m stands for the measured quantity (x or $\hat{\theta}$ in these notes); m_t stands for the true value (μ or θ here); $P(\cdot\,|\,\cdot)$ for $f(\cdot\,|\,\cdot)$.

> *to statements about $P(m \,|\, m_t)$ and never uses a probability density in the variable m_t.*
> *This strict classical approach can be considered to be either a virtue or a flaw, but I think that both critics and adherents commonly make a mistake in describing coverage from the narrow point of view which I described in the preceding paragraph. As Neyman himself pointed out from the beginning, the concept of coverage is not restricted to the idea of an ensemble of hypothetical nearly-identical experiments. Classical confidence intervals have a much more powerful property: if, in an ensemble of real, different, experiments, each experiment measures whatever observables it likes, and construct a 68% C.L. confidence interval, then in the long run 68% of the confidence intervals cover the true value of their respective observables. This is directly applicable to real life, and is the real beauty of classical confidence intervals."*

I think that the reader can judge for himself whether this approach seems reasonable. From the Bayesian point of view, the full answer is provided by $P(m_t \,|\, m)$, to use the same notation of Ref. [110]. If this evaluation has been carried out under the requirement of coherence, from $P(m_t \,|\, m)$ one can evaluate a probability for m_t to lie in the interval (m_1, m_2). If this probability is for instance 68%, it implies:

- one believes 68% that m_t is in that interval;
- one is ready to place a $\approx 2:1$ bet on m_t being in that interval and a $\approx 1:2$ bet on m_t being elsewhere;
- if one imagines n situations in which one has similar conditions and thinks of the relative frequency with which one expects that this statement will be true (f_n), logic applied to the basic rules of probability implies that, with the increasing n, one considers more and more improbable that f_n will differ much from 68% (Bernoulli theorem, see Sec. 7.3).

So, the intuitive concept of 'coverage' is naturally included in the Bayesian result and it is expressed in intuitive terms (probability of true value and expected frequency). But this result has to depend also on priors, as seen in the previous section and in many other places in this report (see, for example, Sec. 1.7). Talking about coverage independently of prior knowledge (as frequentists do) makes no sense, and leads to contradictions and paradoxes. Imagine, for example, an experiment operated for one hour in a collider having a center of mass energy of 200 GeV and reporting zero candidate events for zirconium-antizirconium production[20] in e^+e^- in the absence of expected background. I do not think that there is a single particle physicist ready to believe that, if the experiment is repeated many times, in only 68% of the cases the 68% C.L. interval [0.00, 1.29] will contain the true value of the 'Poisson signal mean', as a blind use of Table II of Ref. [73] would imply.[21] If this example seems a bit odd, I invite you to think about the many 95% C.L. lower limits on the mass of postulated particles published in the last decades. Do you really believe that in 95% of the cases the mass is above the limit, and in 5% of the cases below the limit? If this is the case, you would bet $5 on a mass value below the limit, and receive $100 if this happened to be true (you should be ready to accept the bet, since, if you believe in frequentistic coverage, you must admit that the bet is fair). But perhaps you will never accept such a bet because you believe much more than 95% that the mass is above the limit, and then the bet is not fair at all; or because you are aware of thousands of lower limits, and a particle has never shown up on the 5% side...[22]

[20]Zirconium, of atomic mass about 91 could be produced in such collisions, from the energy-momentum balance point of view. But, being a complex atom, it will 'never' be produced in e^+e^- annihilations.

[21]One would object that this is, more or less, the result that we could obtain making a Bayesian analysis with a uniform prior. But it was said that this prior assumes a positive attitude of the experimenters, i.e. that the experiment was planned, financed, and operated by rational people, with the hope of observing something (see Secs. 6.7 and 7.4). This topic, together with the issue of reporting experimental results in a prior-free way, is discussed in detail in Chapter 13.

[22]According to one of the authors of the 'unified approach' (Ref. [73]), the reason ***"is because people have been flip-flopping. Had they used a unified approach, this would not have happened"*** [111]. Up to the end of 2002 the "unified approach" has been used for hundreds of results in almost 200 papers. I wonder what the next excuse will be to justify the fact that such confidence intervals are not 'true' as often as expected by frequentistic coverage. In other words, the problem is not only about the meaning of the long term property of frequentistic coverage, but it is a question of internal consistency. According to the ideas at the basis of coverage, a statistical method could give an 'absurd' region (like a null interval) in 5% of the cases and a 'tautological' interval (like $[-\infty, +\infty]$) in 95% of the cases, without being accused of inconsistency, since, in the long term, the intervals cover the true value 95% of the times (see details in Ref. [12]). Here the problem is that, besides what the intervals of the 'unified approach' might mean, experience shows that they just do not do 'their job'.

10.7.1 *Orthodox teacher versus sharp student - a dialogue by George Gabor*

As a last comment about frequentistic ideas related to confidence intervals and coverage I would like to add here a nice dialogue, which was circulated via internet on February 1999, with an introduction and comment by the author, the statistician George Gabor [112] of Dalhousie University (Halifax, N.S., Canada). It was meant as a contribution to a discussion triggered by D.A. Berry (that of Refs. [13, 97]) a few days before.

> *"Perhaps a Socratic exchange between an ideally sharp, i.e. not easily bamboozled student (S.) of a typical introductory statistics course and his prof (P.) is the best way to illustrate what I think of the issue. The class is at the point where confidence interval (CI) for the normal mean is introduced and illustrated with a concrete example for the first time.*
>
> **P.** *...and so a 95% CI for the unknown mean is (1.2, 2.3).*
>
> **S.** *Excuse me sir, just a few minutes ago you emphasized that a CI is some kind of random interval with certain coverage properties in REPEATED trials.*
>
> **P.** *Correct.*
>
> **S.** *What, then, is the meaning of the interval above?*
>
> **P.** *Well, it is one of the many possible realizations from a collection of intervals of a certain kind.*
>
> **S.** *And can we say that the 95 collective, is somehow carried over to this particular realization?*
>
> **P.** *No, we can't. It would be worse than incorrect; it would be meaningless for the probability claim is tied to the collective.*
>
> **S.** *Your claim is then meaningless?*
>
> **P.** *No, it isn't. There is actually a way, called Bayesian statistics, to attribute a single-trial meaning to it, but that is beyond the scope of this course. However, I can assure you that there is no numerical difference between the two approaches.*
>
> **S.** *Do you mean they always agree?*
>
> **P.** *No, but in this case they do provided that you have no reason, prior to obtaining the data, to believe that the unknown mean is in any particularly narrow area.*
>
> **S.** *Fair enough. I also noticed sir that you called it 'a' CI, instead of 'the' CI. Are there others then?*
>
> **P.** *Yes, there are actually infinitely many ways to obtain CI's which all have the same coverage properties. But only the one above is a Bayesian interval (with the proviso above added, of course).*
>
> **S.** *Is Bayesian-ness the only way to justify the use of this particular one?*
>
> **P.** *No, there are other ways too, but they are complicated and they operate with concepts that draw their meaning from the collective (except the so called likelihood interval, but then this strange guy*

does not operate with probability at all).

...

It could be continued ad infinitum. Assuming sufficiently more advanced students one could come up with similar exchanges concerning practically every frequentist concept orthodoxy operates with (sampling distribution of estimates, measures of performance, the very concept of independence, etc.). The point is that orthodoxy would fail at the first opportunity had students been sufficiently sharp, open minded, and inquisitive. That we are not humiliated repeatedly by such exchanges (in my long experience not a single one has ever taken place) says more about... well, I don't quite know about what — the way the mind plays tricks with the concept of probability? The background of our students? Both?

Ultimately then we teach the orthodoxy not only because of intellectual inertia, tradition, and the rest; but also because, like good con artists, we can get away with it. And that I find very disturbing. I must agree with Basu's dictum that nothing in orthodox statistics makes sense unless it has a Bayesian interpretation. If, as is the case, the only thing one can say about frequentist methods is that they work only in so far as they don't violate the likelihood principle; and if they don't (and they frequently do), they numerically agree with a Bayesian procedure with some flat prior - then we should go ahead and teach the real thing, not the substitute. (The latter, incidentally, can live only parasitically on an illicit Bayesian usage of its terms. Just ask an unsuspecting biologist how he thinks about a CI or a P-value.)

One can understand, or perhaps follow is a better word, the historical reasons orthodoxy has become the prevailing view. Now, however, we know better."

10.8 Why do frequentistic hypothesis tests 'often work'?

The problem of classifying hypotheses according to their credibility is natural in the Bayesian framework. Let us recall briefly the following way of drawing conclusions about two hypotheses in the light of some data:

$$\frac{P(H_i\,|\,\mathrm{Data})}{P(H_j\,|\,\mathrm{Data})} = \frac{P(\mathrm{Data}\,|\,H_i)}{P(\mathrm{Data}\,|\,H_j)} \cdot \frac{P_\circ(H_i)}{P_\circ(H_j)}\,. \tag{10.6}$$

This form is very convenient, because:

- it is valid even if the hypotheses H_i do not form a complete class [a necessary condition if, instead, one wants to give the result in the standard form of Bayes' theorem given by formula (3.11)];

- it shows that the Bayes factor is an unbiased way of reporting the result (especially if a different initial probability could substantially change the conclusions);
- the Bayes factor depends only on the likelihoods of observed data and not at all on unobserved data (contrary to what happens in conventional statistics, where conclusions depend on the probability of all the configurations of data in the tails of the distribution[23]). In other words, Bayes' theorem applies in the form (10.6) and not as

$$\underbrace{\frac{P(H_i\,|\,\text{Data+Tail})}{P(H_j\,|\,\text{Data+Tail})} = \frac{P(\text{Data+Tail}\,|\,H_i)}{P(\text{Data+Tail}\,|\,H_j)} \cdot \frac{P_\circ(H_i)}{P_\circ(H_j)}}_{?};$$

- testing a single hypothesis does not make sense: one may talk of the probability of the Standard Model (SM) only if one is considering an Alternative Model (AM), thus getting, for example,

$$\frac{P(\text{AM}\,|\,\text{Data})}{P(\text{SM}\,|\,\text{Data})} = \frac{P(\text{Data}\,|\,\text{AM})}{P(\text{Data}\,|\,\text{SM})} \cdot \frac{P_\circ(\text{AM})}{P_\circ(\text{SM})}:$$

  $P(\text{Data}\,|\,\text{SM})$ can be arbitrarily small, but if there is not a reasonable alternative one has only to accept the fact that some events have been observed which are very far from the expectation value;
- repeating what has been said several times, in the Bayesian scheme the conclusions depend only on observed data and on previous knowledge; in particular, they do not depend on
  - how the data have been combined;
  - data not observed and considered to be even rarer than the observed data;
  - what the experimenter was planning to do before starting to take data. (I am referring to predefined fiducial cuts and the stopping rule, which, according to the frequentistic scheme should be defined in the test protocol. Unfortunately I cannot discuss this matter here in detail and I recommend the reading of Ref. [13].)

At this point we can finally reply to the question: "why do commonly-used methods of hypothesis testing usually work?" (see Secs 1.8 and 1.9).

[23]The necessity of using integrated distributions is due to the fact that the probability of observing a particular configuration is always very small, and a frequentistic test would reject the null hypotheses.

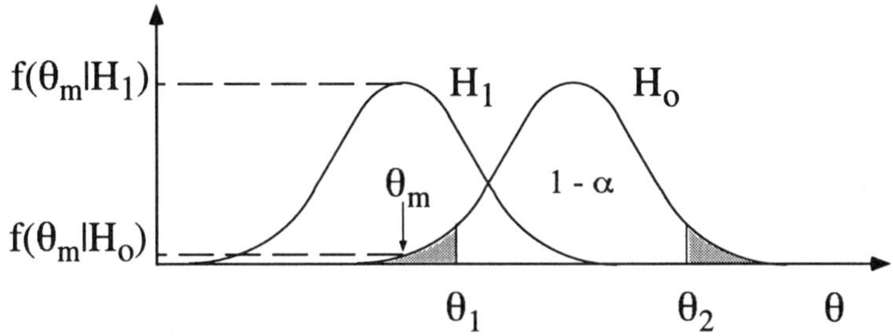

Fig. 10.2 Testing a hypothesis $H_\circ$ implies that one is ready to replace it with an alternative hypothesis.

By reference to Fig. 10.2 (imagine for a moment the figure without the curve H_1), the argument that θ_m provides evidence against $H_\circ$ is intuitively accepted and often works, not (only) because of probabilistic considerations of θ in the light of $H_\circ$, but because it is often reasonable to imagine an alternative hypothesis H_1 that

(1) maximizes the likelihood $f(\theta_m \mid H_1)$ or, at least

$$\frac{f(\theta_m \mid H_1)}{f(\theta_m \mid H_\circ)} \gg 1\,;$$

(2) has a comparable prior $[P_\circ(H_1) \approx P_\circ(H_\circ)]$, such that

$$\frac{P(H_1 \mid \theta_m)}{P(H_\circ \mid \theta_m)} = \frac{f(\theta_m \mid H_1)}{f(\theta_m \mid H_\circ)} \cdot \frac{P_\circ(H_1)}{P_\circ(H_\circ)} \approx \frac{f(\theta_m \mid H_1)}{f(\theta_m \mid H_\circ)} \longrightarrow \gg 1\,. \tag{10.7}$$

As counter-examples in which there is no correspondence between the probability of the tail and the Bayes factor, let us consider the observed data point ($x_m = 5$) and the three hypotheses of Fig. 10.3 to be compared. The probabilities of the tails, i.e. $P(X \geq x_m \mid H_i)$ are equal to 9%, 13% and 4% for $i = 1, 2, 3$. Therefore H_3 fails the 5% significance threshold, while H_1 passes the 5% threshold, but fails the 10%. The 'best agreement' seems performed by H_2. But this is certainly absurd, because $f(x_m \mid H_2)$ is equal to zero, i.e. this hypothesis cannot produce that observation, and therefore it should be falsified without any doubt! This is the effect of including the probability of non-observed data (the tails) in the inference. Instead, the Bayesian answer depends only on x_m and the priors, as it logically should

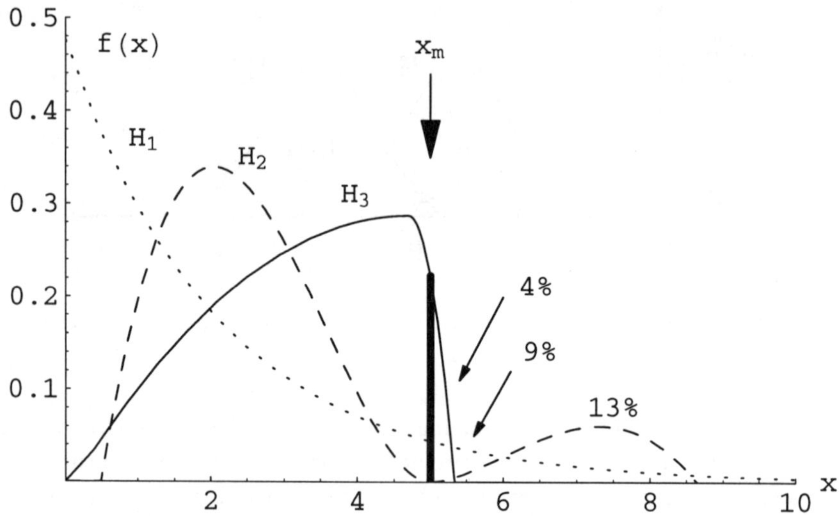

Fig. 10.3 Which of these three hypotheses is favored by the observation $x = 5$?.

be. The Bayes factors can be evaluated from

$$f(x_m \,|\, H_1) : f(x_m \,|\, H_2) : f(x_m \,|\, H_3) = 0.044 : 0 : 0.22\,. \tag{10.8}$$

Hypothesis H_2 is ruled out, while the odds for H_3 versus H_1 increase by a factor five.

Figure 10.4 shows an example in which the experimental observation cannot modify our beliefs, because the likelihood is the same for all hypotheses. This is the essence of the Likelihood Principle, a highly desirable property of frequentistic methods, but often missed (see e.g. Ref. [11]).

Summing up, even though there is no objective or logical reason why the frequentistic scheme should work, the reason why it often does work is that in many cases the test is made when one has serious doubts about the null hypothesis and reasonable (belivable alternatives) are easily conceivable. But a peak appearing in the middle of a distribution, or any excess of events, is not, in itself, a hint of new physics (Fig. 10.5 is an invitation to meditation...). My recommendations are therefore the following.

- Be very careful when drawing conclusions from χ^2 tests, '$3\,\sigma$ golden rule', and other 'bits of magic';
- Do not pay too much attention to fixed rules suggested by statistics 'experts', supervisors, and even Nobel laureates, taking also into account

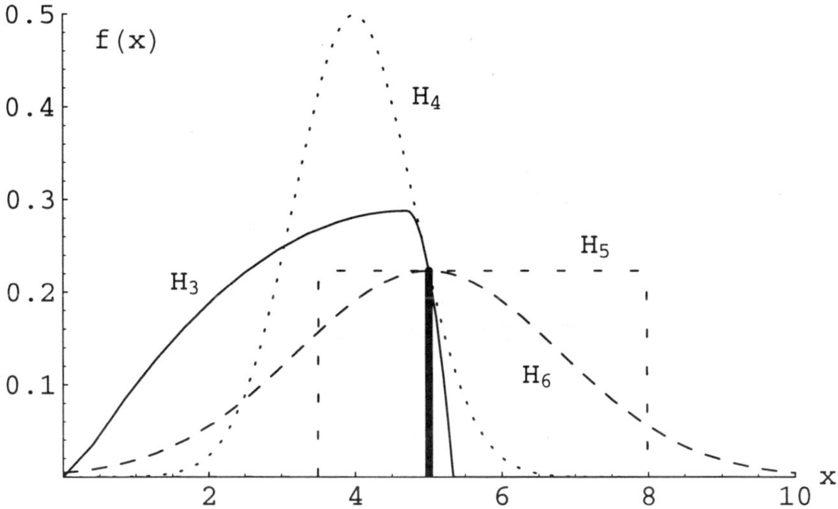

Fig. 10.4 The observation $x = 5$ cannot rationally update the reative beliefs on the four hypotheses which could have caused it.

that

- they usually have permanent positions and risk less than PhD students and postdocs who do most of the real work;
- they have been 'miseducated' by the exciting experience of the glorious 1950s to 1970s: as Giorgio Salvini says, *"when I was young, and it was possible to go to sleep at night after having added within the day some important brick to the building of the elementary particle palace. We were certainly lucky."* [114]. Especially when they were hunting for resonances, priors were very high, and the 3–4 σ rule was a good guide.

- Fluctuations exist. There are millions of frequentistic tests made every year in the world. And there is no probability theorem ensuring that the most extreme fluctuations occur to a precise Chinese student, rather than to a large HEP collaboration (this is the same reasoning of many Italians who buy national *lotteria* tickets in Rome or in motorway restaurants, because 'these tickets win more often'...).

As a conclusion to these remarks, and to invite the reader to take with much care the assumption of equiprobability of hypothesis (a hidden assumption in many frequentistic methods), I would like to add this quotation

Fig. 10.5 Experimental obituary (courtesy of Alvaro de Rujula [113]).

by Poincaré [8]:

> *"To make my meaning clearer, I go back to the game of écarté mentioned before.* [See Sec. 1.6] *My adversary deals for the first time and turns up a king. What is the probability that he is a sharper? The formulae ordinarily taught give 8/9, a result which is obviously rather surprising. If we look at it closer, we see that the conclusion is arrived at as if, before sitting down at the table, I had considered that there was one chance in two that my adversary was not honest. An absurd hypothesis, because in that case I should certainly not have played with him; and this explains the absurdity of the conclusion. The function on the à priori probability was unjustified, and that is why the conclusion of the à posteriori probability led me into an inadmissible result. The importance of this preliminary convention is obvious. I shall even add that if none were made, the problem of the à posteriori probability would have no meaning. It must be always made either explicitly or tacitly."*

10.9 Comparing 'complex' hypotheses – automatic Ockham' Razor

The comparison of hypotheses we have seen so far, recalled by Eq. (10.6) in the previous section, applies when we have models which are 'simple', in the sense that the likelihood does not depend on parameters of the models. In general, each model has a different number of parameters and the situation becomes a bit more complex. Let us consider, for example, models $\mathcal{M}_A$ characterized by n_A parameters $\boldsymbol{\alpha}$, and $\mathcal{M}_B$ with n_B parameters $\boldsymbol{\beta}$. Using probability rules, the Bayes factor (see Sec. 3.7) now becomes

$$\frac{P(\text{Data}\,|\,\mathcal{M}_A, I)}{P(\text{Data}\,|\,\mathcal{M}_B, I)} = \frac{\int P(\text{Data}\,|\,\mathcal{M}_A, \boldsymbol{\alpha}, I)\, f_0(\boldsymbol{\alpha}\,|\,I)\, \mathrm{d}\boldsymbol{\alpha}}{\int P(\text{Data}\,|\,\mathcal{M}_B, \boldsymbol{\beta}, I)\, f_0(\boldsymbol{\beta}\,|\,I)\, \mathrm{d}\boldsymbol{\beta}} \tag{10.9}$$

$$= \frac{\int \mathcal{L}_A(\boldsymbol{\alpha};\, \text{Data})\, f_0(\boldsymbol{\alpha})\, \mathrm{d}\boldsymbol{\alpha}}{\int \mathcal{L}_B(\boldsymbol{\beta};\, \text{Data})\, f_0(\boldsymbol{\beta})\, \mathrm{d}\boldsymbol{\beta}}, \tag{10.10}$$

where $f_0(\boldsymbol{\alpha}\,|\,I)$ and $f_0(\boldsymbol{\beta}\,|\,I)$ are the parameter priors. The inference depends, then, on the ***integrated likelihood***

$$\int \mathcal{L}_{\mathcal{M}}(\boldsymbol{\theta};\, \text{Data})\, f_0(\boldsymbol{\theta})\, \mathrm{d}\boldsymbol{\theta}\,, \tag{10.11}$$

where $\mathcal{M}$ and $\boldsymbol{\theta}$ stand for the generic model and its parameters. This integrated likelihood is sometimes called *evidence* in statistical jargon. Note that $\mathcal{L}_M(\boldsymbol{\theta};\, \text{Data})$ has its largest value around the maximum likelihood point $\boldsymbol{\theta}_{ML}$, but the evidence takes into account all prior possibilities of the parameters. Thus, it is not enough that the best fit of one model is superior to its alternative, in the sense that, for instance,

$$\mathcal{L}_A(\boldsymbol{\alpha}_{ML};\, \text{Data}) > \mathcal{L}_B(\boldsymbol{\beta}_{ML};\, \text{Data})\,, \tag{10.12}$$

and hence, assuming Gaussian models (see Sec. 8.1),

$$\chi^2_A(\boldsymbol{\alpha}_{min\,\chi^2};\, \text{Data}) < \chi^2_B(\boldsymbol{\beta}_{min\,\chi^2};\, \text{Data})\,, \tag{10.13}$$

in order to prefer model $\mathcal{M}_A$ to model $\mathcal{M}_B$. In this case there are not only the model priors which matter, but also the *space of possibilities* of the parameters, i.e. the *adaptation capability* of each model. It is well understood that we do not choose an $n-1$ order polynomial as the best description — 'best' in inferential terms — of the n experimental points, though such a model always offers a 'perfect fit'. Similarly, we are much more impressed by, and we tend *a posteriori* to believe more in, a theory that absolutely predicts an experimental observation, within a reasonable error, than another theory which performs similarly or even better after having adjusted a couple of parameters.

This intuitive reasoning is expressed formally in Eqs. (10.9)–(10.10). The 'evidence' is given integrating the product $\mathcal{L}(\boldsymbol{\theta})$ and $f_0(\boldsymbol{\theta})$ over the parameter space. So, the more $f_0(\boldsymbol{\theta})$ is concentrated around $\boldsymbol{\theta}_{ML}$ the greater is the 'evidence' in favor of that model. Instead, a model with a volume of the parameter space much larger than the one selected by $\mathcal{L}(\boldsymbol{\theta})$ gets disfavored.[24] The extreme limit is that of a hypothetical model with so many parameters it can describe perfectly whatever we observe. We would never take such a model seriously.

This effect is highly welcome, and follows the *Ockham's Razor* scientific rule of discarding unnecessarily complicated models (*"entities should not be multiplied unnecessarily"*). This rule comes out automatically in the Bayesian approach. A nice introduction to the connection between Ockham's Razor and Bayesian reasoning, with examples from physics, can be found in Ref. [115], while Refs. [116, 117, 118] offer recent examples of application of the method in cosmology, providing also a detailed presentation of underlying Bayesian ideas. Oher useful information and examples can be found in Ref. [119].

Although Eq. (10.9) has the nice properties which relate it to Ockham's Razor, its use requires deeper thought than was necessary in the parametric inference seen in Chapters 6-7. This is due to the fact that we cannot use improper priors (see Sec. 6.5). In fact, unless the models do not depend on the same number of parameters, defined in the same ranges, improper priors do not simplify in the numerator and the denominator of Eq. (10.9). Therefore, some care has to be taken to choose proper priors depending on the problem (see Refs. [115, 116, 117, 118]). Reference [119] shows an example of how this method 'chooses' the degree of a polynomial which describes at best the data. Finally, as recent example of model comparison applied to real data, and written also with didactic intent see Ref. [120].

[24] Another way to understand the source of the effect is to consider the integrated likelihood, or 'evidence', Eq. (10.11) as a average likelihood, i.e. average of the likelihood $\mathcal{L}_{\mathcal{M}}(\boldsymbol{\theta};\,\text{Data})$ weighted with $f_0(\boldsymbol{\theta})$. If there are 'many' values of $\boldsymbol{\theta}$ (more precisely, large regions of the parameter space, since we are dealing with continuous quantities) for which $\mathcal{L}_{\mathcal{M}}(\boldsymbol{\theta};\,\text{Data})$ vanishes, these 'many' zero's make the average likelihood small.

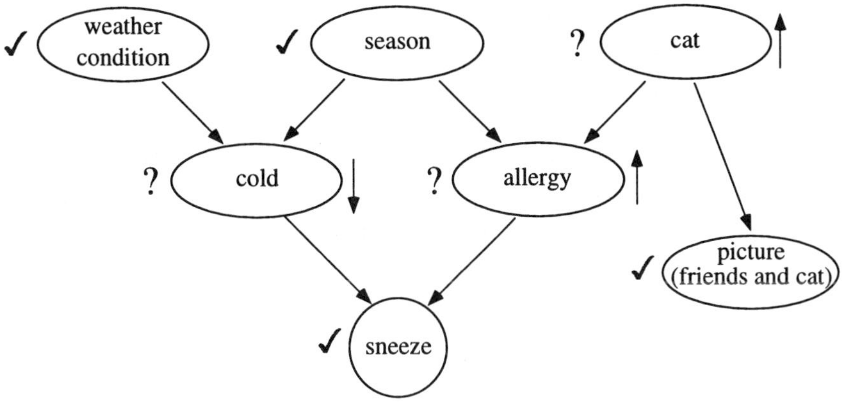

Fig. 10.6 An example of belief network.

10.10 Bayesian networks

10.10.1 *Networks of beliefs – conceptual and practical applications*

In Sec. 10.4 I mentioned the network of beliefs which give the perceived status of objectivity to consolidated science. In fact, belief networks, also called Bayesian networks, are not only an abstract idea useful in epistemology. They represent one of the most promising applications of Bayesian inference and they have generated a renewed interest in the field of artificial intelligence, where they are used for expert systems, decision makers, etc. [120] and even in forensic science [121].

Although, to my knowledge, there are not yet specific physics applications of these methods, I would like to give a rough idea of what they are and how they work, with the help of a simple example. You are visiting some friends, and, minutes after being in their house, you sneeze. You know you are allergic to pollen and to cats, but it could also be a cold. What is the cause of the sneeze? Figure 10.6 sketches the problem. There are some facts about which you are sure (the sneeze, the weather conditions and the season), but you don't know if the sneeze is a symptom of a cold or of an allergy. In particular, you don't know if there is a cat in the house.

Then, you see a picture of your friend with a cat. This could be an indication that they have a cat, but it is just an indication. Nevertheless, this indication increases the probability that there is a cat around, and then the probability that the cause of the sneeze is cat's hair allergy increases, while the probability of any other potential cause decreases. If you then establish with certainty the presence of the cat, the cause of the allergy also becomes practically certain.

The idea of Bayesian networks is to build a network of causes and effects. Each event, generally speaking, can be certain or uncertain. When there is a new piece of evidence, this is transmitted to the whole network and all the beliefs are updated. The research activity in this field consists of the most efficient way of doing the calculation, using Bayesian inference, graph theory, and numerical approximations.

If one compares Bayesian networks with other ways of pursuing artificial intelligence their superiority is rather clear: they are close to the natural way of human reasoning, the initial beliefs can be those of experts (avoiding the long training needed to set up, for example, neural networks, infeasible in practical applications), and they learn by experience as soon as they start to receive evidence.

10.10.2 *The gold/silver ring problem in terms of Bayesian networks*

As a simple Bayesian network, let us reconsider the gold/silver ring problem of Chapter 3. The diagram is framed in the *JavaBayes* [122] interface. The three upper bubbles ('nodes') of the left diagram of Fig. 10.8 correspond to the variables of the problem. The variable *Box* has three equiprobable states of value GG, GS and SS. The 'child' variable *Obs_1* has two possible states, G and S which might be true or false, with probabilities which depend on the 'parent' Box. The conditional beliefs $P(\text{Obs\_1}\,|\,\text{Box})$ are given by a 2×3 matrix

$P(\text{Obs\_1}\,|\,\text{Box})$

	Box		
Obs_1	GG	GS	SS
G	1	1/2	0
S	0	1/2	1

also shown in a *JavaBayes* dialog box (Fig. 10.7). The second observation *Obs_2* depends on the box composition and the first result — it has two parents — and needs a $2 \times 3 \times 2$ array for its representation:

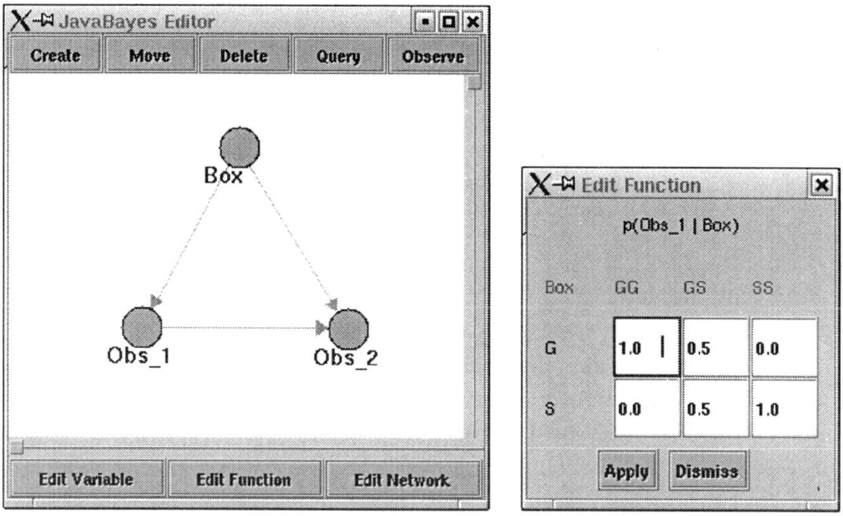

Fig. 10.7 Bayesian network to model the gold/silver ring problem implemented in *JavaBayes* [122]. The screenshot on the right shows the *JavaBayes* dialog box for entering $P(\text{Obs\_1} \mid \text{Box})$.

$P(\text{Obs\_2} \mid \text{Obs\_1}, \text{Box})$

	Obs_1 = G			Obs_1 = S		
	Box			Box		
Obs_2	GG	GS	SS	GG	GS	SS
G	1	0	0	0	1	0
S	0	1	0	0	0	1

Note that the probabilities of both outcomes have been set to zero if the conditions are impossible (in general, the sum of all probabilities given the same conditions have to sum up to 1). From these tables we can calculate the joint probability function[25] $P(\text{Obs\_2, Obs\_1, Box})$. This is done 'factorizing' conditional probabilities, using the well-known 'chain rule':

$$P(\text{Obs\_2, Obs\_1, Box}) = P(\text{Box}) \cdot P(\text{Obs\_1} \mid \text{Box}) \cdot P(\text{Obs\_2} \mid \text{Obs\_1, Box}) \qquad (10.14)$$

[25]Note the extension of the concept of probability function to include *states* and not only values of a quantity.

$P(\text{Obs\_2}, \text{Obs\_1}, \text{Box})$

	Obs_1 = G			Obs_1 = S		
	Box			Box		
Obs_2	GG	GS	SS	GG	GS	SS
G	1/3	0	0	0	1/6	0
S	0	1/6	0	0	0	1/3

Note that the numbers in this table sum up to 1. Instead, the sum of the numbers of the two rows gives the probability of getting gold or silver in the second observation (the 'marginals'). They are both equal to 1/2, as to be expected by symmetry. From this table we can get all other probability function of interest, like $P(\text{Obs\_2}, \text{Obs\_1})$ and $P(\text{Obs\_2}, \text{Box})$, $P(\text{Obs\_1})$, and so on. In particular, using Bayes' theorem, we get the probability of the second observation conditioned by the first observation. Here are the tables of $P(\text{Obs\_2}\,|\,\text{Obs\_1})$ and $P(\text{Obs\_2}, \text{Obs\_1})$:

$P(\text{Obs\_2}, \text{Obs\_1})$

	Obs_1	
Obs_2	G	S
G	1/3	1/6
S	1/6	1/3

$P(\text{Obs\_2}\,|\,\text{Obs\_1})$

	Obs_1	
Obs_2	G	S
G	2/3	1/3
S	1/3	2/3

The answer to our problems is the first element of table $P(\text{Obs\_2}\,|\,\text{Obs\_1})$, i.e. $P(\text{Obs\_2} = G\,|\,\text{Obs\_1} = G) = 2/3$.

Anyone who arrived more or less intuitively at the solution might think this procedure is no more than a formal complication. However we only need to make our network a bit more complicated and it becomes obvious that intuition is of little help. Imagine we cannot observe the extracted ring directly. The observations are mediated by persons and the possibility exists that these persons could lie. They could lie in different ways, and even react differently to gold and silver. This is the meaning of the two bottom nodes in the diagram of Fig. 10.8. The situation seems bizarre, but the analogy with detectors which lie (we say 'err') gives an idea of the scientific relevance of this modelling.

Let us assign some values to the probability functions $P(\text{Rep\_1}\,|\,\text{Obs\_1})$ and $P(\text{Rep\_2}\,|\,\text{Obs\_2})$

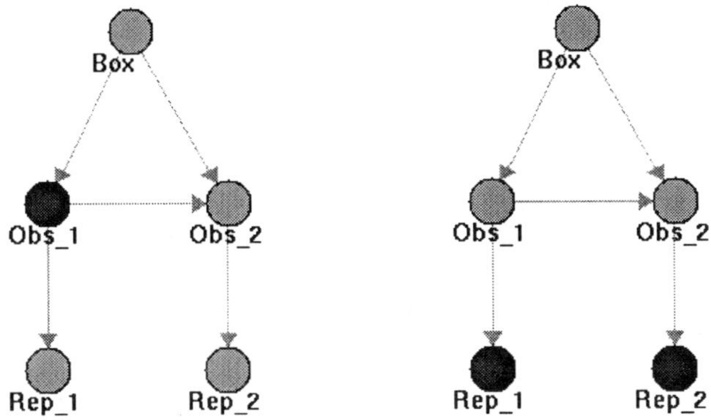

Fig. 10.8 Bayesian network to model the gold/silver ring problem, complicated by the possibility that whoever observes the ring might report something different from what he sees. The two diagrams show the graphical representation of evidence on Obs_1 or on Rep_1 and Rep_2 (darker bubbles).

P(Rep_1 | Obs_1)

	Obs_1	
Rep_1	G	S
G	0.7	0.2
S	0.3	0.8

P(Rep_2 | Obs_2)

	Obs_2	
Rep_2	G	S
G	0.75	0.1
S	0.25	0.9

and try to answer the following questions: a) what is the probability that the first observer has seen Gold if he says Gold? b) what is the probability that the second observer says Gold if the first observer said Gold? c) What is the probability that the second observation was really Gold if both observers say Gold? d) What is the probability that the box content is SS if both observers said Gold? Bayesian networks offer two kinds of help in answering these questions: conceptual help comes from the graphical representation of the problem; practical help, essential when problems become complicated, comes from the powerful mathematical and computational tools that have become available in recent years. For example, this minimal variation of the basic problem is easily solved using *JavaBayes*. The details are given in Ref. [123]. Let us give here the answers to only some of the questions:

- Probability that the observer has seen gold if he says gold:

$P(\text{Obs\_1}\,|\,\text{Rep\_1})$

	Rep_1	
Obs_1	G	S
G	0.78	0.27
S	0.22	0.73

$P(\text{Obs\_2}\,|\,\text{Rep\_2})$

	Rep_2	
Obs_2	G	S
G	0.88	0.22
S	0.12	0.78

- Probability of the second observation given the first report, and probability of the second report given the first report

$P(\text{Obs\_2}\,|\,\text{Obs\_1})$

	Obs_1	
Obs_2	G	S
G	0.59	0.42
S	0.41	0.58

$P(\text{Rep\_2}\,|\,\text{Rep\_1})$

	Rep_1	
Rep_2	G	S
G	0.49	0.38
S	0.51	0.62

- Probability of the box content given the two reports:

$P(\text{Box}\,|\,\text{Rep\_1}, \text{Rep\_2})$

Box	Rep_1 = G Rep_2 = G	Rep_1 = G Rep_2 = S	Rep_1 = S Rep_2 = G	Rep_1 = S Rep_2 = S
GG	0.80	0.36	0.25	0.07
GS	0.17	0.51	0.49	0.23
SS	0.03	0.13	0.26	0.70

More details can be found in Ref. [123], where the reasoning is extended to include the treatment of uncertainty in measurement. In the cited web site the simple version of the problem is solved 'by hand' using a *Mathematica* notebook. This gives an idea of the complex calculations needed when the problem becomes complicated. When many nodes and continuous node states (i.e. continuous random variables) are involved, the exact calculations of all summations (i.e. integrals) becomes prohibitive and Monte Carlo methods are needed. The most powerful techniques to perform numerical calculations associated to Bayesian networks are based on Markov Chain Monte Carlo (MCMC). Many references and applications can be found on the BUGS [124] web site (where also free software is available), while a starting point to search for MCMC literature is Ref. [125].

Chapter 11

Combination of experimental results: a closer look

"Every theory should be tested against experiment.
Every experiment should be tested against theory"

We provide here a practical example of how to model prior knowledge in order to solve an often debated problem in physics: how to combine data in the presence of 'outliers'. Before tackling this problem we recall the standard combination rule and its conditions of validity. The proposed method was applied to what was considered a hot problem in frontier physics at the end of the past century, i.e. the determination of the *direct CP violation parameter* ϵ'/ϵ (see Ref. [126] for the latest experimental and references on measurements and theory).

11.1 Use and misuse of the standard combination rule

Every physicist knows the rule for combining several experimental results, which we have derived in Sec. 6.3, and rewrite here, for the reader convenience, in the following form:

$$\mu = \frac{\sum_i d_i/s_i^2}{\sum_i 1/s_i^2} \tag{11.1}$$

$$\sigma(\mu) = \left(\sum_i 1/s_i^2\right)^{-\frac{1}{2}}, \tag{11.2}$$

where μ refers to (the best estimate of) the true value and $d_i \pm s_i$ stands for the individual data point (the use of s_i, instead of the usual σ_i, for the "errors" by the experiments will become clear later).

It is a matter of fact that these popular formulae, in many cases learned as kind of practical formulae in laboratory courses, are often used a bit blindly. Only when the result are manifestly odd, one is forced to think that there is something strange going on. In some cases the radicated habit to use uncritically these formulae leads to the misleading conclusion that are the individual results to be 'wrong'. Two examples of this kind have been shown already in Sec. 8.11. It is worth recalling here a case similar to that discussed in Sec. 8.11 was reported by the PDG statistics experts [79] as argument against Bayesian statistics.

Let us take another numerical example of this kind. Three independent counting experiments performed during equal observation time report the following results: $4 \pm \sqrt{4}$, $7 \pm \sqrt{7}$, and $10 \pm \sqrt{10}$. The combination according to Eqs. (11.1)–(11.2) gives 6.0 ± 1.4. The result look suspicious, because one would think of the three experiments as a single one running the triple of time of the individual experiment. The result would then be $21 \pm \sqrt{21}$, equivalent to 7.0 ± 1.5 for a single experiment. Also in this case the *standard combination rule* (11.1)–(11.2) has been misused. In fact, as shown in Sec. 6.3, this rule is based on some important assumptions:

i) all measurements refer to the same quantity;
ii) the measurements are independent;
iii) the probability distribution of d_i around μ is described by a Gaussian distribution with standard deviation given by $\sigma_i = s_i$.

If one, or several, of these hypotheses are not satisfied, the result of formulae (11.1)–(11.2) is questionable. In the case under study, the solution is simple: assumption *iii)* fails, and we have to solve the problem in the most general way to infer the Poisson parameter λ associated with the quantity of interest. Following the procedure often used in Chapters 6 and 7, we get (with obvious meaning of the symbols)

$$f(\lambda \,|\, \boldsymbol{x}) \propto e^{-n\lambda} \lambda^{\sum_{i=1}^{n} x_i} f_\circ(\lambda)\,, \tag{11.3}$$

a result already shown in Eq. (7.65). Using the numbers of our example and a uniform prior, we get a maximum of belief at $\lambda = 7$, in agreement with the intuitive considerations. The result in terms of expected value and standard deviation is, instead, 7.3 ± 1.6. The reason of an expected value slightly larger than 7 has already been discussed in Sec. 7.4, and it is consistent with assumed prior knowledge. What is mostly remarkable is that we would get exactly the same result (p.d.f. and all other summaries) if we considered an experiment having observed 21 counts.

11.2 'Apparently incompatible' experimental results

The case seen in the previous section was an easy one. More complicated is the situation in which some data point 'seem to be incompatible.' Now we are confronted with the problem that we are never absolutely sure if the hypotheses behind the standard combination rule are true or not. If we were absolutely convinced that the hypotheses were correct, there would be no reason to hesitate to apply Eqs. (11.1)–(11.2), no matter how 'apparently incompatible' the data points might appear. But we know from experience that unrecognized sources of systematic errors might affect the results, or that the uncertainty associated with the recognized sources might be underestimated (but we also know that, often, this kind of uncertainty is prudently overstated...).

As is always the case in the domain of uncertainty, there is no 'objective' method for handling this problem; neither in deciding if the data are in mutual disagreement, nor in arriving at a universal solution for handling those cases which are judged to be troublesome. Only good sense gained by experience can provide some guidance. Therefore, all automatic 'prescriptions' should be considered *cum grano salis*. For example, the usual method for checking the hypothesis that 'the data are compatible with each other' is to make a χ^2 test. The hypothesis is accepted if, generally speaking, the χ^2 does not differ too much from the expected value. As a strict rule, the χ^2 test is not really logically grounded (Sec. 1.8) although it does 'often work', due to implicit hypotheses which are external to the standard χ^2 test scheme (see Sec. 10.8), but which lead to mistaken conclusions when the unstated hypotheses are not reasonable (Sec. 1.9). Therefore, I shall not attempt here to quantify the degree of suspicion. I shall assume a situation in which experienced physicists, faced with a set of results, tend to be uneasy about the mutual consistency of the picture that those data offer.

As a real life example, which was considered a hot topic a couple of years ago [127], let us consider the results of Tab. 11.1, which are also reported in a graphical form in Fig. 11.1. Figure 11.2 shows also the combined result obtained using Eqs. (11.1)–(11.2), as well as some combinations of subsamples of the results. These results have not been chosen as the best example of disagreeing data, but because of the physics interest, and

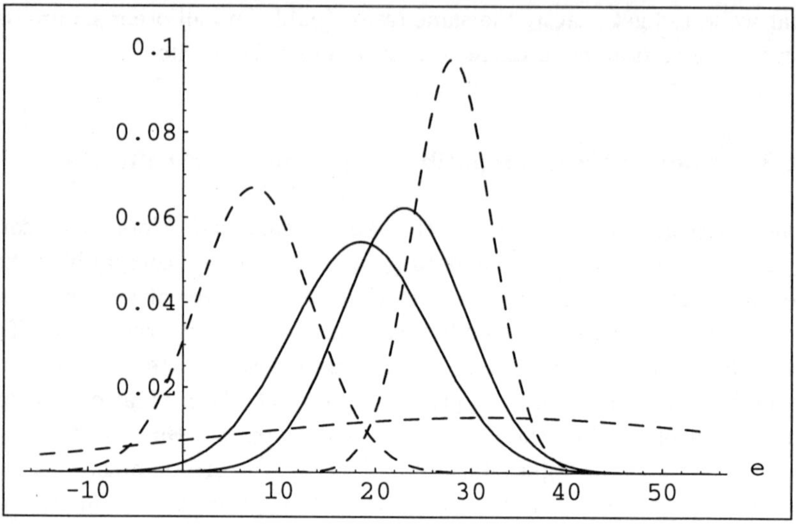

Fig. 11.1 Results on $\mathrm{Re}(\epsilon'/\epsilon)$ obtained at CERN (solid line) and Fermilab (dashed line), where $e = \mathrm{Re}(\epsilon'/\epsilon) \times 10^4$.

also because the situation is at the edge of where one starts worrying. The impression of uneasiness was not only because the mutual agreement among the experimental results is not at the level one would have wished, but also because the value of $\mathrm{Re}(\epsilon'/\epsilon)$ around which the experimental results cluster

Table 11.1 Published results on $\mathrm{Re}(\epsilon'/\epsilon)$ (values in units of 10^{-4}). Data points indicated by $\surd$ have been used for quantitative evaluations. Owing to correlations between the 1988 and 1993 uncertainties of NA31, only the combined value published in 1993 is used (see Ref. [127] for details and references).

	Experiment	Central value	$\pm\sigma_{stat} \pm \sigma_{syst}$	σ_{tot}
$\surd$	E731 (1988)	32	$\pm 28 \pm 12$	30
	NA31 (1988)	33	$\pm 6.6 \pm 8.3$	11
$\surd$	E731 (1993)	7.4	$\pm 5.2 \pm 2.9$	5.9
	NA31 (1993)	20	$\pm 4.3 \pm 5.0$	7
$\surd$	NA31 (1988+1993)	23.0	$\pm 4 \pm 5$	6.5
$\surd$	KTeV (1999)	28.0	$\pm 3.0 \pm 2.8$	4.1
$\surd$	NA48 (1999)	18.5	$\pm 4.5 \pm 5.8$	7.3

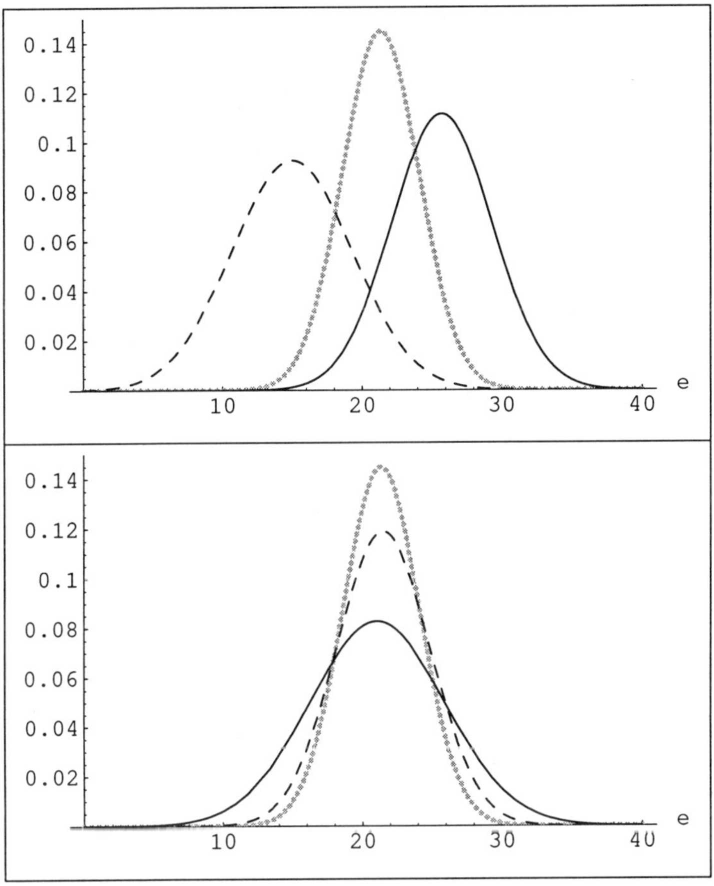

Fig. 11.2 Some combinations of the experimental results obtained using the standard combination rule of Eqs. (11.1)–(11.2). Upper plot: old results (dashed line), 1999 results (solid line), overall combination (dotted grey line). Lower plot: CERN experiments (solid line), Fermilab experiments (dashed), overall combination (dotted grey line).

was somewhat far from the theoretical evaluations (see e.g. Ref. [128] and references therein). Now, it is clear that experimentalists should not be biased towards theoretical expectations, and the history of physics teaches us about wrong results published to please theory. But we are also aware of unexpected results (either claims of new physics, or simply a quantitative disagreement with respect to the global scenario offered by other results within the framework of the Standard Model) which finally turn out to be false alarms. In conclusion, given the picture at that time of theory versus experiments about ϵ'/ϵ, there was plenty of room for doubt: Doubt about theory, about individual experiments, and about experiments as a whole.

In this situation, drawing conclusions based on a blind application of Eqs. (11.1)–(11.2) seems a bit naïve. For example, a straightforward conclusion of the standard combination rule leads to a probability that $\mathrm{Re}(\epsilon'/\epsilon)$ is smaller than zero of the order of 0.5×10^{-14}, and I don't think that experienced physicists would have shared without hesitation beliefs of this order of magnitude. As a matter of fact, at that time the question if CP symmetry was directly violated in kaon decay was still highly debated, and sophisticate experiments were still run to get evidence on value and size of ϵ'/ϵ.

11.3 Sceptical combination of experimental results

Once we have understood what is behind the simple combination rule, it is possible to change one of the hypotheses entering Eqs. (11.1)–(11.2). Obviously, the problem has no unique solution.[1] This depends to a great extent on the status of knowledge about the experiments which provided the results. For example, if one has formed a personal idea concerning the degree of reliability of the different experimental teams, one can attribute different weights to different results, or even disregard results considered unreliable or obsolete (for example their corrections for systematic effects could depend on theoretical inputs which are now considered to be obsolete).[2]

Wishing to arrive at a solution which, with all the imaginable limitations a general solution may have, is applicable to many situations without an inside, detailed knowledge of each individual experiment, we have to make some choices. First, we decide that *our sceptic is democratic*, i.e. 'he' has no a priori preference for a particular experiment. Second, the easiest way of modelling his scepticism, keeping the mathematics simple, is to consider

---

[1]The method discussed here follows the basic ideas of Ref. [129]. See Ref. [130] for an alternative approach.

[2]For example, it is known that Millikan applied quality-of-measurement rating to his electron charge determination, or selected in a pure subjective way the values to calculate the mean value. The method might seem 'not scientific', but it is amazing that the average obtained using 58 data points selected from a total of 140 gave a value which is still in excellent agreement with the currently accepted value of e (see e.g. Ref. [34]).

the likelihood still Gaussian,

$$f(d_i \mid \mu, \sigma_i) = \frac{1}{\sqrt{2\pi}\,\sigma_i} \exp\left[-\frac{(d_i-\mu)^2}{2\,\sigma_i^2}\right], \tag{11.4}$$

but with a standard deviation which might differ from that quoted by the experimentalists by a factor r_i which is not exactly known:

$$r_i = \frac{\sigma_i}{s_i}\,. \tag{11.5}$$

The uncertainty about r_i can be described by a p.d.f. $f(r_i)$. This uncertainty changes the likelihood (11.4), as it can be evaluated by the probability rules:

$$f(d_i \mid \mu) = \int f(d_i \mid \mu, r_i, s_i)\, f(r_i)\, \mathrm{d}r_i\,, \tag{11.6}$$

with

$$f(d_i \mid \mu, r_i, s_i) = \frac{1}{\sqrt{2\pi}\, r_i\, s_i} \exp\left[-\frac{(d_i-\mu)^2}{2\, r_i^2\, s_i^2}\right]. \tag{11.7}$$

If one believes that all r_i are exactly one, i.e. $f(r_i) = \delta(r_i - 1)\ \forall i$, the standard combination rule is recovered. Because of our basic assumption of democracy, the mathematical expression of the p.d.f. of r_i will not depend on i, therefore we shall talk hereafter, generically, about r and $f(r)$.

A solution to the problem of finding a parametrization of $f(r)$ such that this p.d.f. is acceptable to experienced physicists, even though the integral (11.6) still has a closed form, has been proposed by Dose and von der Linden [129]; an improved version of it will be used here. Following Ref. [129], we choose initially the variable $\omega = 1/r^2 = s_i^2/\sigma_i^2$, and consider it to be described by a gamma distribution:

$$f(\omega) = \frac{\lambda^\delta\, \omega^{\delta-1}\, e^{-\lambda\,\omega}}{\Gamma(\delta)}, \tag{11.8}$$

where λ and δ are the so-called scale and shape parameters, respectively. As a function of these two parameters, expected value and variance of ω are $\mathrm{E}(\omega) = \delta/\lambda$ and $\mathrm{Var}(\omega) = \delta/\lambda^2$. Using probability calculus we get the p.d.f of r:

$$f(r \mid \lambda, \delta) = \frac{2\,\lambda^\delta\, r^{-(2\,\delta+1)}\, e^{-\lambda/r^2}}{\Gamma(\delta)}, \tag{11.9}$$

where the parameters have been written explicitly as conditionands for the probability distribution. Expected value and variance of r are:

$$\mathrm{E}(r) = \frac{\sqrt{\lambda}\,\Gamma(\delta - 1/2)}{\Gamma(\delta)} \tag{11.10}$$

$$\mathrm{Var}(r) = \frac{\lambda}{\delta - 1} - \frac{\lambda\,\Gamma^2(\delta - 1/2)}{\Gamma^2(\delta)}\,, \tag{11.11}$$

existing simultaneously if $\lambda > 0$ and $\delta > 1$.

The individual likelihood, integrated over the possible values of r, is obtained by inserting Eqs. (11.7) and (11.9) in Eq. (11.6):

$$f(d_i \,|\, \mu, s_i) = \frac{\lambda^\delta}{\sqrt{2\pi}s_i}\,\frac{\Gamma(\delta + 1/2)}{\Gamma(\delta)}\left(\lambda + \frac{(d_i - \mu)^2}{2\,s_i^2}\right)^{-(\delta+1/2)} . \tag{11.12}$$

Using a uniform prior distribution for μ, and remembering that we are dealing with independent results, we have finally:

$$f(\mu \,|\, \boldsymbol{d}, \boldsymbol{s}) \propto f(\boldsymbol{d} \,|\, \boldsymbol{s}, \mu) \propto \prod_i \left(\lambda + \frac{(d_i - \mu)^2}{2\,s_i^2}\right)^{-(\delta+1/2)} , \tag{11.13}$$

where $\boldsymbol{s} = \{s_1, s_2, \ldots, s_n\}$. The normalization factor can be determined numerically. Equation (11.13) should be written, to be precise, as $f(\mu \,|\, \boldsymbol{d}, \boldsymbol{s}, \lambda, \delta)$, to remind us that the solution depends on the choice of λ and δ, and teaches us how to get a solution which takes into account all reasonable choices of the parameters:

$$f(\mu \,|\, \boldsymbol{d}, \boldsymbol{s}) = \int f(\mu \,|\, \boldsymbol{d}, \boldsymbol{s}, \lambda, \delta)\, f(\lambda, \delta)\, \mathrm{d}\lambda \mathrm{d}\delta\,, \tag{11.14}$$

where $f(\lambda, \delta)$ quantifies the confidence on each possible pair of parameters.[3]

A natural constraint on the values of the parameters comes from the request $\mathrm{E}(r) = 1$, modelling the assumption that the σ's agree, on average, with the stated uncertainties. The standard deviation of the distribution gives another constraint. Conservative considerations suggest $\sigma(r)/\mathrm{E}(r) \approx \mathcal{O}(1)$. The condition $\mathrm{E}(r) = \sigma(r) = 1$ is obtained for $\lambda \approx 0.6$ and $\delta \approx 1.3$. The resulting p.d.f. of r is shown as the continuous line of Fig. 11.3. One can see that the parametrization of $f(r)$ corresponds qualitatively to intuition: the barycenter of the distribution is 1; values below $r \approx 1/2$ are considered practically impossible; on the other hand, very large values of r

[3] λ and δ are the same for all experiments as we are modelling a democratic scepticism. In general they could depend on the experiment, thus changing Eq. (11.13).

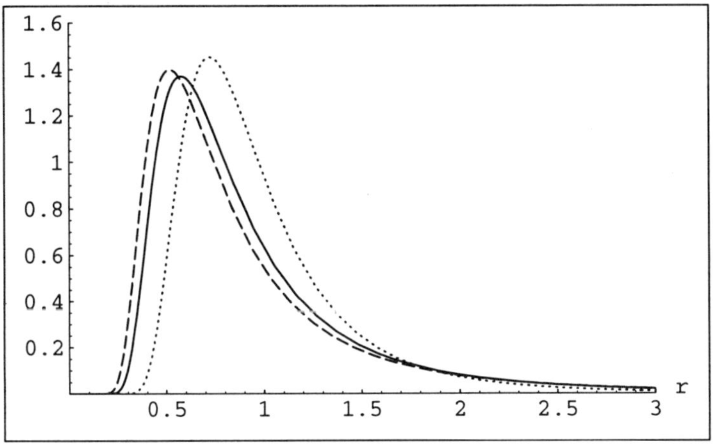

Fig. 11.3 Distribution of the rescaling factor $r = \sigma_{\text{true}}/\sigma_{\text{est}}$ using the parametrizations of Eq. (11.9) for several values of the set of parameters (λ, δ); the solid line corresponds to what will be taken as the reference distribution in this paper, yielding $\mathrm{E}(r) = \sigma(r) = 1$, and it is obtained for $\lambda \approx 0.6$ and $\delta \approx 1.3$. Dotted and dashed lines show the p.d.f. of r yielding $\sigma(r) = 0.5$ and 1.5, respectively.

are conceivable, although with very small probability, indicating that large overlooked systematic errors might occur. Anyway, we feel that, besides general arguments and considerations about the shape of $f(r)$ (to which we are not used), what matters is how reasonable the results look. Therefore, the method has been tested with simulated data, shown in the left plots of Fig. 11.4.

For simplicity, all individual results are taken to have the same standard deviation (note that the upper left plot of Fig. 11.4 shows the situation of two identical results). The solid curve of the right-hand plots shows the combined result obtained using Eq. (11.13) with $\lambda = 0.6$ and $\delta = 1.3$, yielding $\mathrm{E}(r) = \sigma(r) = 1$. For comparison, the dashed lines show also the result obtained by the standard combination. The method described here, with parameters chosen by general considerations, tends to behave in qualitative agreement with the expected point of view of a sceptical experienced physicist. As soon as the individual results start to disagree, the combined distribution gets broader than the standard combination, and might become multi-modal if the results cluster in several places. However, if the agreement is somehow 'too good' (first and last case of Fig. 11.4) the combined distribution becomes narrower than the standard result.

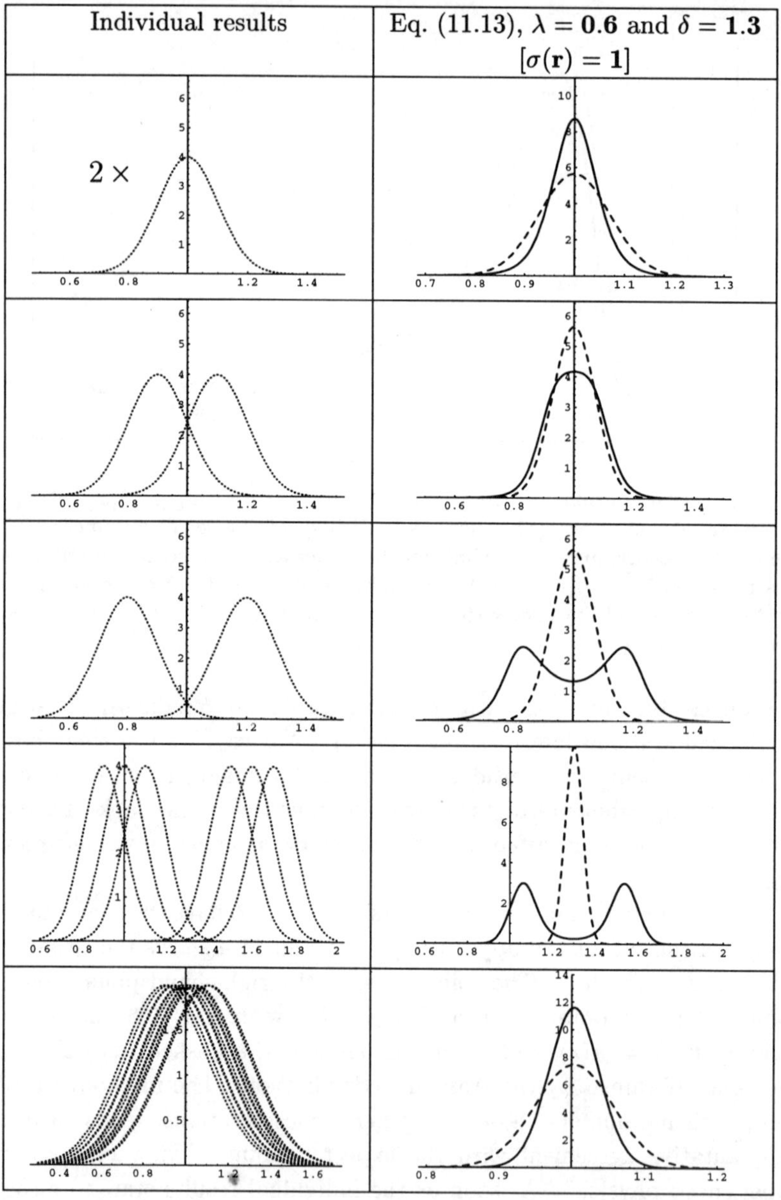

Fig. 11.4 Examples of sceptical combination of results. The plots on the left-hand side show the individual results (in the upper plot the two results coincide). The plots on the right-hand side show the combined result obtained using Eq. (11.13) with the constraint $\mathrm{E}(r) = \sigma(r) = 1$ (continuous lines), compared with the standard combination (dashed lines).

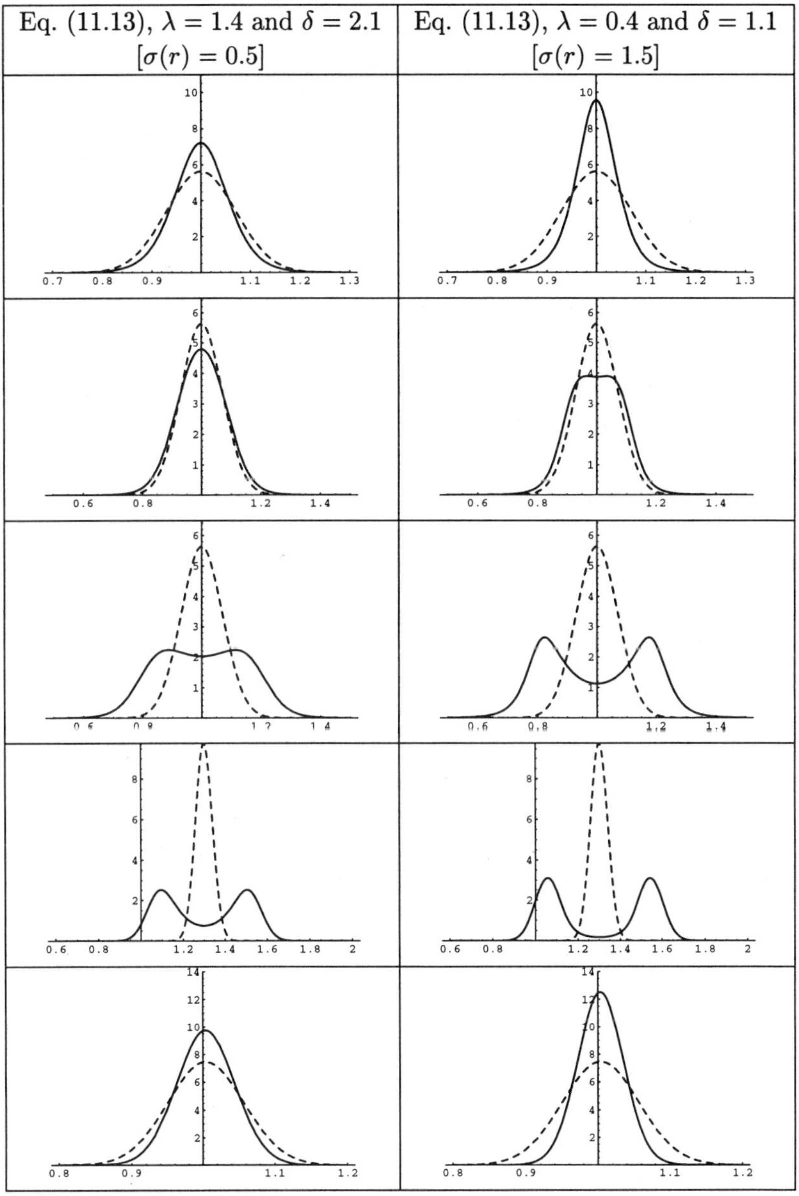

Fig. 11.5 Combination of results obtained by varying the parameters of the sceptical combination, in order to hold $\mathrm{E}(r)$ to one and change $\sigma(r)$ by $\pm 50\%$.

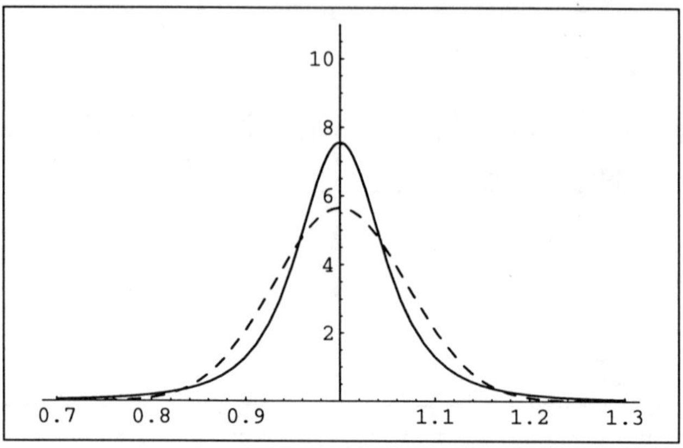

Fig. 11.6 Sceptical perception of a single measurement having a standard deviation equivalent to the standard combination of the top of Fig. 11.4. Note how the result differs from the combination of the individual results.

In order to get a feeling about the sensitivity of the results from the choice of the parameters, two other sets of parameters have been tried, keeping the requirement $\mathrm{E}(r) = 1$, but varying $\sigma(r)$ by $\pm 50\,\%$: $\sigma(r) = 0.5$ is obtained for $\lambda \approx 1.4$ and $\delta \approx 2.1$; $\sigma(r) = 1.5$ is obtained for $\lambda \approx 0.4$ and $\delta \approx 1.1$. The resulting p.d.f.'s of r are shown in Fig. 11.3. The results obtained using these two sets of parameters on the simulated data of Fig. 11.4 are shown in Fig. 11.5. We see that, indeed, the choice $\mathrm{E}(r) = \sigma(r) = 1$ seems to be an optimum, and the $\pm 50\%$ variations of $\sigma(r)$ give results which are at the edge of what one would consider to be acceptable. Therefore, we shall take the parameters providing $\mathrm{E}(r) = \sigma(r) = 1$ as the reference ones.

Another interesting feature of Eq. (11.13) is its behavior for a single experimental result, as shown in Fig. 11.6. For comparison, we have taken a result having a stated standard deviation equal to $1/\sqrt{2}$ of each of those of Fig. 11.4. Figure 11.6 has to be compared with the upper right plots of Fig. 11.4. The sceptical combination takes much more seriously two independent experiments, each reporting in an uncertainty σ, than a single experiment performing $\sigma/\sqrt{2}$. On the contrary, the two situations are absolutely equivalent in the standard combination rule. In particular, the tails of the p.d.f. obtained by the sceptical combination vanish more slowly than in the Gaussian case, while the belief in the central value is higher. The result models the qualitative attitude of sceptical physicists, according to whom a single experiment is never enough to establish a value, no

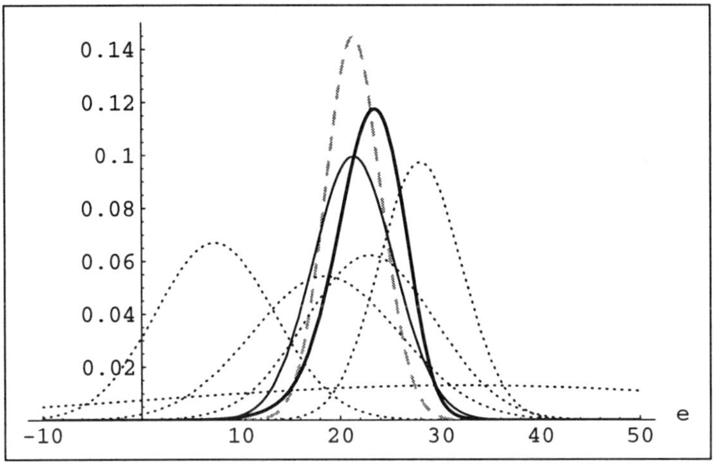

Fig. 11.7 Individual results compared with the standard combination (grey dashed), the PDG-rescaled combination (solid thin) and the sceptical combination as described in this paper (solid thick).

matter how precise the result may be, although the true value might have more chance to be within one standard deviation than the probability level calculated from a Gaussian distribution.

11.3.1 *Application to ϵ'/ϵ*

The combination rule based on Eq. (11.13) has been applied to the results about $\mathrm{Re}(\epsilon'/\epsilon)$ shown in Tab. 11.1. As discussed above, our reference parameters are $\lambda = 0.6$ and $\delta = 1.3$, corresponding to $\mathrm{E}(r) \approx \sigma(r) \approx 1$. The resulting p.d.f. for $e = \mathrm{Re}(\epsilon'/\epsilon) \times 10^4$ is shown as the thick continuous line of Fig. 11.7, together with the individual results (dotted lines). For comparison, we also give the result obtained using the combination rules commonly applied in particle physics. The grey-dashed line of Fig. 11.7 is obtained with the standard combination rule [Eqs. (11.1) and (11.2)]. The thin continuous line has been evaluated using the PDG 'prescription' [79]. According to this rule, the standard deviation (11.2) is enlarged by a factor given by $\sqrt{\chi^2/(N-1)}$, where χ^2 is the chi-2 of the data with respect to the average (11.1) and N is the number of independent results.

We see that although the PDG rule gives a distribution wider than that obtained by the standard rule, the barycenters of the distributions coincide, thus not taking into account that one of the results is quite far from where

Table 11.2 Comparison of the different methods of combining the results.

Combination	Mean	σ	Median	Mode	$P[\mathrm{Re}(\epsilon'/\epsilon < 0)]$
Standard	21.4	2.7	21.4	21.4	5×10^{-15}
PDG rule [79]	21.4	4.0	21.4	21.4	5×10^{-8}
Sceptical	22.7	3.5	23.0	23.5	1.5×10^{-6}

the others seem to cluster. Moreover, the p.d.f. is assumed to be Gaussian, independently of the configuration of experimental points. Instead, the sceptical combination takes into account better the configuration of the data points. The peak of the distribution is essentially determined by the three results which appear more consistent with each other. Nevertheless, there is a more pronounced tail for small values of $\mathrm{Re}(\epsilon'/\epsilon)$, to take into account that there is indeed a result providing evidence in that region, and that cannot be ignored.

A quantitative comparison of the different methods is given in Tab. 11.2, where the most relevant statistical summaries are provided, together with some probability intervals. It is worth recalling that each of these summaries gives some information about the distribution, but, when the uncertainty of this result has to be finally propagated into other results (together with other uncertainties), it is the average and standard deviation which matter. The standard 'error propagation' is based on linearization, on the property of expected value and variance under a linear combination and on central limit theory (the result of several contributions will be roughly Gaussian). Therefore, propagating mode (or median) and 68% probability intervals does not make any sense, unless the input distributions are Gaussian.

An interesting comparison is given by the probability that $\mathrm{Re}(\epsilon'/\epsilon)$ is negative. The sceptical combination gives the largest value, but still at the level of one part per million, indicating that, even in this conservative analysis, a positive value of the direct CP violation parameter was 'practically' established already with those measurements.

The sensitivity of the result on the parameters of the combination formula can be inferred from Fig. 11.8, where the results obtained changing $\sigma(r)$ by $\pm 50\%$ are shown. The combined result is quite stable. This is particularly true if one remembers that these extreme values of parameters

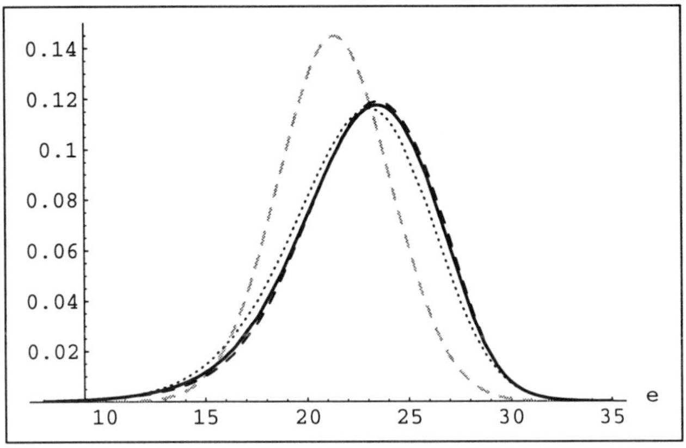

Fig. 11.8 Dependence of the sceptical combination on the choice of the parameters. Continuous, dotted and dashed lines are, respectively: $\lambda = 0.6$ and $\delta = 1.3$ [$\sigma(r) = 1$)]; $\lambda = 0.4$ and $\delta = 1.1$ [$\sigma(r) = 0.5$)]; $\lambda = 1.4$ and $\delta = 2.1$ [$\sigma(r) = 0.5$)]. The grey-dashed line gives, for comparison, the result of the standard combination.

are quite at the edge of what one would accept as reasonable, as can be seen in Fig. 11.5. Note that if one would like to combine the results taking also into account the uncertainty about the parameters, one would apply Eq. (11.14). It is reasonable to think that, since the variations of the p.d.f. from that obtained for the reference value of the parameters are not very large, the p.d.f. obtained as weighted average over all the possibilities will not be much different from the reference one.

Figure 11.9 shows the results subdivided into CERN and Fermilab. In these cases the difference between the standard combination and the sceptical combination becomes larger, and, again, the outcome of the sceptical combination follows qualitatively the intuitive one of experienced physicists. The sceptical combination of the CERN results alone is better than that given by the standard one, thus reproducing formally the instinctive suspicion that the uncertainties could have been overestimated. For the Fermilab ones the situation is reversed. In any case, both partial combinations tend to establish strongly the picture of a positive and sizeable $\mathrm{Re}(\epsilon'/\epsilon)$ value. Finally, note that the $\pm 50\%$ variations in $\sigma(r)$ produce in the partial combinations a larger effect (although not relevant for the conclusions) than in the overall combination. This is due to the fact that the variations produce opposite effects on the two subsets of data in the region of $\mathrm{Re}(\epsilon'/\epsilon)$ around 20×10^{-4}.

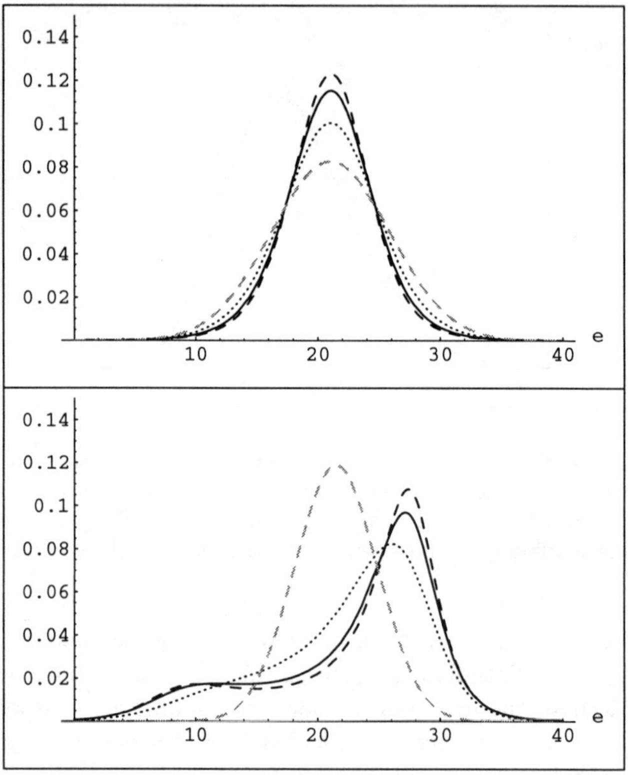

Fig. 11.9 Sceptical combination of CERN and Fermilab results (upper and lower plot, respectively). The continuous line shows the result obtained by Eq. (11.13) and reference parameters. The dashed and dotted lines are the results obtained by varying the standard deviation of $r = \sigma/s$ by +50% and −50%, respectively. The grey-dashed line shows the results obtained by the standard combination rule.

11.3.2 *Posterior evaluation of* σ_i

An interesting by-product of the method illustrated above is the posterior evaluation of the various σ_i, or, equivalently, of the various r_i. Again, we can make use of Bayes' theorem, obtaining

$$f(\boldsymbol{r} \,|\, \boldsymbol{d}, \boldsymbol{s}, \mu) = \frac{f(\boldsymbol{d} \,|\, \boldsymbol{r}, \boldsymbol{s}, \mu)\, f_\circ(\boldsymbol{r} \,|\, \boldsymbol{s}, \mu)}{\int f(\boldsymbol{d} \,|\, \boldsymbol{r}, \boldsymbol{s}, \mu)\, f_\circ(\boldsymbol{r} \,|\, \boldsymbol{s}, \mu)\, \mathrm{d}\boldsymbol{r}}\,, \tag{11.15}$$

where $\boldsymbol{r} = \{r_1, r_2, \ldots, r_n\}$. Since the initial status of knowledge is such that values of r_i are independent of each other, and they are independent of μ and $\boldsymbol{s}$, we obtain

$$f_\circ(\boldsymbol{r}\,|\,\boldsymbol{s},\mu) = f_\circ(\boldsymbol{r}) = \prod_i f_\circ(r_i) \equiv \prod_i f(r_i\,|\,\lambda,\delta) = \prod_i \frac{2\,\lambda^\delta\, r_i^{-(2\,\delta+1)}\, e^{-\lambda/r_i^2}}{\Gamma(\delta)}\,, \tag{11.16}$$

having used Eq. (11.9). As a shorthand for Eq. (11.16), we shall write in the following simply $f_\circ(\boldsymbol{r}) = \prod_i f_\circ(r_i)$.

Since the experimental results are also considered independent, we can rewrite Eq. (11.15) as

$$\begin{aligned} f(\boldsymbol{r}\,|\,\boldsymbol{d},\boldsymbol{s},\mu) &= \frac{\prod_i f(d_i\,|\,r_i,s_i,\mu)\, f_\circ(r_i)}{\int \prod_i f(d_i\,|\,r_i,s_i,\mu)\, f_\circ(r_i)\, \mathrm{d}\boldsymbol{r}} \\ &= \frac{\prod_i f(d_i\,|\,r_i,s_i,\mu)\, f_\circ(r_i)}{\prod_i \int f(d_i\,|\,r_i,s_i,\mu)\, f_\circ(r_i)\, \mathrm{d}r_i}\,. \end{aligned} \tag{11.17}$$

The marginal distribution of each r_i, still conditioned by μ (and, obviously, by the experimental values), is obtained by integrating $f(\boldsymbol{r}\,|\,\boldsymbol{d},\boldsymbol{s},\mu)$ over all r_j, with $j \neq i$. As a result, we obtain

$$f(r_i\,|\,\boldsymbol{d},\boldsymbol{s},\mu) = \frac{f(d_i\,|\,r_i,s_i,\mu)\, f_\circ(r_i)}{\int f(d_i\,|\,r_i,s_i,\mu)\, f_\circ(r_i)\, \mathrm{d}r_i}\,. \tag{11.18}$$

Making use of Eqs. (11.7), (11.9) and (11.12) we get:

$$f(r_i\,|\,\boldsymbol{d},\boldsymbol{s},\mu) = \frac{\frac{1}{\sqrt{2\,\pi}\, r_i\, s_i} \exp\left[-\frac{(d_i-\mu)^2}{2\, r_i^2\, s_i^2}\right] \frac{2\,\lambda^\delta\, r_i^{-(2\,\delta+1)}\, e^{-\lambda/r_i^2}}{\Gamma(\delta)}}{\frac{\lambda^\delta}{\sqrt{2\,\pi}\, s_i}\, \frac{\Gamma(\delta+1/2)}{\Gamma(\delta)} \left(\lambda + \frac{(d_i-\mu)^2}{2\, s_i^2}\right)^{-(\delta+1/2)}}\,. \tag{11.19}$$

The final result is obtained by eliminating, in the usual way, the condition μ, i.e.

$$f(r_i\,|\,\boldsymbol{d},\boldsymbol{s}) = \int f(r_i\,|\,\boldsymbol{d},\boldsymbol{s},\mu)\, f(\mu\,|\,\boldsymbol{d},\boldsymbol{s})\, \mathrm{d}\mu\,. \tag{11.20}$$

Making use of Eq. (11.13), and neglecting in Eq. (11.19) all factors not depending on r_i and μ, we get the unnormalized result

$$\begin{aligned} f(r_i\,|\,\boldsymbol{d},\boldsymbol{s}) &\propto r_i^{-(2\,\delta+2)}\, e^{-\lambda/r_i^2} \\ &\times \int \mathrm{d}\mu\, \exp\left[-\frac{(d_i-\mu)^2}{2\, r_i^2\, s_i^2}\right] \prod_{j\neq i} \left(\lambda + \frac{(d_j-\mu)^2}{2\, s_j^2}\right)^{-(\delta+1/2)}\,. \end{aligned} \tag{11.21}$$

This formula is clearly valid for $n \geq 2$. If this is not the case, the product over $j \neq i$ is replaced by unity, and the integral is proportional to r_i. Equation (11.21) then becomes $f(r_1 \,|\, d_1, s_1) \propto r_1^{-(2\,\delta+1)}\, e^{-\lambda/r_1^2}$, i.e. we have recovered the initial distribution (11.9). In fact, if we have only one data point, there is no reason to change our beliefs about r. Only the comparison with other results can induce us to change our opinion.

Once we have got $f(r_i \,|\, \boldsymbol{d}, \boldsymbol{s})$ we can give posterior estimates of r_i in terms of average and standard deviations, and they can be compared with the prior assumption $\mathrm{E}(r) = \sigma(r) = 1$, to understand which uncertainties have been implicitly rescaled by the sceptical combination.[4] Convenient formulae to evaluate numerically first and second moments of the posterior distribution of r_i are given by

$$\mathrm{E}(r_i) = \frac{\Gamma(\delta)}{\Gamma(\delta+1/2)} \cdot \frac{\int \left(\lambda + \frac{(d_i-\mu)^2}{2\,s_i^2}\right)^{1/2} \prod_j \left(\lambda + \frac{(d_j-\mu)^2}{2\,s_j^2}\right)^{-(\delta+1/2)} \mathrm{d}\mu}{\int \prod_j \left(\lambda + \frac{(d_j-\mu)^2}{2\,s_j^2}\right)^{-(\delta+1/2)} \mathrm{d}\mu} \tag{11.22}$$

$$\mathrm{E}(r_i^2) = \frac{\Gamma(\delta-1/2)}{\Gamma(\delta+1/2)} \cdot \frac{\int \left(\lambda + \frac{(d_i-\mu)^2}{2\,s_i^2}\right) \prod_j \left(\lambda + \frac{(d_j-\mu)^2}{2\,s_j^2}\right)^{-(\delta+1/2)} \mathrm{d}\mu}{\int \prod_j \left(\lambda + \frac{(d_j-\mu)^2}{2\,s_j^2}\right)^{-(\delta+1/2)} \mathrm{d}\mu} . \tag{11.23}$$

Note that, since $\prod_j(\ldots)$ of the integrands are proportional to $f(\mu \,|\, \boldsymbol{d}, \boldsymbol{s})$, Eqs. (11.22)–(11.23) can be written in the compact form

$$\mathrm{E}(r_i) = \frac{\Gamma(\delta)}{\Gamma(\delta+1/2)} \cdot \mathrm{E}_\mu \left[\left(\lambda + \frac{(d_i-\mu)^2}{2\,s_i^2}\right)^{1/2}\right] \tag{11.24}$$

$$\mathrm{E}(r_i^2) = \frac{\Gamma(\delta-1/2)}{\Gamma(\delta+1/2)} \cdot \mathrm{E}_\mu \left[\lambda + \frac{(d_i-\mu)^2}{2\,s_i^2}\right], \tag{11.25}$$

where $\mathrm{E}_\mu(\cdot)$ indicates expected values over the p.d.f. of μ.

At this point it is important to anticipate the objection of those who think that it is incorrect to infer $n+1$ quantities (μ and $\boldsymbol{r}$) starting from n data points. Indeed, there is nothing wrong in doing so. But, obviously, the results are correlated, and they depend also on the prior distribution

[4]Note that it is incorrect to feed again into the procedure the rescaled uncertainties, as they come from this analysis. The procedure has already taken into account all possible rescaling factors in the evaluation of $f(\mu \,|\, \boldsymbol{d}, \boldsymbol{s})$.

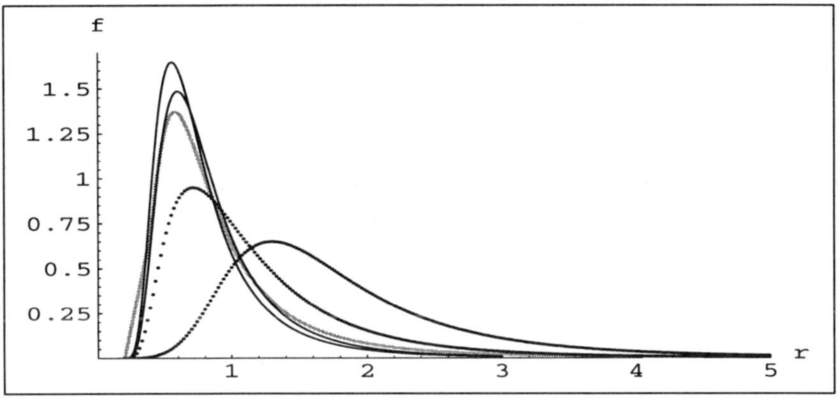

Fig. 11.10 Final distributions of r corresponding to the four most precise results on $\mathrm{Re}(\epsilon'/\epsilon)$, compared with the reference prior one (grey line). The continuous lines refer to the CERN results, dotted lines to the Fermilab ones.

of r_i, which acts as a constraint. In fact we have seen above that for $n = 1$ the result on r is trivial.

Figure 11.10 gives the final distributions of $r_i = \sigma_i/s_i$ for the four most precise determinations of $\mathrm{Re}(\epsilon'/\epsilon)$, compared with the reference initial distribution having $\sigma(r) = 1$ (grey line in the plot). The distributions relative to the CERN results are shown with continuous lines, the Fermilab ones by dots. In particular, the one that has a substantial probability mass above 1 is the 1993 E731 result. Average and standard deviations of the distributions are given in Tab. 11.3, also showing the values that one would obtain with the other sets of parameters that we have considered to be edge ones.

Once more, the results are in qualitative agreement with intuition: The

Table 11.3 Posterior estimation of $r = \sigma_i/s_i$ starting from identical priors having $\mathrm{E}_\circ(r) = 1$ and $\sigma_\circ(r) = 0.5$, 1.0 and 1.5. The individual results are given by $d_i \pm s_i$ to be consistent with the notation used throughout this section (see Ref. [127] for details and references) .

Experiment	$d_i \pm s_i$	Posterior $\mathrm{E}(r_i)$ $(\sigma(r_i))$		
		$\sigma_\circ(r) = 0.5$	$\sigma_\circ(\mathbf{r}) = \mathbf{1}$	$\sigma_\circ(r) = 1.5$
E731 (1988)	32 ± 30	0.9 (0.4)	0.8 (0.5)	0.7 (0.5)
E731 (1993)	7.4 ± 5.9	1.6 (0.7)	1.9 (1.2)	2.1 (1.5)
NA31 (1988+1993)	23.0 ± 6.5	0.9 (0.4)	0.8 (0.5)	0.8 (0.6)
KTeV (1999)	28.0 ± 4.1	1.2 (0.6)	1.2 (0.9)	1.2 (1.0)
NA48 (1999)	18.5 ± 7.3	0.9 (0.4)	0.9 (0.5)	0.9 (0.6)

CERN curves are slightly squeezed below $r = 1$, as the uncertainty evaluation seems to be a bit conservative. The Fermilab ones show instead some drift towards large r. In particular, figure and table make one suspect that some contribution to the error has been overlooked in the E731 data. Note that in this case the average value of the rescaling factor is smaller than one could expect from alternative procedures which require the overall χ^2 to equal the number of degrees of freedom. The reason is the shape of the initial distribution of r, which protects us against unexpectedly large values of the rescaling factors.

Chapter 12

Asymmetric uncertainties and nonlinear propagation

"Two and two equal four. The trouble is,
in that world of shadows and distorting mirrors
what may or may not appear to be two, when
multiplied by a factor that may or may not be two,
could possibly come out at four but probably will not."
(Frederick Forsyth – The fist of God)

Measurement results are frequently published in the form

$$\text{best value}\ ^{+\Delta_+}_{-\Delta_-}\ .$$

It is interesting to try to understand how experimentalists arrive at these asymmetric uncertainties and how this information is commonly used in subsequent analyses. We shall see that the present practice is far from being acceptable and, indeed, could bias the value of important physics quantities. The alternative is, of course, to use Bayesian reasoning. The general strategy is briefly outlined and some approximate formulae, useful for avoiding complex calculations, are derived.

12.1 Usual combination of 'statistic and systematic errors'

The combination in quadrature of uncertainties due to systematic effects has become quite standard practice in physics. It is also common practice to add these uncertainties in quadrature to those from random effects ('statistic and systematic errors', in physics jargon — see remarks in Sec. 5.1). Usually the two kinds of uncertainties are given separately, and the systematic-effect uncertainties are listed individually (at least for the most relevant ones) in order to show the potential of further measurements made with the same apparatus. This combination rule has arisen as a kind of pragmatic procedure [30], in analogy to the combination of standard deviations in probability theory, although cannot justifiably be termed within 'conventional' statistics. The same is true for the use of the covariance matrix to handle correlated uncertainties.

There is less agreement when the uncertainties due to systematic effects are asymmetric and/or they produce asymmetric shifts in the final quantity of interest due to nonlinear propagation of uncertainty. As a numerical example of the latter case, take a quantity Y depending on three 'influence quantities' X_1, X_2 and X_3, which could be calibration constants, environment quantities or theoretical parameters. Suppose that, for the reference values of the X's, the analysis procedure gives (in arbitrary units) $Y = 1.000 \pm 0.050$, where the uncertainty associated with the result is that due to random effects (in this chapter a notation different from that of Chapter 8 is used, to come closer to what the reader is familiar when dealing with uncertainty propagation problems). Consider now that by 'varying reasonably the quantities X_i' (the expression is intentionally left vague for the moment) the following deviations from the central values occur: $\Delta Y_{1\pm} = {}^{+0.060}_{-0.090}$, $\Delta Y_{2\pm} = {}^{+0.098}_{-0.147}$, and $\Delta Y_{3\pm} = {}^{+0.104}_{-0.156}$. An often-used practice[1] is to combine in quadrature separately positive and negative deviations, obtaining the following result: $Y = 1.00 \pm 0.05\,(\text{stat.})\,{}^{+0.15}_{-0.23}\,(\text{syst.})$, subsequently summarized as $Y = 1.00^{+0.16}_{-0.24}$.

Now we are faced with the problem that the result of this *ad hoc* procedure has no theoretical justification. Hence the uncertainty content of the statement (i.e. its probabilistic meaning) is unclear and, as a consequence, it is not obvious how to make use of this information in further analyses, even in the simple case in which the data points are uncorrelated. As a matter of fact, most people remove the asymmetry in further analysis of the results, getting something equivalent to a standard deviation to be used in χ^2 fits. This 'standard deviation' is evaluated either by taking the largest value between Δ_+ and Δ_-, or by averaging the two values (some use the arithmetic, others the geometric average). The result is that in both procedures the uncertainty is symmetrized and the result is considered as if it were described, for all practical purposes, by a Gaussian model around the published best estimate.[2]

[1]There is also who combines left and right deviations linearly, obtaining for this example $Y = 1.00 \pm 0.05\,(\text{stat.})\,{}^{+0.26}_{-0.39}\,(\text{syst.})$.

[2]A more complicated 'prescription' is described by the PDG [79], which we report

The main worry is not that the combined uncertainties will be incorrect (we anticipate that the arithmetic average of Δ_+ and Δ_- gives indeed the correct uncertainty in most cases of practical interest), but rather that the result itself can be biased with respect to what one could get using consistently the best knowledge concerning the input quantities, as will be shown in the sequel.

12.2 Sources of asymmetric uncertainties in standard statistical procedures

12.2.1 *Asymmetric χ^2 and '$\Delta\chi^2 = 1$ rule'*

We saw in Sec. 8.1 that minimum χ^2 fits can be considered special cases of Bayesian methods. In particular, if the final distribution is approximately (multivariate-)Gaussian, i.e. the χ^2 function is parabolic around its minimum, the covariance matrix of the parameter can be evaluated from the curvature of the function (the "Hessian"). Under the same hypotheses, the "$\Delta\chi^2 = 1$" rule holds [and the equivalent "$\Delta(-\ln\mathcal{L}) = 1/2$" rule, due to Eq. (8.3)]. But in real life, more complicated situations can happen, as shown in the one-dimensional examples of Fig. 12.1.

- If the χ^2 is perfectly parabolic (Fig. 12.1, frame 1A) the final distribution, proportional to $\exp[-\chi^2/2]$ is Gaussian (frame 1B). The $\Delta\chi^2 = 1$ rule and Hessian provide the standard deviation to be associated to the uncertainty about μ, and define a 68% probability interval (which we could also call 'confidence interval', in the sense that we are 68% confident that μ is inside it).
- In the case of 'minimal deviation from a parabola' (Fig. 12.1, frame 2A) the $\Delta\chi^2 = 1$ rule and Hessian give different results. The practice is to prefer the asymmetric result provided by $\Delta\chi^2 = 1$ rule, 5^{+1}_{-2} in this case, as that which provides a 68% probability around the 'best

here for the convenience of the reader: *"When experimenters quote asymmetric errors $(\delta x)^+$ and $(\delta x)^-$ for a measurement x, the error that we use for that measurement in making an average or a fit with other measurements is a continuous function of these three quantities. When the resultant average or fit $\overline{x}$ is less than $x - (\delta x)^-$, we use $(\delta x)^-$; when it is greater than $x + (\delta x)^+$, we use $(\delta x)^+$. In between, the error we use is a linear function of x. Since the errors we use are functions of the result, we iterate to get the final result. Asymmetric output errors are determined from the input errors assuming a linear relation between the input and the output quantities."* This rule does not seem to be applied by others than the PDG. As examples of other ad hoc procedures, see Refs. [131, 132, 133].

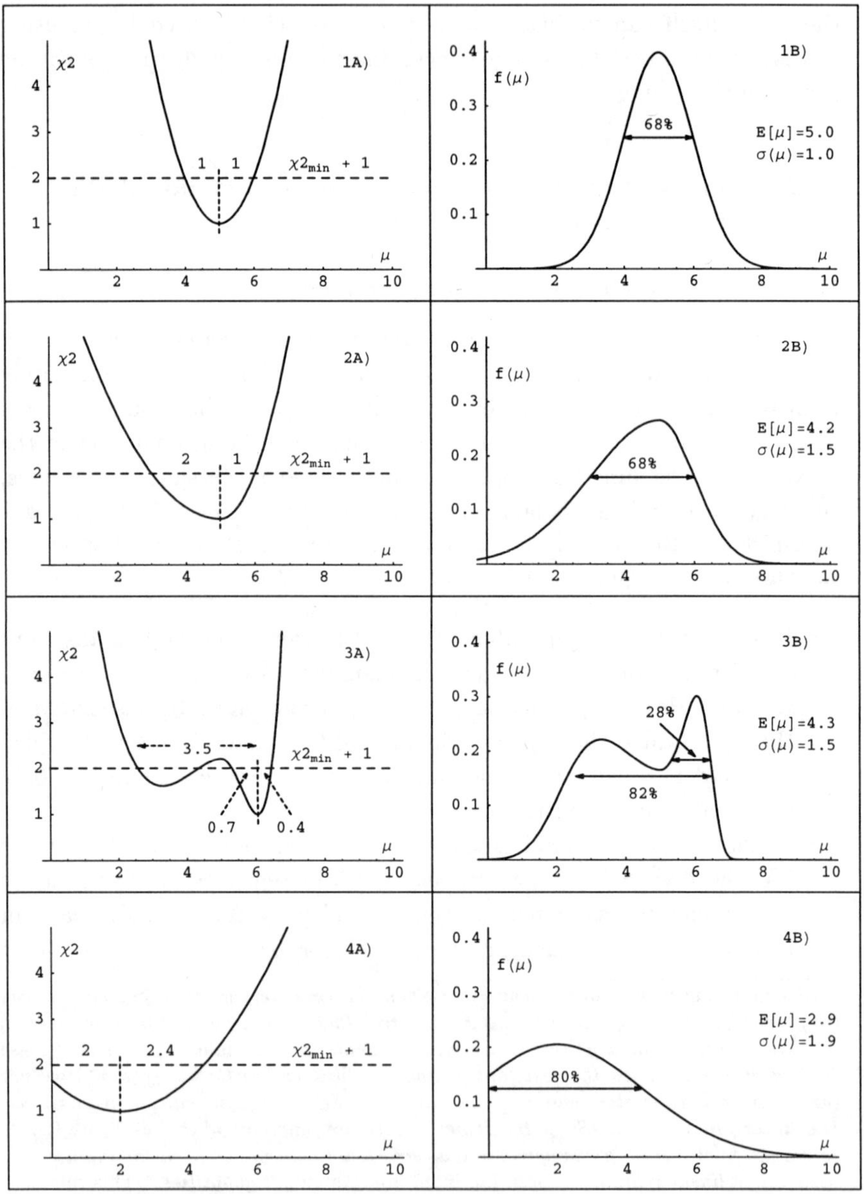

Fig. 12.1 Some common shapes of χ^2 functions. Results based on the $\chi^2_{min}+1$ 'prescription' are compared with final p.d.f.'s based on uniform prior, i.e. $f(\mu\,|\,\text{data}) \propto \exp[-\chi^2/2]$.

estimate' of μ. The usual justification is that one might think of a monotonic transformation $\mu \to \mu'$, such that the χ^2 is parabolic in μ' and the $\Delta\chi^2(\mu') = 1$ rule defines a 68% probability interval (assuming a uniform prior in μ'!). Then, since probability intervals are maintained in monotonic transformations, the condition $\Delta\chi^2(\mu') = 1$ defines a probability interval for μ too. Some comments on this prescription are in order.

- The Bayesian analysis (Fig. 12.1, frame 2B) provides the same 68% probability interval, but without having to swap the variable of interest μ with a virtual one, and having to assume, implicitly or explicitly, a uniform prior on this 'mythical' μ'.[3] The standard deviation is in general different from the $\Delta\pm$ that result from the $\Delta\chi^2 = 1$ rule.
- The $\Delta\chi^2 = 1$ rule does not provide expected value and standard deviation of the quantity of interest, but only a 'best value' (coinciding with the mode) and asymmetric uncertainty. The problem is that this best value is not suitable for subsequent propagations of uncertainty, as already discussed in Sec. 12.1.

• When the χ^2 function becomes a bit more complicated,[4] like in the examples of Fig. 12.1, frames 3A and 4A, the reasoning based on the swapping of variables also fails, Hessian and $\Delta\chi'^2 = 1$ rule results diverge

[3] Frequentists will complain that "the procedure does not yield a probability interval" but a "confidence interval", that "they do not use priors", and so on (and that I have never understood their methods, which is in some sense true, since they have 'no meaning'...). My intention is to review the results of their methods, trying to understand when and why they are reasonable.

[4] When systematic effects, too, are included 'ad hoc' and 'minimum χ^2 fits' are performed minimizing objects which have little to do with the χ^2 of probability theory, the situation becomes surreal, like in the cases of the following table, [134] which shows $\Delta\chi^2$ rule arbitrarily modified to obtain the desired result:

Collaboration	Rule	Value of α_s (adimensional)
CTEQ6	$\Delta\chi^2 = 100$	0.1165 ± 0.0065
ZEUS	$\Delta\chi^2_{eff} = 50$	$0.1166 \pm 0.0008(\text{uncor}) \pm 0.0032(\text{corr})$ $\pm 0.0036(\text{norm}) \pm 0.0018(\text{model})$
MRST	$\Delta\chi^2 = 20$	$0.119 \pm 0.002(\text{exp}) \pm 0.003(\text{theory})$
H1	$\Delta\chi^2 = 1$	$0.115 \pm 0.0017(\text{exp}) \pm 0.005(\text{theory})$

(α_s is the QCD coupling constant, evaluated at the Z^0 mass square scale). As the authors of Ref. [134] remark, *"The values obtained are consistent, and the errors not too dissimilar given the wide variation in $\Delta\chi^2$ used. This is largely because each group has chosen a method which gives a reasonable and believable error."* No further comment is needed to emphasize how 'objective' many non-subjective methods can be.

and the latter can have multiple solutions. Essentially, the numbers resulting from these rules lose any unequivocal and reasonable probabilistic meaning. Instead, we can see that the Bayesian approach still provides acceptable and consistent results (see frames 3B and 4B), not dominated by local minima or numerical fluctuations, of clear meaning and valid under well stated conditions.

12.2.2 *Systematic effects*

The source of asymmetry described in the previous sub-section is related to what is usually defined as 'statistical analysis'. Others arise when from systematic effects are taken into account, typically in the following way. Variations are applied to the values of physics, simulation and analysis parameters. Deviations from the nominal result are registered and used to assess the 'systematic error'. Needless to say, this procedure requires much "scientific judgment" [5] [5], and I think that reasoning in terms of ISO type B uncertainty would help a lot (see Chapter 8). Let us see what is done in practice and what are the potential problems.

Before going into details, let me make a general comment. Without guidance of the kind provided by the ISO recommendations, or more in general by subjective probability, there is no common agreement on *what these variations should mean* (which is a different problem from 'what they really are' [6]). Sometimes they are just one standard deviation (when the

[5]A related problem, which also requires much subjective commitment by the experimentalists is when to stop making systematic checks and adding contributions to the overall uncertainty. As eloquently said [135], *"one could correlate a particle physics result with the phase of the Moon or the position of Jupiter, and find most likely no significant effect, with some uncertainty; but certainly we don't want to take care of this uncertainty."* (But if you are patient enough searching, you can also find 'significative' effects...) Only contributions which are in principle relevant should be considered in the uncertainty evaluation. Even if the effect is 'statistically significant', one should try to understand if it can physically influence the result, before including it in the list of contribution to the overall uncertainty.

[6]To make this distinction clear, think of the following example. If somebody quotes a 50% probability interval for a parameter, I will take this information to be what that person believes. Perhaps other experts would quote different intervals at the same probability level, but this is not a problem. At least I have got the information that somebody I trust has expressed that degree of confidence. The real problem comes when I get a number and nobody tells me what meaning he attaches to the number., When somebody tells me that me that this is a frequentistic CL I really don't know what to think, unless I analyze in detail the procedure used. In fact, I know by experience that sometimes they are equivalent, under reasonable but unstated hypotheses, to probabilities. But other times, almost always when results for search of rare phenomena are reported, they

information about that parameter comes from a 'statistical method') sometimes they mean a kind of maximum conceivable variations, other times they are just intervals of "high probability". In other cases they are $\pm 50\%$ variations of the nominal value, or a 'factor-of-two' variation (i.e. $\times 2$ and $\times 1/2$).

12.2.2.1 *Asymmetric beliefs on systematic effects*

A direct asymmetry in the systematic effect comes when the interval in which the parameter could be is assessed in a non-symmetric way around the best estimate. For example, the parameter of a theory could have more chance of being larger than/smaller than the 'accepted' reference value. Often this happens because that parameter produces nonlinear (e.g. logarithmic) effects to the observables which depend on it. We have already given in Sec. 8.10 the example of a calibration constant believed to be $1.0^{+0.1}_{-0.0}$. Also 'factor of two' variations of parameters produce asymmetric uncertainty ($a^{+a}_{-a/2}$, where a is the nominal value).

12.2.2.2 *Nonlinear propagation of uncertainties*

The most common source of asymmetric uncertainty due to systematics is nonlinear dependence of the true value Y on the influence quantity X_i. In fact, the procedure of obtaining $\Delta Y_\pm$ deviations of the output value from $\pm\Delta X_i$ of the input value is nothing but a numerical estimate of the derivative $\partial Y/\partial X_i$ around $\mathrm{E}(\boldsymbol{X})$. Nonlinearity makes ΔY_- differ from ΔY_+. Some examples are given in Fig. 12.2 for the generic input variables X_i and output variables Y_i.

12.3 General solution of the problem

From a Bayesian perspective, the solution is in principle simple: model the probability functions of all quantities of interest and apply probability theory to propagate the uncertainty of each *input quantity* into the *output quantities.* Some useful p.d.f.'s to model uncertainty on input quantities are shown in Fig. 8.1, while Fig. 12.2 shows how some of these p.d.f.'s are transformed by a nonlinear propagation. Figure 12.3 shows another example of uncertainty propagations, based on the variables X_i and Y_i of

are not. An important physics case (the dreamed Higgs particle!) in which the same experimental teams report 95% CL upper and lower limits which do not have the same limit is reported in Ref. [59].

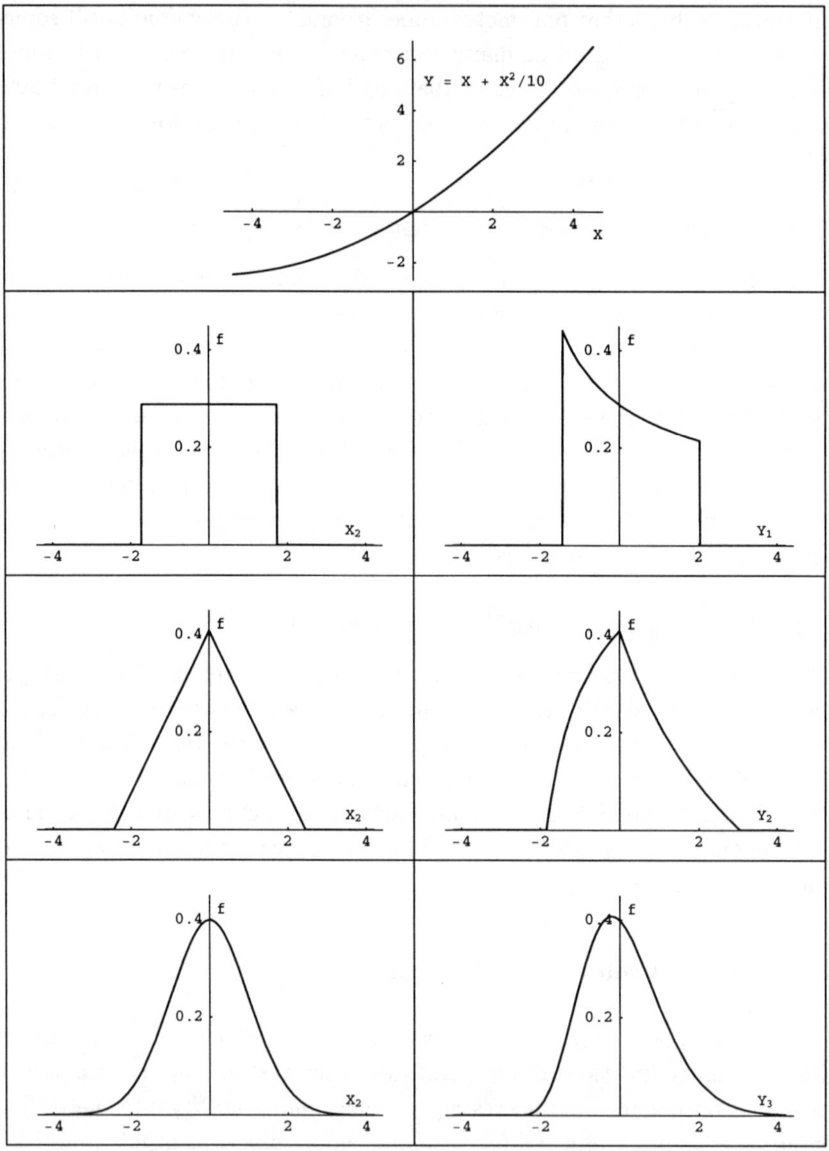

Fig. 12.2 Propagation of uniform, triangular and Gaussian distribution under a nonlinear transformation. $f(Y_i)$ were obtained analytically using Eq. (4.95).

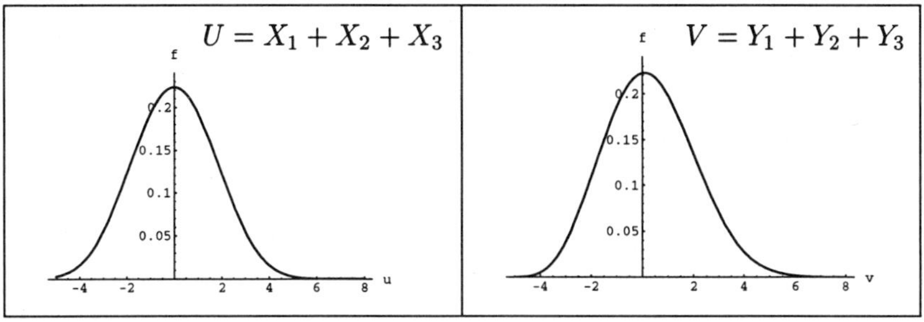

Fig. 12.3 Probability density functions of the sum of the quantities X_i and of their nonlinear transformations Y_i defined in Fig. 12.2.

Fig. 12.2. In all these examples analytical calculations have been performed. In many practical cases the computational part can be challenging and Monte Carlo methods are used — but this is just a technical detail.[7]

The result will be given by the joint p.d.f. of all final quantities, and can be summarized with the usual statistics objects: expected value, standard deviation, mode, median, probability intervals, and so on. In particular, an important summary is the correlation matrix if several output quantities are involved. It is important to stress, once more in this text, the most important objects for further propagations of uncertainty are expected value, standard deviation and correlation matrix. In fact, there is no simple rule for the combination of the mode and probability intervals, as discussed in Sec. 4.4 (see in particular Fig. 4.3).

12.4 Approximate solution

We have already seen in Chapters 4 and 8 that the approximate solution of uncertainty propagation, and hence of treatment of uncertainty due to systematic effects, is achieved via linearization. When nonlinear effects are sizable, at least second order approximation has to be considered. In this section, we first review the linear approximation and then analyze the second order approximation. The first part follows what was done in

[7] In my opinion, many papers place too much emphasis on the details of the Monte Carlo, instead of telling in simpler words what the aim was. In many cases, it would be enough to state that *"the integral has been done by Monte Carlo"*. Similarly, I note a misuse of the expression 'getanken experiment' (of nobler origin) just to refer to these technicalities, i.e. to say that a likelihood, meant as response of the apparatus, has been estimated by Monte Carlo.

Sec. 8.6, but with more details and a different notation, close to that of Sec. 4.4.

12.4.1 *Linear expansion around E(X)*

Let us call X_i and Y_j the input and output variables, respectively, and with $\boldsymbol{X}$ and $\boldsymbol{Y}$ their ensemble. $\boldsymbol{X}$ and $\boldsymbol{Y}$ are related by functions $Y_j = Y_j(\boldsymbol{X})$. The first-order expansion of $Y_i(\boldsymbol{X})$ around the expected values of X_i gives

$$Y_j \approx Y_j(\mathrm{E}[\boldsymbol{X}]) + \sum_i \left.\frac{\partial Y_j}{\partial X_i}\right|_{\mathrm{E}[\boldsymbol{X}]} (X_i - \mathrm{E}[X_i]) \tag{12.1}$$

$$\approx k + \sum_i \left.\frac{\partial Y_j}{\partial X_i}\right|_{\mathrm{E}[\boldsymbol{X}]} X_i\,, \tag{12.2}$$

where the derivatives are evaluated for $\boldsymbol{x} = \mathrm{E}(\boldsymbol{X})$ (this will be implicit hereafter). The second formula is very convenient to calculate the variance, having put in k all terms which do not contain X_i. Evaluating the expected values from Eq. (12.1), and variances and covariances from Eq. (12.2), we get (the symbol '≈' has been replaced by '=' to indicate that there are no further approximations other than linearization):

$$\mathrm{E}(Y_j) = Y_j(\mathrm{E}[\boldsymbol{X}])\,, \tag{12.3}$$

$$\sigma^2(Y_j) = \sum_i \left(\frac{\partial Y_j}{\partial X_i}\right)^2 \sigma_i^2 + \left\{2\sum_{l<m} \left(\frac{\partial Y_j}{\partial X_l}\right)\left(\frac{\partial Y_j}{\partial X_m}\right) \rho_{lm}\,\sigma_l\,\sigma_m\right\}. \tag{12.4}$$

$$\mathrm{Cov}(Y_j, Y_k) = \sum_i \left(\frac{\partial Y_j}{\partial X_i}\right)\left(\frac{\partial Y_k}{\partial X_i}\right) \sigma_i^2 + \left\{2\sum_{l<m} \left(\frac{\partial Y_j}{\partial X_l}\right)\left(\frac{\partial Y_k}{\partial X_m}\right) \rho_{lm}\,\sigma_l\,\sigma_m\right\}, \tag{12.5}$$

where σ_i are shorthand for $\sigma(X_i)$ and ρ_{lm} are the correlation coefficients between X_l and X_m. The terms within $\{\cdot\}$ vanish if the input quantities are uncorrelated, as is often the case when relevant systematic effects are considered.

In complex real-life cases the derivatives are not performed analytically. Instead, the effects of the input values on the output values are evaluated numerically, often by Monte Carlo techniques by $\pm\Delta x_i$ variations around the 'best estimates'. In order to obtain sensible approximate formulae, it is convenient to take $\Delta x_i = \sigma_i$ and to consider the variation around the expected value. Calling [8] $\Delta_{\pm_{ji}}$ the variation of Y_j due to a variation of X_i of $\pm 1\,\sigma_i$ around $\mathrm{E}[X_i]$, linearity implies that

$$\frac{\partial Y_j}{\partial X_i} \approx \frac{\Delta_{+_{ji}}}{\sigma_i} \approx \frac{\Delta_{-_{ji}}}{\sigma_i}. \tag{12.6}$$

Since in the linear approximation $\Delta_{+_{ji}}$ and $\Delta_{-_{ji}}$ are practically equal, we call either of them Δ_{ji} (taking the average of the two if there are small differences; the case of large differences, a hint that there are nonlinear effects, will be discussed in the next section). We get, finally, the following practical formulae for the elements of the covariance matrix:

$$\sigma^2(Y_j) = \sum_i \Delta_{ji}^2 + \left\{ 2 \sum_{l<m} \rho_{lm}\, \Delta_{jl}\, \Delta_{jm} \right\}, \tag{12.7}$$

$$\mathrm{Cov}(Y_j, Y_k) = \sum_i \Delta_{ji}\, \Delta_{ki} + \left\{ 2 \sum_{l<m} \rho_{lm}\, \Delta_{jl}\, \Delta_{km} \right\}. \tag{12.8}$$

In the simple case of independent input quantities, Eqs. (12.7)–(12.8) reduce to

$$\sigma^2(Y_j) = \sum_i \Delta_{ji}^2 \tag{12.9}$$

$$\mathrm{Cov}(Y_j, Y_k) = \sum_i \Delta_{ji}\, \Delta_{ki} \tag{12.10}$$

$$= \sum_i \mathrm{Cov}_i(Y_j, Y_k) = \sum_i s_{ijk}\, |\Delta_{ji}|\, |\Delta_{ki}|, \tag{12.11}$$

where $\mathrm{Cov}_i(Y_j, Y_k)$ stands for the contribution to the covariance from the ith input quantity, and s_{ijk} indicate the product of the signs of the absolute increments of Y_j and Y_k for a variation of X_i ($|\Delta_{ji}|$ have the meaning of standard uncertainty of Y_j due to X_i alone).

At this point, we have to remember that μ_r defined in Sec. 8.6 is considered as one of the input quantities, and that in the most general case there will be many μ_{r_j}, each associated with one and only one output quantity

[8] The following notation is used: $\Delta_+ = Y(\mathrm{E}[X]+\sigma_X) - Y(\mathrm{E}[X])$ and $\Delta_- = Y(\mathrm{E}[X]) - Y(\mathrm{E}[X]-\sigma_X)$. Therefore, for monotonic functions around $\mathrm{E}[X]$ the increments Δ_+ and Δ_- have the same sign.

Y_j. The resulting covariance matrix will be equal to the sum of the covariance matrix of the μ_{r_j} (they can be correlated because they could come from fitting procedures, unfolding, or other statistical techniques) and the covariance matrix due to the systematic effects. Let us write down, as an easy and practical example, the formulae for the case when we have N values μ_{r_j} and the influence quantities are uncorrelated:

$$\sigma^2(Y_j) = \sigma^2_{r_j} + \sum_{i>N} \Delta^2_{ji}\,, \tag{12.12}$$

$$\mathrm{Cov}(Y_j, Y_k) = \mathrm{Cov}(\mu_{r_j}, \mu_{r_k}) + \sum_{i>N} s_{ijk}\, |\Delta_{ji}|\, |\Delta_{ki}|\,, \tag{12.13}$$

where we have taken into account that the Δ_{ji} associated with μ_{r_i} are given by $\Delta_{ji} = \sigma_i\, \delta_{ij}$, where δ_{ij} is the Kronecker symbol. In fact, the derivatives of Y_j with respect to μ_{r_i}, evaluated at the point of best estimate of $\boldsymbol{X}$, are equal to 1 if $i = j$, and equal to 0 otherwise.

12.4.2 *Small deviations from linearity*

Let us consider now nonlinearity effects, which are mostly responsible for the published asymmetric uncertainties due to systematics. Nonlinearity in fact makes $\Delta_{+_{ji}}$ and $\Delta_{-_{ji}}$ differ considerably. We treat here only second-order effects. Figure 12.2 shows an example of the transformation of some important p.d.f.'s, all characterized by $\mathrm{E}(X) = 0$ and $\sigma(X) = 1$, while Fig. 12.3 shows the probability distribution of two variables based on those of Fig. 12.2. One can see that indeed the p.d.f. of the sum of both the original and the transformed quantities can be described by a Gaussian for the practical purposes of interest in uncertainty evaluations (see Fig. 4.5 for another striking example).

In order to simplify the formulae, let us consider first the case of only one input quantity and one output quantity (see Appendix B of Ref. [136] for the general case). Taking the second-order expansion, we have

$$Y = Y(\mathrm{E}[X]) + \frac{\partial Y}{\partial X}\,(X - \mathrm{E}[X]) + \frac{1}{2}\frac{\partial^2 Y}{\partial X^2}\,(X - \mathrm{E}[X])^2\,. \tag{12.14}$$

Expected value and variance of Y are then

$$\mathrm{E}(Y) = Y(\mathrm{E}[X]) + \frac{1}{2}\frac{\partial^2 Y}{\partial X^2}\,\sigma^2(X)\,, \tag{12.15}$$

$$\begin{aligned}\sigma^2(Y) = {} & \left(\frac{\partial Y}{\partial X}\right)^2 \sigma^2(X) + \frac{\partial Y}{\partial X}\frac{\partial^2 Y}{\partial X^2}\,\mathrm{E}\left[(X-\mathrm{E}[X])^3\right] \\ & + \frac{1}{4}\left(\frac{\partial^2 Y}{\partial X^2}\right)^2 \left[\mathrm{E}[(X-\mathrm{E}[X])^4] - \sigma^4(X)\right]\,. \end{aligned} \tag{12.16}$$

These formulae can be transformed into more practical ones if the derivatives are replaced by their numerical evaluations from the $\pm 1\,\sigma$ variations of X around $\mathrm{E}(X)$, which produce variations $\Delta_\pm$ in Y. The approximate derivatives evaluated in $E[X]$ are

$$\frac{\partial Y}{\partial X} \approx \frac{1}{2}\left(\frac{\Delta_+}{\sigma(X)} + \frac{\Delta_-}{\sigma(X)}\right) = \frac{\Delta_+ + \Delta_-}{2\,\sigma(X)}\,, \tag{12.17}$$

$$\frac{\partial^2 Y}{\partial X^2} \approx \frac{1}{\sigma(X)}\left(\frac{\Delta_+}{\sigma(X)} - \frac{\Delta_-}{\sigma(X)}\right) = \frac{\Delta_+ - \Delta_-}{\sigma^2(X)}\,. \tag{12.18}$$

The formula of the variance, Eq. (12.16), can be simplified using skewness ($\mathcal{S}$) and kurtosis ($\mathcal{K}$), defined as

$$\mathcal{S}(X) = \frac{\mathrm{E}\left[(X-\mathrm{E}[X])^3\right]}{\sigma^3(X)} \tag{12.19}$$

$$\mathcal{K}(X) = \frac{\mathrm{E}\left[(X-\mathrm{E}[X])^4\right]}{\sigma^4(X)}\,. \tag{12.20}$$

We get finally

$$\mathrm{E}(Y) = Y(\mathrm{E}[X]) + \delta\,, \tag{12.21}$$

$$\sigma^2(Y) = \overline{\Delta}^2 + 2\,\overline{\Delta}\cdot\delta\cdot\mathcal{S}(X) + \delta^2\cdot[\mathcal{K}(X) - 1]\,, \tag{12.22}$$

where δ is the semi-difference of the two shifts and $\overline{\Delta}$ is their average:

$$\delta = \frac{\Delta_+ - \Delta_-}{2} \tag{12.23}$$

$$\overline{\Delta} = \frac{\Delta_+ + \Delta_-}{2}\,. \tag{12.24}$$

The interpretation of Eq. (12.21) is simple and corresponds to a procedure that some might have already guessed: Asymmetric uncertainties produce a shift in the best estimate of the quantities. In the case that the dependence between Y and X is linear, δ is ≈ 0 and we recover the result given in Sec. 12.4.1. Note also that the second term of Eq. (12.22) disappears if the distribution describing the uncertainty on X is symmetric around $\mathrm{E}(X)$ (skewness is an indicator of asymmetry), and that the third term plays a minor role, since the difference between Δ_+ and Δ_- is usually smaller than their sum, and $\mathcal{K}(X)$ is around 2 or 3 for the distributions of interest (see Fig. 8.1, which gives standard deviation, skewness and kurtosis of some distributions important to model uncertainty in measurement).

The extension to several independent input quantities is straightforward, as one only needs to add together the individual contributions to expected value and the variance. Considering the most common case in which the second and third terms of the r.h.s. of Eq. (12.22) are negligible[9], we obtain the following simple practical formulae:

$$\mathrm{E}(Y) \approx Y(\mathrm{E}[\boldsymbol{X}]) + \sum_i \delta_i \,, \tag{12.25}$$

$$\sigma^2(Y) \approx \sum_i \overline{\Delta}_i^2 \,. \tag{12.26}$$

Averaging positive and negative deviations is indeed a good practice, but the shift of the central value should not be neglected. For the separation of input quantities into μ_{r_i} and influence factor, see Eqs. (12.12)–(12.13). The formulae for the more general case of several output quantities and of correlations among input quantities can be found in Ref. [136].

12.5 Numerical examples

Let us go back to the numerical example at the beginning of this chapter. Those numbers were indeed simulated from a quadratic dependence of Y on the influence quantities, each having a slightly different functional form and a different model to describe its uncertainty. Including also μ_r as X_0, we can write the dependence of Y on X_i in the following explicit form:

$$Y = \sum_{i=0}^{3} \alpha_i X_i + \beta_i X_i^2 \,, \tag{12.27}$$

where α_i and β_i are given in Tab. 12.1, in which the uncertainty model is also indicated. As stated in Sec. 12.1, the expression 'reasonable variation

[9]For symmetric distributions the skewness is zero, while the kurtosis is around 3 for the distributions of interest and enters with δ^2.

Table 12.1 Parameters of the input quantities used in the numerical example of the text. X_0 is identified with the value μ_r, obtained when X_{1-3} are equal to their expected values.

Interpretation 1: 'reasonable variations' = $\pm 1\,\sigma$ for all X_i							
Input/Output	Model p.d.f.	E[X]	$\sigma(X)$	α	β	ΔY_-	ΔY_+
$X_0(\equiv \mu_r)$	Gaussian	1	0.05	1	0	+0.050	+0.050
X_1	Gaussian	0	0.3	0.25	-0.167	+0.090	+0.060
X_2	Triangular $[-1,1]$	0	0.41	0.30	-0.147	+0.147	+0.098
X_3	Uniform $[-1,1]$	0	0.58	0.225	-0.078	+0.156	+0.104
Y	$\approx$ **Gaussian**	**0.93**	**0.20**				

Interpretation 2: 'reasonable variations' = $\pm 1\,\sigma$ for μ_r and X_1; $\pm\Delta x$ for others							
Input/Output	Model p.d.f.	E[X]	$\sigma(X)$	α	β	ΔY_- (rescaled† at $1\,\sigma_{X_i}$)	ΔY_+
$X_0(\equiv \mu_r)$	Gaussian	1	0.05	1	0	+0.050	+0.050
X_1	Gaussian	0	0.3	0.25	-0.167	+0.090	+0.060
X_2	Triangular $[-1,1]$	0	0.41	0.123	-0.0245	+0.054	+0.046
X_3	Uniform $[-1,1]$	0	0.58	0.130	-0.026	+0.084	+0.066
Y	$\approx$ **Gaussian**	**0.97**	**0.13**				

† *The rescaling is applied to input quantities not described by Gaussian models, i.e. X_2 and X_3.*

of the parameters' was intentionally left vague. We consider the two cases in which the variations of non-Gaussian quantities correspond to $\pm 1\,\sigma$ or to $\pm$ half-interval, respectively ('interpretation 1' and 'interpretation 2' in Tab. 12.1). The details of the first evaluation are

$$\begin{aligned} \mathrm{E}(Y) &= 1.00 + \sum_i \delta_i = 1.00 + (0 - 0.015 - 0.026 - 0.0245) \\ &= 0.9345\,, \end{aligned} \tag{12.28}$$

$$\begin{aligned} \sigma^2(Y) &= \sigma_r^2(Y) + \sigma_{sys}^2(Y) \\ &= (0.05)^2 + (0.1983)^2 = (0.2046)^2\,. \end{aligned} \tag{12.29}$$

(see Ref. [136] for further details). This result can be summarized as $Y^{(1)} = 0.93 \pm 0.05 \pm 0.20 = 0.93 \pm 0.20$. The result given in Eqs. (12.28)–(12.29) is in perfect agreement with $\mathrm{E}(Y) = 0.9344$ and $\sigma(Y) = 0.2046$ obtained directly from the p.d.f. of Y estimated by Monte Carlo with 10^6 extractions. In contrast, the result obtained combining separately positive and negative deviations in quadrature (see Sec. 12.1) shows a bias which amounts to 35% of σ. Assuming the second interpretation we would get

$$\begin{aligned} \mathrm{E}(Y) &= 1.00 + \sum_i \delta_i = 1.00 + (0 - 0.015 - 0.004 - 0.009) \\ &= 0.972\,, \end{aligned} \tag{12.30}$$

$$\begin{aligned} \sigma^2(Y) &= \sigma_r^2(Y) + \sigma_{sys}^2(Y) \\ &= (0.05)^2 + (0.1173)^2 = (0.1275)^2\,, \end{aligned} \tag{12.31}$$

i.e. $Y^{(2)} = 0.97 \pm 0.05 \pm 0.12 = 0.97 \pm 0.13$.

12.6 The non-monotonic case

Sometimes a variation of $\pm 1\sigma$ of an influence parameter might produce values of Y which are both above or both below the value obtained with the reference value, i.e. Δ_+ and Δ_- have opposite signs in that case. This result indicates that the function is not monotonic, and this situation has to be treated with some care. In fact, although the formulae derived in this paper do not depend on whether the functions are monotonic or not, the transformed distribution can be very different from those of Fig. 12.2 and can bring a large non-Gaussian contribution to the overall distribution. As an example, let us consider Fig. 12.4, which describes an input quantity normally distributed around 0 with $\sigma = 0.3$, a parabolic dependence of

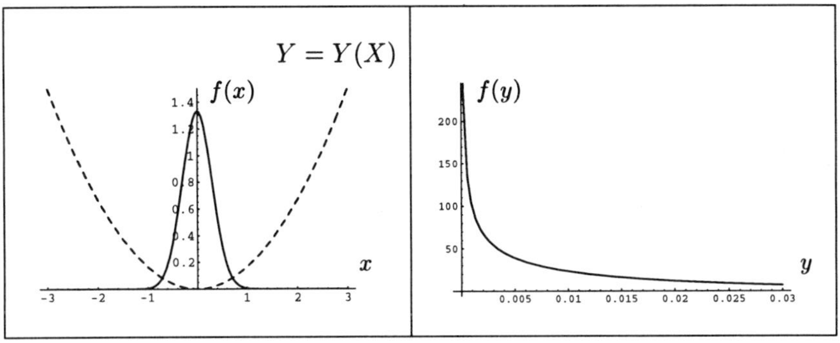

Fig. 12.4 Example of non-monotonic relation between input and output quantity. The left plot show the parabolic dependence of Y on X (dashed line) and the Gaussian p.d.f. of X (solid line). The right plot shows the p.d.f. of Y.

Y on X given by $Y = 0.167\,X^2$ (i.e. like X_1 of Tab. 12.1, but with the $\alpha = 0$ and β reversed in sign, just for graphical convenience). The $\pm 1\,\sigma$ variations are $\Delta_+ = +0.015$ and $\Delta_- = -0.015$, but certainly one would not quote 0 as the expected value of Y, nor 0.015 its standard deviation. $\mathrm{E}(X)$ being at the minimum of the distribution, the p.d.f. of X ends sharply at zero, and is very asymmetric. In fact it is easy to recognize in $f(y)$ a scale transformation of the χ^2 with one degree of freedom, namely $Y = 0.015 \times \chi^2_1$. Expected values and standard deviation are then $\mathrm{E}(Y) = 0.015$ and $\sigma(Y) = 0.015 \times \sqrt{2} = 0.021$. We can compare the result with what we get from Eqs. (12.21)–(12.22):

$$\begin{cases} \delta = 0.015 \\ \overline{\Delta} = 0 \end{cases} \Longrightarrow \begin{cases} \mathrm{E}(Y) = 0 + 0.015 = 0.015 \\ \sigma^2(Y) = 0 + 0 + 0.015^2 \times 2 = (0.021)^2 \,. \end{cases} \tag{12.32}$$

The result is exactly the same, as it should be, since in this example the function is parabolic and, therefore, there are no approximations in Eqs. (12.21)–(12.22). We see that in this case only the quadratic terms appear. Similarly, it would be wrong to consider the best estimate of Y as equal 0, with an uncertainty equal to the deviation: The result would have a standard deviation smaller by $\sqrt{2}$, and the best estimate would have a bias of -140% of the reported standard deviation.

Chapter 13

Which priors for frontier physics?

"Sometimes I dream of Higgs."
(Tom Stoppard, The real inspector Hound)

"Ogni limite ha una pazienza"
(Totò, Totò a colori)

The role of priors in inference has been discussed at several places in this book. In foregoing chapters I have tried to convince the reader, using general reasoning, formulae and examples, that in many circumstances priors are practically irrelevant. But I hope I have not given the impression that one can always forget about priors, or that Bayesian methods are usable only when the results are prior-independent. This would give a wrong sense of the Bayesian spirit, and would reduce much of the power of the approach for solving complex problems. In this last chapter I shall examine some cases in which the role of priors is so crucial that it is better to avoid reporting probabilistic results altogether. We shall see what the formal – and physical – origin of this problem is, and shall arrive at the useful classification of *closed* and *open* likelihood. Finally some practical recommendations will be given for reporting search results in the most efficient and unbiased way.

13.1 Frontier physics measurements at the limit to the detector sensitivity

There are important experiments in physics, that I like to classify with the label *frontier*, which have two things in common.

- Researchers are highly uncertain about the result of the experiment, and do not usually share the same opinion about the underlying phe-

nomenology.
- The experimental conditions are extreme, in the sense that what is being looked for are very tiny signals in the presence of background processes (usually further complicated by the fact that even these background processes are not precisely known).

Under these circumstances the lucky case in which the experiment ends with a spectacular effect, convincing all experts of a uniform interpretation of the result is extremely rare. Instead, in many cases, there is agreement that the experiment does not provide hints of the searched for signal. Unfortunately, situations arise which fall somewhere between these two extreme cases and researchers may be in serious doubt about what to do, scared and undecided between losing the chance of a discovery or losing their reputation ...

13.2 Desiderata for an optimal report of search results

Let us specify an optimal report of a search result in terms of some desired properties.

- The way of reporting the result should not depend on whether the experimental team is more or less convinced they have found the signal they were looking for.
- The report should allow an easy, consistent and efficient combination of all pieces of information which could come from several experiments, search channels and running periods. By efficient I mean the following: if many independent data sets each provides a little evidence in favor of the searched-for signal, the combination of all data should enhance that hypothesis; if, instead, the indications provided by the different data are incoherent, their combination should result in stronger constraints on the intensity of the postulated process (a higher mass, a lower coupling, etc.).
- Even results coming from low sensitivity (and/or very noisy) data sets should be included in the combination, without them spoiling the quality of the result obtainable by the clean and high-sensitivity data sets alone. If the poor-quality data carry the slightest piece of evidence, this information should play its correct role and slightly increase the global evidence.
- The presentation of the result (and its meaning) should not depend on the particular application (Higgs search, scale of contact-interaction,

proton decay, etc.).

- The result should be stated in such a way that it cannot be misleading. This requires that it should easily map onto the natural categories developed by the human mind for uncertain events.
- Uncertainties due to systematic effects of uncertain size should be included in a consistent and (at least conceptually) simple way.
- Subjective contributions of the persons who provide the results should be kept to a minimum. These contributions cannot vanish, in the sense that we always have to rely on the "understanding, critical analysis and integrity" [5] of the experimenters, but at least the dependence on the believed values of the quantity should be minimal.
- The result should summarize the experiment in the most complete way, and no extra items of information (luminosity, cross-sections, efficiencies, expected number of background events, observed number of events) should be required for further analyses.
- The result should be ready to be turned into probabilistic statements, needed to form one's opinion about the quantity of interest or to take decisions.
- The result should not lead to paradoxical conclusions.

13.3 Master example: Inferring the intensity of a Poisson process in the presence of background

As a guiding example to be referred to throughout the rest of this chapter, let us consider a case which often happens in frontier physics. We assume that a physics process, *believed to exist*, produces events modelled with a Poisson distribution, whose expected value is proportional to the observation time T, i.e. $X \sim \mathcal{P}_{rT}$. Our aim is to infer the process intensity r on the basis of the observed number of events. Unfortunately, the assumed process is not the only cause of the observed events. Other processes, generally speaking *noise*, could produce similar observations. A practical example of this type of experimental scenario is the difficult task of measuring the rate of gravitational wave (g.w.) bursts above a certain threshold (for details see Ref. [137], on which this chapter is mostly based). To fix our ideas in physics terms we shall refer through this chapter to this example and use the expressions 'intensity of Poisson process' and 'g.w. burst rate' as synonyms. Since in this kind of search, a coincidence of at least two g.w. detectors is required in order to reduce the background, we shall often refer to the 'events' as *coincidences*, and indicate them by n_c (corresponding the the generic x used elsewhere in this book).

13.4 Modelling the inferential process

Now that the inferential scheme has been set up, let us rephrase our problem in the language of Bayesian statistics.

- The physical quantity of interest, with respect to which we are in a state of great uncertainty, is the g.w. burst rate r.
- We feel quite confident[1] about the expected rate of background events r_b (but not about the number which will actually be observed).
- What is certain is the number n_c of coincidences which have been observed (stating that the observed number of coincidences is $n_c \pm \sqrt{n_c}$ does not make any sense!), although we do not know how many of these events have to be attributed to background and how many (if any) to g.w. bursts.

For a given hypothesis r the number of coincidence events which can be observed in the observation time T is described by a Poisson process having an intensity which is the sum of that due to background and that due to signal. Therefore the likelihood is

$$f(n_c \,|\, r, r_b) = \frac{e^{-(r+r_b)\,T}((r + r_b)\,T)^{n_c}}{n_c!}\,, \tag{13.1}$$

and, making use of Bayes' theorem, we get

$$f(r \,|\, n_c, r_b) \propto \frac{e^{-(r+r_b)\,T}((r + r_b)\,T)^{n_c}}{n_c!} f_\circ(r)\,. \tag{13.2}$$

13.5 Choice of priors

At this point we are faced with the problem of what $f_\circ(r)$ to choose. The best way of understanding why this choice can be troublesome is to illustrate the problem with numerical examples. Let us consider T as unit time (e.g. one month), a background rate r_b such that $r_b \times T = 1$, and the following hypothetical observations: $n_c = 0$; $n_c = 1$; $n_c = 5$.

---

[1] Uncertainty about r_b can be handled easily in our scheme, as seen in Chapters 6–8.

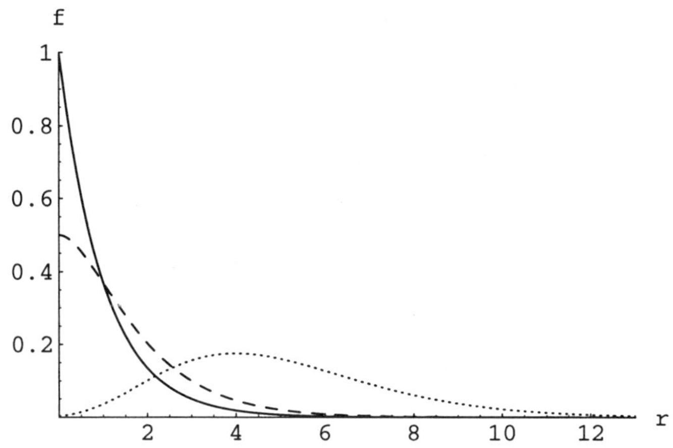

Fig. 13.1 Distribution of the values of the rate r, in units of events/month, inferred from an expected rate of background events $r_b = 1$ event/month, an initial uniform distribution $f_\circ(r) = k$ and the following numbers of observed events: 0 (solid); 1 (dashed); 5 (dotted).

13.5.1 *Uniform prior*

One might think that a good 'democratic' choice would be a uniform distribution in r, i.e. $f_\circ(r) = k$. Inserting this prior in Eq. (13.2) and normalizing the final distribution we get, using Eq. (7.75),

$$f(r \,|\, n_c, r_b, f_\circ(r) = k) = \frac{e^{-r\,T}((r + r_b)\,T)^{n_c}}{n_c! \sum_{n=0}^{n_c} \frac{(r_b\,T)^n}{n!}}\,. \tag{13.3}$$

The resulting final distributions are shown in Fig. 13.1. For $n_c = 0$ and 1 the distributions are peaked at zero, while for $n_c = 5$ the distribution appears so neatly separated from $r = 0$ that it seems a convincing proof that the postulated physics process searched for does exist. In the cases $n_c = 0$ and 1 researchers usually present the result with an upper limit (typically 95 %) on the basis that $f(r)$ seems compatible with no effect, as suggested by Fig. 13.1. For example, in the simplest and well-known case of $n_c = 0$ the 95 % C.L. upper limit is 3 events/month (see Sec. 7.4.1). The usual meaning [30] one attributes to the limit is that, if the physics process of interest exists, then there is a 95 % probability that its rate is below 3 events/month, resulting from the following equation

$$\int_0^3 f(r \,|\, n_c = 0, r_b = 1, f_\circ(r) = k)\, \mathrm{d}r = 0.95\,. \tag{13.4}$$

But there are other infinite probabilistic statements that can be derived from $f(r \,|\, r_b = 1, n_c = 0)$. For example, $P(r > 3\,\text{events/month}) = 5\,\%$, $P(r > 0.1\,\text{events/month}) = 90\,\%$, $P(r > 0.01\,\text{events/month}) = 99\,\%$, and so on. Without doubt, researchers will not hesitate to publish the 95 % upper limit, but they would feel uncomfortable stating that they believe 99 % that, if the g.w. bursts exist at all, then the rate is above 0.01 events/month. The reason for this uneasiness can be found in the uniform prior, which might not correspond to the prior knowledge that researchers really have. Let us, then, examine more closely the meaning of the uniform distribution and its consequences. Saying that $f_\circ(r) = k$, means that $\mathrm{d}P/\mathrm{d}r = k$, i.e. $P \propto \Delta r$; for example,

$$P(0.1 \le r \le 1) = \frac{1}{10}\,P(1 \le r \le 10) = \frac{1}{100}\,P(10 \le r \le 100)\ldots\,, \tag{13.5}$$

and so on. But, taken literally, this prior is hardly ever reasonable, at least for the physics case of gravitational wave detection. The problem is not due to the divergence for $r \to \infty$ which makes $f_\circ(r)$ not normalizable, i.e. 'improper' (see Sec. 6.5). This mathematical nuisance is automatically cured when $f_\circ(r)$ is multiplied by the likelihood, which, for a finite number of observed events, vanishes rapidly enough for $r \to \infty$. A much more serious problem is related to the fact that the uniform distribution assigns to all the infinite orders of magnitude below 1 a probability which is only 1/9 of the probability of the decade between 1 and 10, or 1 % of the probability of the first two decades, and so on. This is the reason why, even if no coincidence events have been observed, the final distribution obtained from zero events observed (solid curve of Fig. 13.1) implies that $P(r \ge 1\,\text{event/month}) = 37\,\%$.

13.5.2 *Jeffreys' prior*

A prior distribution alternative to the uniform can be based on the observation that what often seems uniform is not the probability per unit of r, but rather the probability per decade of r, i.e. researchers may feel equally uncertain about the orders of magnitudes of r, namely

$$P(0.1 \le r \le 1) = P(1 \le r \le 10) = P(10 \le r \le 100)\ldots\,. \tag{13.6}$$

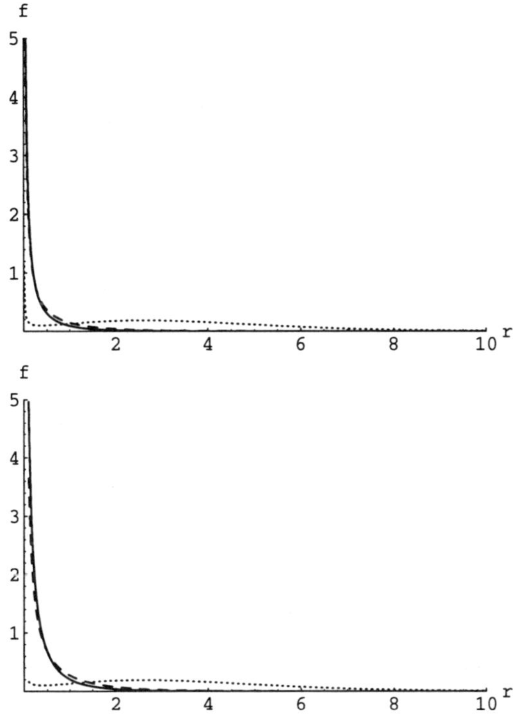

Fig. 13.2 Final distributions for the same experimental configuration of Fig. 13.1, but with a Jeffreys' prior with cut-off at $r_{min} = 0.01$ events/month (upper plot) and $r_{min} = 0.1$ events/month (lower plot).

This implies that $\mathrm{d}P/\mathrm{d}\ln r = k$, or $\mathrm{d}P/\mathrm{d}r \propto 1/r$. This prior is known as Jeffreys' prior [49] (see also Ref. [46], and it is very interesting indeed, at least from a very abstract point of view (though it tends to be misused, as is discussed in Ref. [33]). If we take Jeffreys' prior literally, it does not work in our case either. In fact, when inserted in Eq. (13.2), it produces a divergence for $r \to 0$. This is due to the infinite orders of magnitude below 1, to each of which we give equal prior probability, and to the fact that the likelihood (13.1) goes to a constant for $r \to 0$. Therefore, for any $r_\circ > 0$, we have $P(r < r_\circ)/P(r > r_\circ) = \infty$. To get a finite result we need a cut-off at a given r_{min}.

As an exercise, just to get a feeling of both the difference with respect to the case of the uniform distribution, and the dependence on the cut-off, we report in Fig. 13.2 the results obtained for the same experimental conditions as Fig. 13.1, but with a Jeffreys' prior truncated at $r_{min} = 0.1$ and 0.01. One can see that the final distributions conditioned by 0 or 1 events observed are pulled towards $r = 0$ by the new priors, while the case of $n_c = 5$ is more robust, although it is no longer nicely separated from zero.

13.5.3 *Role of priors*

The strong dependence of the final distributions on the priors shown in this example should not be considered a bad feature, as if it were just an artifact of Bayesian inference. Putting it the other way round, the Bayesian inference reproduces, in a formal way, what researchers already have clear in their minds as a result of intuition and experience. In the numerical examples we are dealing with, the dependence of the final distributions on the priors is just a hint of the fact that the experimental data are not so strong as to lead every scientist to the same conclusion (in other words, the experimental and theoretical situation is far from being the well-established one upon which intersubjectivity is based). The possibility that scientists might have distant and almost non-overlapping priors, such that agreement is reached only after a huge amount of very convincing data, should not be overlooked, as this is, in fact, the typical situation in frontier research. For this reason, one should worry, instead, about statistical methods which advertise 'objective' probabilistic results in such a critical situation.

When the experimental situation is more solid, as for example in the case of five events observed out of only 0.1 expected from background, the conclusions become very similar, virtually independent of the priors (see Fig. 13.3), unless the priors reflected really widely differing opinions.

13.5.4 *Priors reflecting the positive attitude of researchers*

Having clarified the role of priors in the assessment of probabilistic statements about true values, and their critical influence on frontier-research results, it is clear that, in our opinion, "reference priors do not exist" [33, 107]. However, I find that the "concept of a 'minimal informative' prior specification - appropriately defined!" [27] can sometimes be useful, if the practitioner is aware of the assumptions behind the specification.

We can now ask ourselves what kind of prior would be shared by rational and responsible people who have planned, financed and operated a frontier-type experiment. This is what I like to call 'positive attitude of researchers'

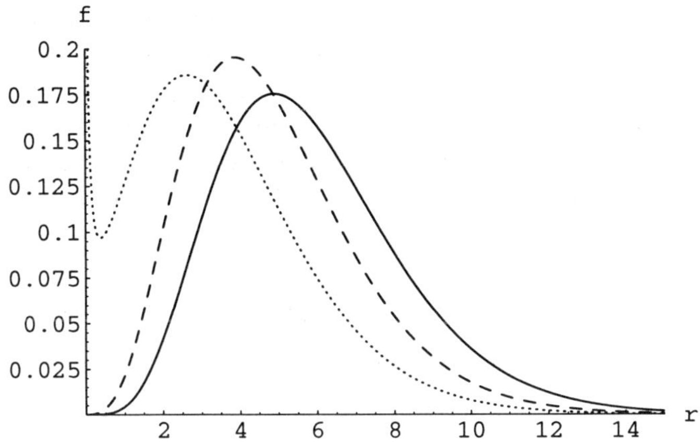

Fig. 13.3 Distribution of the values of the rate r, in units of events/month, inferred from five observed events, an expected rate of background events $r_b = 0.1$ events/month, and the following priors: uniform distribution $f_\circ(r) = k$ (solid); Jeffreys' prior truncated at $r_{min} = 0.01$ (dashed). The case of the Jeffreys' priors is also reported for $r_b = 1$ event/month (dotted).

(see Sec. 6.7). Certainly, the researchers believed there was a good chance, depending on the kind of measurement, that they would end up with a number of candidate events well above the background; or that the physical quantity of interest was well above the experimental resolution; or that a certain rate would be in the region of sensitivity.[2] One can show that the results obtained with reasonable prior distributions, chosen to model this positive attitude, are very similar to those obtainable by an improper uniform prior and, in particular, the upper/lower bounds obtained are very stable (see Secs. 6.7 and 7.7.1).

Let us apply this idea to this chapter guiding example: 0, 1 or 5 events observed over a background of 1 event (Fig. 13.1). Searching for a rare

[2]In some cases researchers are aware of having very little chance of observing anything, but they pursue the research to refine instrumentation and analysis tools in view of some positive results in the future. A typical case is gravitational wave search. In this case it is not scientifically correct to provide probabilistic upper limits from the current detectors, and the honest way to provide the result is that described here [138]. However, some could be tempted to use a frequentistic procedure which provided an 'objective' upper limit 'guaranteed' to have a 95% coverage. This behavior is irresponsible since these researchers are practically sure that the true value is below the limit. Loredo shows in Sec. 3.2 of Ref. [139] an instructive real-life example of a 90% C.I. which certainly does not contain the true value (the web site [139] contains several direct comparisons between frequentistic versus Bayesian results).

process with a detector having a background of 1 event/month, for an exposure time of one month, a positive attitude would be to think that signal rates of several events per month are quite possible. On the other hand, the fact that the process is considered to be rare implies that one does not expect a very large rate (i.e. large rates would contradict previous experimental information), and also that there is some belief that the rate could be very small, virtually zero. Let us assume that the researchers are almost sure that the rate is below 30 events/month. We can consider, for instance, the following prior distributions.

- A uniform distribution between 0 and 30:

$$f_\circ(r) = 1/30 \qquad (0 \leq r \leq 30). \tag{13.7}$$

- A triangular distribution:

$$f_\circ(r) = \frac{1}{450}\,(30 - r) \qquad (0 \leq r \leq 30). \tag{13.8}$$

- A half-Gaussian distribution of $\sigma_\circ = 10$

$$f_\circ(r) = \frac{2}{\sqrt{2\pi}\,\sigma_\circ}\exp\left[-\frac{r^2}{2\,\sigma_\circ^2}\right] \qquad (r \geq 0). \tag{13.9}$$

The last two functions model the fact that researchers might believe that small values of r are more possible than high values, as is often the case. Moreover, the half-Gaussian distribution also describes the more realistic belief that rates above 30 events/month are not excluded, although they are considered very unlikely.[3] The three priors are shown in the upper plot of Fig. 13.4. The resulting final distributions are shown in the lower plot of the same figure. The three solutions are practically indistinguishable, and, in particular, very similar to the results obtained by an improper uniform distribution (Fig. 13.1). This suggests that the improper uniform prior represents a practical and easy way of representing the prior specification for this kind of problem if one assumes what we have called the positive attitude of the researchers. Therefore, this prior could represent a way of reporting conventional probabilistic results, if one is aware of the limits of the convention. Seeking a truly objective probabilistic result — I stress yet again — is an illusory dream.

[3]As discussed in Ref. [72], realistic priors can be roughly modelled by a log-normal distribution. With parameters chosen to describe the positive attitude we are considering, this distribution would give results practically equivalent to the three priors we are using now.

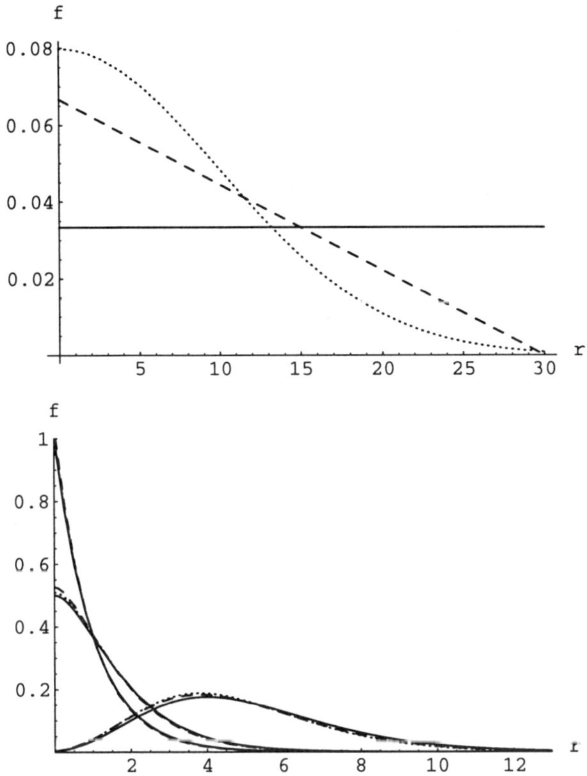

Fig. 13.4 The upper plot shows some reasonable priors reflecting the positive attitude of researchers: uniform distribution (solid); triangular distribution (dashed); half-Gaussian distribution (dotted). The lower plot shows how the results of Fig. 13.1, obtained starting from an improper uniform distribution, (do not!) change if, instead, the priors of the upper plot are used.

13.6 Prior-free presentation of the experimental evidence

At this point, I want to reassure the reader that it is possible to present data in an 'objective' way, on the condition that all thoughts of providing probabilistic results about the measurand are abandoned.

Let us again take Bayes' theorem, which we rewrite here in terms of the uncertain quantities of interest

$$f(r \,|\, n_c, r_b) \propto f(n_c \,|\, r, r_b) \cdot f_\circ(r)\,, \tag{13.10}$$

and consider only two possible values of r, let them be r_1 and r_2. From Eq. (13.10) it follows that

$$\frac{f(r_1 \,|\, n_c, r_b)}{f(r_2 \,|\, n_c, r_b)} = \underbrace{\frac{f(n_c \,|\, r_1, r_b)}{f(n_c \,|\, r_2, r_b)}}_{\textit{Bayes factor}} \cdot \frac{f_\circ(r_1)}{f_\circ(r_2)} \,. \tag{13.11}$$

This is a common way of rewriting the result of the Bayesian inference for a couple of hypotheses, keeping the contributions due to the experimental evidence and to the prior knowledge separate. The ratio of likelihoods is known as the Bayes factor and it quantifies the ratio of evidence provided by the data in favor of either hypothesis. The Bayes factor is considered to be practically objective because likelihoods (i.e. probabilistic description of the detector response) are usually much less critical than priors about the physics quantity of interest.

The Bayes factor can be extended to a continuous set of hypotheses r, considering a function which gives the Bayes factor of each value of r with respect to a reference value r_{REF}. The reference value could be arbitrary, but for our problem the choice $r_{REF} = 0$, giving

$$\mathcal{R}(r; n_c, r_b) = \frac{f(n_c \,|\, r, r_b)}{f(n_c \,|\, r = 0, r_b)} \,, \tag{13.12}$$

is very convenient for comparing and combining the experimental results [140, 38]. The function $\mathcal{R}$ has nice intuitive interpretations which can be highlighted by reordering the terms of Eq. (13.11) in the form

$$\frac{f(r \,|\, n_c, r_b)}{f_\circ(r)} \Big/ \frac{f(r = 0 \,|\, n_c, r_b)}{f_\circ(r = 0)} = \frac{f(n_c \,|\, r, r_b)}{f(n_c \,|\, r = 0, r_b)} = \mathcal{R}(r; n_c, r_b) \tag{13.13}$$

(valid for all possible a priori r values). $\mathcal{R}$ has the probabilistic interpretation of *relative belief updating ratio*, or the geometrical interpretation of *shape distortion function* of the probability density function. $\mathcal{R}$ goes to 1 for $r \to 0$, i.e. in the asymptotic region in which the experimental sensitivity is lost: As long as $\mathcal{R}$ stays at 1, the shape of the p.d.f. (and therefore the relative probabilities in that region) remains unchanged. Instead, in the limit $\mathcal{R} \to 0$ (for large r) the final p.d.f. vanishes, i.e. the beliefs go to zero no matter how strong they were before. In the case of the Poisson process we are considering, the relative belief updating factor becomes

$$\mathcal{R}(r; n_c, r_b, T) = e^{-r\,T} \left(1 + \frac{r}{r_b}\right)^{n_c} \,, \tag{13.14}$$

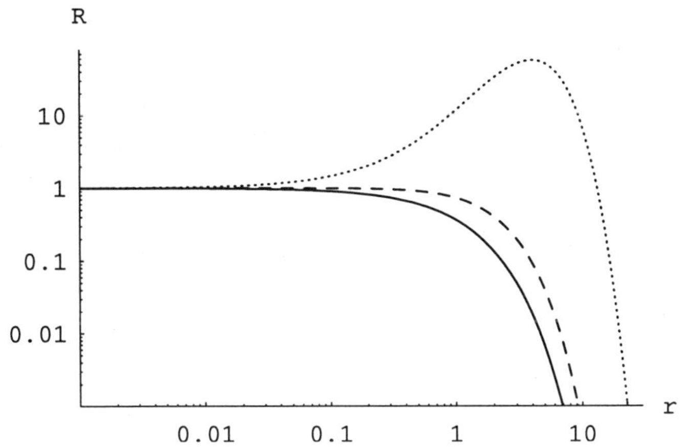

Fig. 13.5 Relative belief updating ratio $\mathcal{R}$ for the Poisson intensity parameter r for the cases of Fig. 13.1.

with the condition[4] $r_b > 0$ if $n_c > 0$.

Figure 13.5 shows the $\mathcal{R}$ function for the numerical examples considered above. The abscissa has been drawn in a log scale to make it clear that several orders of magnitude are involved. These curves transmit the result of the experiment immediately and intuitively:

- whatever one's beliefs on r were before the data, these curves show how one must change them;
- the beliefs one had for rates far above 20 events/month are killed by the experimental result;
- if one believed strongly that the rate had to be below 0.1 events/month, the data are irrelevant;
- the case in which no candidate events have been observed gives the strongest constraint on the rate r;
- the case of five candidate events over an expected background of one produces a peak of $\mathcal{R}$ which corroborates the beliefs around 4 events/month only if there were sizable prior beliefs in that region.

[4]The case $r_b = n_c = 0$ yields $\mathcal{R}(r) = e^{-r}$, obtainable starting directly from Eq. (13.12), defining $\mathcal{R}$, and from Eq. (13.1), giving the likelihood. Also the case $r_b \to \infty$ has to be evaluated directly from the definition of $\mathcal{R}$ and from the likelihood, yielding $\mathcal{R} = 1\ \forall\, r$; finally, the case $r_b = 0$ and $n_c > 0$ makes $r = 0$ impossible, thus prompting a claim for discovery – and it no longer makes sense for the $\mathcal{R}$ function defined above to have that nice asymptotic behavior in the insensitivity region.

Moreover there are some technical advantages in reporting the $\mathcal{R}$ function as a result of a search experiment.

- One deals with numerical values which can differ from unity only by a few orders of magnitude in the region of interest. Instead, the values of the likelihood can be extremely low. For this reason, the comparison between different results given by the $\mathcal{R}$ function can be perceived better than if these results were published in terms of likelihood.
- Since $\mathcal{R}$ differs from the likelihood only by a factor, it can be used directly in Bayes' theorem, which does not depend on constants, whenever probabilistic considerations are needed.[5] In fact,

$$f(r \mid n_c\, r_b) \propto \mathcal{R}(r; n_c, r_b) \cdot f_\circ(r)\,. \tag{13.15}$$

- The combination of different independent results on the same[6] quantity r can be done straightforwardly by multiplying individual $\mathcal{R}$ functions:

$$\mathcal{R}(r; \text{all data}) = \Pi_i \mathcal{R}(r; \text{data}_i)\,. \tag{13.16}$$

- Finally, one does not need to decide a priori if one wants to make a 'discovery' or an 'upper limit' analysis as conventional statistics teaches (see e.g. criticisms in Ref. [13]): the $\mathcal{R}$ function represents the most unbiased way of presenting the results and everyone can draw their own conclusions.

13.7 Some examples of $\mathcal{R}$-function based on real data

The case study we have been dealing with is based on a toy model simulation. To see how the proposed method provides the experimental evidence in a clear way we show in Figs. 13.6 and 13.7 $\mathcal{R}$-functions based on real data. The first is a reanalysis of Higgs search data at LEP [38]; the second comes from the search for contact interactions at HERA made by ZEUS [140]. The extension of Eq. (13.12) to the most general case is

$$\mathcal{R}(\mu; \text{data}) = \frac{f(\text{data} \mid \mu)}{f(\text{data} \mid \mu_{\text{ins}})}\,, \tag{13.17}$$

[5] Note that, although it is important to present prior-free results, at a certain moment a probability assessment about r can be important, for example, in forming one's own idea about the most likely range of r, or in taking decisions about planning and financing of future experiments.

[6] See comments about the choice of the energy threshold in Ref. [72].

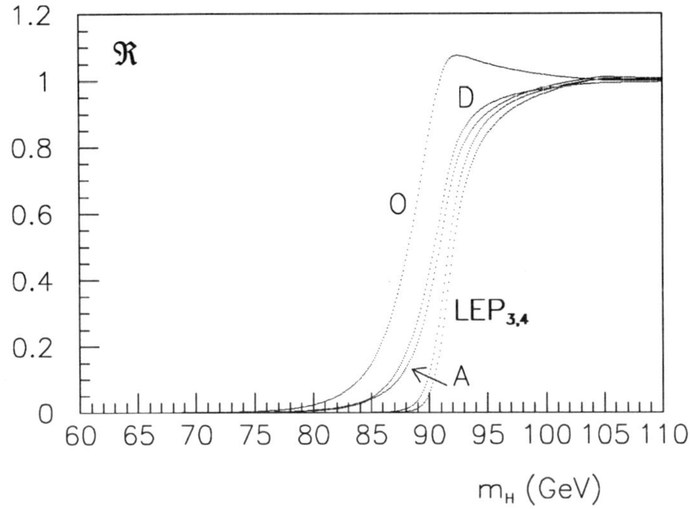

Fig. 13.6 $\mathcal{R}$-function reporting results on Higgs direct search from the reanalysis performed in Ref. [38]. A, D and O stand for ALEPH, DELPHI and OPAL experiments. Their combined result is indicated by LEP_3. The full combination (LEP_4) was obtained by assuming for L3 experiment a behavior equal to the average of the others experiments.

where μ_{ins} stands for the asymptotic insensitivity value (0 or ∞, depending on the physics case) of the generic quantity μ. Figures 13.6 and 13.7 show clearly what is going on, namely which values are practically ruled out and which ones are inaccessible to the experiment. This method has also been used recently to report results of gravitational wave bursts [138].

13.8 Sensitivity bound versus probabilistic bound

It is rather evident from Figs. 13.5, 13.6 and 13.7 how we can summarize the result with a single number which gives an idea of an upper or lower bound. In fact, although the $\mathcal{R}$-function represents the most complete and unbiased way of reporting the result, it might also be convenient to express with just one number the result of a search which is considered by the researchers to be unfruitful. This number can be any value chosen by convention in the region where $\mathcal{R}$ has a transition from 1 to 0. This value would then delimit (although roughly) the region of the values of the quantity which are definitively excluded from the region in which the

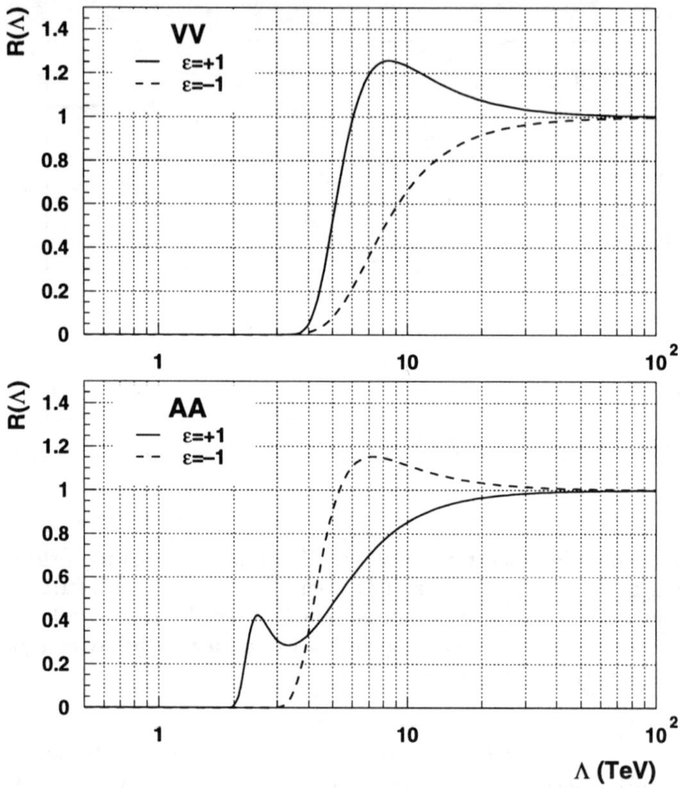

Fig. 13.7 $\mathcal{R}$-functions reporting results on search for contact interactions [140]. The ZEUS paper contains the detailed information to obtain these curves.

experiment can say nothing. The meaning of this bound is not that of a probabilistic limit, but of a *wall*[7] which separates the region in which we 'are', and where we see nothing, from the region we cannot see. We may take as the conventional position of the wall the point where $\mathcal{R}(r_B)$ equals 50%, 5% or 1% of the insensitivity plateau. What is important is not to call this value a bound at a given probability level (or at a given confidence level – the perception of the result by the user will be the same! [30]). A possible unambiguous name, corresponding to what this number indeed is, could

[7]In most cases it is not a sharp solid wall. A hedge might be more realistic, and indeed more poetic: *"Sempre caro mi fu quell'ermo colle, / E questa siepe, che da tanta parte / Dell'ultimo orizzonte il guardo esclude"* (Giacomo Leopardi, *L'Infinito*). The exact position of the hedge doesn't really matter, if we think that on the other side of the hedge there are infinite orders of magnitude unaccessible to us.

be *standard sensitivity bound.* As the conventional level, my suggestion is to choose $\mathcal{R} = 0.05$ [72]. This convention has the advantage that it allows recovery of upper 95% probability bounds obtained in the case of no events observed in an experiment characterized by Poisson likelihood and assuming a uniform prior. Anyhow, choosing a different convention does not change the substance of the result, as we can see from Figs. 13.5, 13.6 and 13.7.

Note that it does not make much sense to give the standard sensitivity bound with many significant digits. The reason becomes clear by observing Figs. 13.5–13.7, in particular Fig. 13.7. I don't think that there will be a single physicist who, judging from the figure, believes that there is a substantial difference concerning the scale of a postulated contact interaction for $\epsilon = +1$ and $\epsilon = -1$. Similarly, looking at Fig. 13.5, the observation of 0 events, instead of 1 or 2, should not produce a significant modification of our opinion about g.w. burst rates. What really matters is the order of magnitude of the bound or, depending on the problem, the order of magnitude of the difference between the bound and the kinematic threshold. A limit should be considered on the same footing as an uncertainty, not as a true value. Sometimes I get the impression that when some people talk about a '95% confidence limit', they think as if they were '95% confident about the limit'. It seems to me that for this reason some are disappointed to see upper limits on the Higgs mass fluctuating, in contrast to lower limits which are more stable and in constant increase with the increasing available energy. In fact, as explained in Ref. [59], these two 95% C.L. limits don't have the same meaning. It is quite well understood by experts that most 95% C.L. limits are in practice $\approx$ 100% probability limits (see also Ref. [60]) – and, then, the frequentistic concept of 'exact coverage' is just an illusion! (See Sec. 10.7.)

I can imagine that at this point there are still researchers who would like to give, in addition to $\mathcal{R}$ function and sensitivity bound, probabilistic limits. In my opinion, the most honest way would be for these researchers to use the likelihood information (or the $\mathcal{R}$ function) together with their initial subjective beliefs to get final beliefs. This procedure is at least unambiguous and would allow the reader to have an idea of what these authors were really thinking (which is what matters in practice, assuming these authors are authoritative scientists! – see Sec. 10.4). An alternative way, of presenting probabilistic results (one which would be more a matter of convention, but which, unfortunately, does not correspond to real beliefs of a real person), is to provide probabilistic limits justified with what in Sec. 13.5.4 was called 'positive attitude of researchers'. There it was shown that, no matter how this 'positive attitude' is reasonably modelled, the final p.d.f. is, for the case of g.w. bursts ($\mu_{\text{ins}} = 0$), very similar to that obtained by a uniform distribution. Therefore, a uniform prior could be used to provide some kind of conventional probabilistic upper limits, which could look acceptable to all those who share that kind of positive attitude. But, certainly, it is not possible to pretend that these probabilistic conclusions could be shared by everyone. Note, however, that this reasoning cannot be applied in a straightforward way when $\mu_{\text{ins}} = \infty$, as can be easily understood (though there is no problem with the result presented in terms of the $\mathcal{R}$-function). In such a case one can work on a sensible conjugate variable which has the asymptotic insensitivity limit at 0. For example, in the case of a search for 'contact interactions', one could use the quantity ϵ/Λ^2 in place of Λ [62, 140, 141]. Ref. [62] also contains the basic idea of using a sensitivity bound, though formulated differently in terms of 'resolution power cut-off.' (This was my instinctive way, at that time before I even knew what Bayes' theorem was, for getting rid of the unsatisfactory 'prescriptions' I found in the literature.)

13.9 Open versus closed likelihood

Figures 13.5, 13.6 and 13.7 show clearly the reason that frontier measurements are crucially dependent on priors: the likelihood only vanishes on one side (let us call these measurements *open likelihood*). In other cases the likelihood goes to zero in both sides (*closed likelihood*). Normal routine measurements belong to the second class, and usually they are characterized by a narrow likelihood, meaning high precision. Most physics measurements belong to the class of closed priors. The two classes can be treated differently, though the publication of non trivial (i.e. non-Gaussian) likelihood is recommended in all cases. This does not mean recovering frequentistic 'flip-flop' (see Ref. [73] and references therein), but recognizing the qualitative, not just quantitative, difference between the two cases.

When the likelihood is closed, the sensitivity on the choice of prior is much reduced, and a probabilistic result can easily be given. The better-understood subcase is when the likelihood is a very narrow distribution (possibly Gaussian). Any reasonable prior which models the knowledge of the expert interested in the inference is practically constant in the narrow range around the maximum of the likelihood. Therefore, we get the same result obtained by a uniform prior. However, when the likelihood is not so narrow, there could still be some dependence on the prior used. Again, this problem has no solution if inference is considered as a kind mathematical game [33] and fancy mathematical functions are proposed (but I will always ask the proposer *"do you really think so? are you prepared to gamble, with odds based on your prior?"*). Things are less problematic if one uses physics intuition and experience. The idea is to use a uniform prior on the quantity which is 'naturally measured' by the experiment. This might look like an arbitrary concept, but is in fact an idea to which experienced physicists are accustomed. For example, we say that 'a tracking devise measures $1/p$', 'radiative corrections measure $\log(M_H)$', 'a neutrino mass experiment is sensitive to m^2', and so on. We can see that our intuitive idea of 'the quantity really measured' is related to the quantity which has a linear dependence on the observation(s). When this is the case, random (Brownian) effects occurring during the process of measurement tend to produce a roughly Gaussian distribution of observations. In other words, we are dealing with a roughly Gaussian likelihood. So, a way to state the natural measured quantity is to refer to the quantity for which the likelihood is roughly Gaussian. This is the reason why we do least-square fits choosing the variable in which the χ^2 is parabolic (i.e. the likelihood is normal) and then interpret the result as probability of the true value. I would recommend continuing with the tradition of considering natural the quantity which gives a roughly normal likelihood. For example, this was the original motive for proposing ϵ/Λ^2 to report compositeness result [62]. This uniform-prior/Gaussian-likelihood duality goes back to Gauss himself [68], and has been shown in Sec. 6.12.

When there is no agreement about the natural quantity one can make a sensitivity analysis of the result, as in the exercise of Fig. 13.8, based on Ref. [142]. If one chooses a prior flat in Higgs mass m_H, rather than in $\log(m_H)$, the p.d.f.'s given by the continuous curve change into those given by the dashed curve. Expected value and standard deviation of the distributions (last digits in parentheses) change from $M_H = 0.10(7)$ TeV to $M_H = 0.14(9)$ TeV. Although this is just an academic exercise, since it is rather well accepted that radiative corrections measure $\log(M_H)$, Fig. 13.8 and the above digits show that the result is indeed rather stable, considering the difficulty of the measurement.

If a case is really controversial, one can still show the likelihood. But it is important to understand that a likelihood is not yet the probabilistic result we physicists want. If only the likelihood is published, the risk is too high that this likelihood will be interpreted anyway and somehow as if

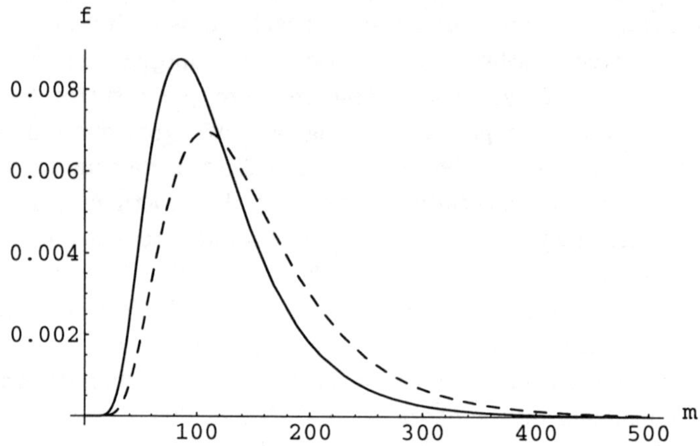

Fig. 13.8 Sensitivity analysis exercise from the indirect Higgs mass determination of Ref. [142]. Solid lines and dashed lines are obtained with priors uniform in $\log(m_H)$ and m_H, respectively.

it were a probabilistic result, as happens now in practice. For this reason I think that, at least in the rather simple case of closed likelihood, those who perform the research should recognize their responsibility and, making use of their best prior knowledge, assess the expected value and standard deviation (plus other information in the case of a strongly non-Gaussian distribution) that they really believe. I do not think that, in most applications, this subjective ingredient would be more influential than the many other subjective choices made during the course of an experimental and that we have to accept anyhow. Adhering strictly to the point of view that one should refrain totally from giving probabilistic results, on the basis of the idealistic principle of avoiding the contribution of personal priors, would halt research. We always rely on somebody else's priors and consult experts. Only a perfect idiot has no priors, and this is not the best person to consult.

Part 4

Conclusion

Chapter 14

Conclusions and bibliography

> *"You take your life in your own hands, and what happens? A terrible thing: no one to blame."*
> *(Erica Jong)*

14.1 About subjective probability and Bayesian inference

I hope to have been able to show that it is possible to build a powerful theory of measurement uncertainty starting from subjective probability and the rules of logics, from which the Bayes' theorem follows. Subjective probability is based on the natural concept of probability, as degree of belief, related to a status of uncertainty, whilst Bayes' theorem is the logical tool to update the probability in the light of new pieces of information.

The main advantages the Bayesian approach has over the others are (in addition to the non-negligible fact that it is able to treat problems on which the others fail):

- the recovery of the intuitive idea of probability as a valid concept for treating scientific problems;
- the simplicity and naturalness of the basic tool;
- the capability of combining prior knowledge and experimental information;
- the automatic updating property as soon as new information is available;

- the transparency of the method which allows the different assumptions on which the inference may depend to be checked and changed;
- the high degree of awareness that it gives to its user.

When employed on the problem of measurement errors, as a special application of conditional probabilities, it allows all possible sources of uncertainties to be treated in the most general way.

When the problems get complicated and the general method becomes too heavy to handle, it is often possible to use approximate methods based on linearization to evaluate average and standard deviation of the distribution, while the central limit theorem makes the final distributions approximately Gaussian. Nevertheless, there are some cases in which the linearization may cause severe problems, as shown in Sec. 8.11. In such cases one needs to go back to the general method or to apply other kinds of approximations which are not just blind use of the covariance matrix.

Many conventional (frequentistic) methods can be easily recovered, like maximum likelihood or χ^2 fitting procedures, as approximation of Bayesian methods, when the (implicit) assumptions on which they are based are reasonable.

14.2 Conservative or realistic uncertainty evaluation?

Finally, I would like to conclude with some remarks about safe (or conservative) evaluation of the uncertainty. The normative rule of coherence requires that all probabilistic statements should be consistent with the beliefs. Therefore, if the uncertainty on a physical quantity is modelled with a Gaussian distribution, and one publishes a result as, for example, $\alpha_s = 0.119 \pm 0.03$, one should be no more nor less sure than 68 % that α_s is in that interval (and one should be 95% sure that the value is within ± 0.06, and so on). If one feels more sure than 68 % this should be explicitly stated, because the normal practice of physics is to publish standard uncertainty in a normal probability model, as also recommended by the ISO Guide [5]. In this respect, the ISO recommendation can be summarized with the following quotation:

> *"This Guide presents a widely applicable method for evaluating and expressing uncertainty in measurement. It provides a realistic rather than a 'safe' value of uncertainty based on the concept that there is no inherent difference between an uncertainty component arising from a random effect and one arising from a correction for a systematic effect.*

The method stands, therefore, in contrast to certain older methods that have the following two ideas in common:

- *The first idea is that the uncertainty reported should be 'safe' or 'conservative' (...) In fact, because the evaluation of the uncertainty of a measurement result is problematic, it was often made deliberately large.*
- *The second idea is that the influences that give rise to uncertainty were always recognizable as either 'random' or 'systematic' with the two being of different nature; (...) In fact, the method of combining uncertainty was often designed to satisfy the safety requirement.*

... When the value of a measurand is reported, the best estimate of its value and the best estimate of the uncertainty of that estimate must be given, for if the uncertainty is to err, it is not normally possible to decide in which direction it should err safe. An understatement of uncertainties might cause too much trust to be placed in the values reported, with sometimes embarrassing and even disastrous consequences. A deliberate overstatement of uncertainty could also have undesirable repercussions."

The examples of the 'undesirable repercussions' given by the ISO Guide are of the metrological type. In my opinion there are other physical reasons which should be considered. Deliberately overstating uncertainty leads to a better (but artificial) agreement between results and 'known' values or results of other experiments. This prevents the identification of possible systematic effects which could have biased the result and which can only be identified by performing the measurement of the same physical quantity with a different instrument, method, etc. (the so-called 'reproducibility conditions' [5]). Behind systematic effects there is always some physics, which can somehow be 'trivial' (noise, miscalibration, row approximations, background, etc.), but also some new phenomenology. If the results of different experiments are far beyond their uncertainty the experimenters could compare their methods, find systematic errors and, finally, the combined result will be of a higher quality. In this respect, a last quotation from Feynman is in order:

"Well, QED is very nice and impressive, but when everything is so neatly wrapped up in blue bows, with all experiments in exact agreement with each other and with the theory - that is when one is learning **absolutely nothing.**"

"On the other hand, when experiments are in hopeless conflict - or when the observations do not make sense according to conventional ideas, or when none of the new models seems to work, in short when the situation

> *is an unholy mess -* **that** *is when one is really making hidden progress and a breakthrough is just around the corner!"* [143]

14.3 Assessment of uncertainty is not a mathematical game

Finally, I would like to conclude with my favorite quotation concerning measurement uncertainty, taken from the ISO Guide [3]:

> *"Although this* Guide *provides a framework for assessing uncertainty, it cannot substitute for critical thinking, intellectual honesty, and professional skill. The evaluation of uncertainty is neither a routine task nor a purely mathematical one; it depends on detailed knowledge of the nature of the measurand and of the measurement. The quality and utility of the uncertainty quoted for the result of a measurement therefore ultimately depend on the understanding, critical analysis, and integrity of those who contribute to the assignment of its value."*

14.4 Bibliographic note

The state of the art of Bayesian theory is summarized in Refs. [27] and [43], where many references can be found. A comprehensive and eloquent presentation of the Bayesian approach in scientific reasoning, covering philosophical, mathematical and statistical aspects is given in Ref. [144], a short account of which can be found in a "Nature" article [10]. Very interesting and insightful philosophical and historical aspects of subjective probability are provided in the introduction of Ref. [69]. Other interesting references to get an idea of what present philosophers think about Bayesian theory see also Refs. [145, 146, 147, 148] and references therein. Eloquent 'defenses of the Bayesian choice' can be found at the end of Refs.[58] and [144]. For an excellent elementary introduction to Bayesian statistics, see Ref. [149]. A clear, concise mathematical presentation of de Finetti subjective probability can be found in the first chapter of Ref. [150]. Reference [29] provides a little formal introduction to physicists, also stressing the importance of teaching subjective probability and Bayesian statistics in the physics curriculum.

As classical books on subjective probability, de Finetti's and Jeffreys's *"Theory of probability"* [16] are a must (same title and no mention of 'Bayesian' in the title!). I found Ref. [151] particularly stimulating and

Ref. [42] very convincing (the latter represents, in my opinion, the only real introductory, calculus-based, textbook on subjective probability and Bayesian statistics available so far, with many examples and exercises). Unfortunately these two books are only available in Italian at the moment. For Italian readers, I also recommend Refs. [152] and [153].

I have consulted Refs. [154] and [155], which also contain many references. References [48, 49, 57, 58, 75, 156, 157, 158, 159, 160] are well-known books among Bayesians. Some literature on Bayesian Networks can be found in Ref. [120], which also contains interesting URLs. Reference [46] is Bayesian book close to the physicist's point of view. For developments on Bayesian theory and practical applications I recommend consulting the proceedings of "Valencia Meetings" [161] and "Maxent Workshops" [93]. An overview of maximum-entropy methods can also be found in Ref. [92], while Ref. [108] is an unpublished cult book for those who adhere to the MaxEnt school. Refs. [55, 92] show some applications of Bayesian reasoning and maximum entropy ideas in statistical mechanics. Ref. [162] show how these ideas can be applied to spectrum analysis and time series (for the latter subject, Ref. [163] is particularly recommended).

Other information on Bayesian literature methods can be found on web sites. As a starting point I would recommend Ref. [164], as well as other sites dedicated to Bayesian networks and artificial intelligence [120]. Physicists will find interesting Tom Loredo's preprints and tutorials, and references therein [165]. Interesting papers for physical and technological applications can be found in Refs. [166, 167].

For an overview about numerical issues in Bayesian analysis Ref. [168] is recommended. When integrals become complicated, the Markov Chain Monte Carlo (MCMC) technique becomes crucial: introductions and applications can be found, for example, in Refs. [75, 125, 169, 170, 171, 172]. A recent application of Bayesian methods in cosmology, which uses MCMC and contains a pedagogical introduction too, can be found in Ref. [173].

Some sources in the history of probability and statistics (from which, for example I have taken the Laplace quote at the beginning of Chapter 7), can be found in Ref. [174].

The applied part of these notes, as well as the critical part, is mostly original. References are given at the appropriate place in the text — only those actually used have been indicated. A concise critical overview of Bayesian reasoning versus frequentistic methods in physics can be found in Ref. [30], whilst Ref. [33] is recommended to those who are still anxious about priors.

As far as measurement uncertainty is concerned, consultation of the ISO Guide [5] is advised. At present the BIPM recommendations are also followed by the American National Institute of Standards and Technology (NIST), whose guidelines [7] are also on the web.

Note: For the reader's convenience, I have added the link to the electronic version of the papers, whenever available, though in a preprint form. In particular, all references in the format `xxx/yymmnnn` (e.g. `hep-ph/9512295` or `physics/9811046`) are available at `http://arxiv.org/abs/xxx/yymmnnn`.

Bibliography

[1] G. D'Agostini, *"Probability and measurement uncertainty in Physics – a Bayesian primer"*, Internal Report N. 1070 of the Dept. of Physics of the Rome University "La Sapienza", and DESY-95-242, December 1995. [hep-ph/9512295].

[2] G. D'Agostini, *"Bayesian reasoning in High Energy Physics – principles and applications"*, CERN Report 99–03, July 1999. [http://www.roma1.infn.it/~dagos/YR.html]

[3] Deutsches Institut für Normung (DIN), *"Grunbegriffe der Messtechnick - Behandlung von Unsicherheiten bei der Auswertung von Messungen"* (DIN 1319 Teile 1–4), Beuth Verlag GmbH, Berlin, Germany, 1985. Only parts 1–3 are published in English. An English translation of part 4 can be requested from the authors of Ref. [36]. Part 3 is going to be rewritten in order to be made in agreement with Ref. [5] (private communication from K. Weise).

[4] R. Kaarls, *BIPM proc.-Verb. Com. Int. Poids et Mesures* **49** (1981), A1-A2 (in French);
P. Giacomo, Metrologia **17** (1981) 73 (draft of English version; for the official BIPM translation see Refs. [5] or [7]).

[5] International Organization for Standardization (ISO), *"Guide to the expression of uncertainty in measurement"*, Geneva, Switzerland, 1993.

[6] International Organization for Standardization (ISO), *"International vocabulary of basic and general terms in metrology"*, Geneva, Switzerland, 1993.

[7] B.N. Taylor and C.E. Kuyatt, *"Guidelines for evaluating and expressing uncertainty of NIST measurement results"*, NIST Technical Note 1297, September 1994 [http://physics.nist.gov/Pubs/guidelines/outline.html].

[8] H. Poincaré, *"Science and Hypothesis"*, 1905 (Dover Publications, 1952).

[9] H. Poincaré, *"Calcul des probabilités"*, University of Paris, 1893–94.

[10] C. Howson and P. Urbach, *"Bayesian reasoning in science"*, Nature, Vol. 350, 4 April 1991, p. 371.

[11] G. Zech, *"Frequentist and Bayesian confidence limits"*, EPJdirect **C12**

(2002) 1 [http://www.edpsciences.com/articles/epjdirectc/abs/2002/contents.html].

[12] P. Clifford, *"Interval estimation as viewed from the world of mathematical statistics"*, Workshop on Confidence Limits, Geneva, Switzerland, January 2000, CERN Report 2000-005 [http://ep-div.web.cern.ch/ep-div/Events/CLW/papers.html].

[13] J.O. Berger and D.A. Berry, *"Statistical analysis and the illusion of objectivity"*, Am. Scientist **76** (1988) 159.

[14] M.J. Schervish, *"P values: what they are and what they are not"*, Am. Stat. **50** (1996) 203.

[15] G. Cowan, *"Statistical data analysis"*, Clarendon Press, Oxford, 1998.

[16] B. de Finetti, *"Theory of probability"*, J. Wiley & Sons, 1974.

[17] K. Baklawsky, M. Cerasoli and G.C. Rota, *"Introduzione alla Probabilità"*, Unione Matematica Italiana, 1984.

[18] www.desy.de/pr-info/desy-recent-hera-results-feb97_e.html, (*"DESY Science Information on Recent HERA Results"*, Feb. 19, 1997).

[19] DESY '98 - Highlights from the DESY Research Center, *"Throwing 'heads' seven times in a row - what if it was just a statistical fluctuation?"*.

[20] ZEUS Collaboration, J. Breitweg et al., *"Comparison of ZEUS data with Standard Model predictions for $e^+p \to e+X$ scattering at high x and Q^2"*, Z. Phys. **C74** (1997) 207;
H1 Collaboration, C. Adloff et al., *"Observation of events at very high Q^2 in ep collisions at HERA"*, Z. Phys. **C74** (1997) 191.

[21] C. Tully in an interview to Physics Web, September 2000: "Higgs boson on the horizon", by V. Jamieson, http://PhysicsWeb.org/article/news/4/9/2/1.

[22] G. Bunce, in BNL News Release "Physicists announce possible violation of standard model of particle physics", February 2001, http://www.bnl.gov/bnlweb/pubaf/pr/bnlpr020801.htm.

[23] FNAL, Press Pass November 7, 2001, "Neutrino Measurement Surprises Fermilab Physicists", http://www.fnal.gov/pub/presspass/press_releases/NuTeV.html

[24] I. Kant, *"Prolegomena to any future metaphysics"*, 1783.

[25] A. Einstein, *"Autobiographisches"*, in *"Albert Einstein: Philosopher-Scientist"*, P.A. Schilpp ed., Library of Living Philosophers, Tudor, Evanston, Ill., 1949, pp. 2–95.

[26] A. Einstein, *"Über die spezielle und die allgemeine Relativitätstheorie (gemeinverständlich)"*, Vieweg, Braunschweig, 1917. Translation: *"The special and the general Theory. A popular exposition"*, London Methuen 1946.

[27] J.M. Bernardo and A.F.M. Smith, "*Bayesian theory*", John Wiley & Sons, 1994.

[28] D. Hume, *"Enquiry concerning human understanding"* (1748), see, e.g., http://www.utm.edu/research/hume/wri/1enq/1enq-6.htm,.

[29] G. D'Agostini, *"Teaching statistics in the physics curriculum. Unifying*

and clarifying role of subjective probability", Am. J. Phys. **67** (1999) 1260 [physics/9908014].

[30] G. D'Agostini, *"Bayesian reasoning versus conventional statistics in high energy physics"*, Proc. XVIII International Workshop on Maximum Entropy and Bayesian Methods, Garching (Germany), July 1998, V. Dose et al. eds., Kluwer Academic Publishers, Dordrecht, 1999 [physics/9811046].

[31] G. D'Agostini, contribution to the panel discussion at Workshop on Confidence Limits, Geneva, Switzerland, January 2000, CERN Report 2000-005 pp. 285–286 [ep-div.web.cern.ch/ep-div/Events/CLW/QA/PS/clwdiscuss.ps].

[32] G. D'Agostini, *"Role and meaning of subjective probability: some comments on common misconceptions"*, XX International Workshop on Maximum Entropy and Bayesian Methods in Science and Engineering, Gif-sur-Yvette (France), July 2000, A. Mohammad-Djafari, ed, AIP Conference Proceedings, Vol. 568, 2001 [physics/0010064].

[33] G. D'Agostini, *"Overcoming priors anxiety"*, *Bayesian Methods in the Sciences*, J. M. Bernardo Ed., special issue of *Rev. Acad. Cien. Madrid*, Vol. 93, Num. 3, 1999 [physics/9906048].

[34] S.J. Press and J.M. Tanur, *"The subjectivity of scientists and the Bayesian approach"*, John Wiley & Sons, 2001.

[35] K. Weise, private communication, August 1995.

[36] K. Weise, W. Wöger, *"A Bayesian theory of measurement uncertainty"*, Meas. Sci. Technol. **4** (1993) 1.

[37] H. O. Lancaster, "*The Chi-squared Distribution*", John Wiley & Sons, 1969.

[38] G. D'Agostini and G. Degrassi, *"Constraints on the Higgs boson mass from direct searches and precision measurements"*, Eur. Phys. J. **C10** (1999) 633 [hep-ph/9902226].

[39] P.-S. Laplace, *"Théorie Analityque des Probabilités"*, 1812.

[40] B. de Finetti, *"Probabilità"*, entry for Enciclopedia Einaudi, 1980.

[41] E. Schrödinger, *"The foundation of the theory of probability – I"*, Proc. R. Irish Acad. **51A** (1947) 51; reprinted in *Collected papers* Vol. 1 (Vienna 1984: Austrian Academy of Science) 463.

[42] R. Scozzafava, *"La probabilità soggettiva e le sue applicazioni"*, Masson, Editoriale Veschi, Roma, 1993.

[43] A. O'Hagan, *"Bayesian Inference"*, Vol. 2B of Kendall's advanced theory of statistics (Halsted Press, 1994).

[44] E.T. Jaynes, *"Information theory and statistical mechanics"*, Phys. Rev. **106** (1957) 620.

[45] R.T. Cox, *"Probability, Frequency and Reasonable Expectation"* Am. J. Phys. **14** (1946) 1.

[46] D.S. Sivia, *"Data analysis - a Bayesian tutorial"*, Clarendon Press, Oxford University Press, 1997.

[47] F.H. Fröhner *"Evaluation and Analysis of Nuclear Resonance Data"*, JEFF Report **18** (Nuclear Energy Agency and Organization for Eco-

nomic Co-operation and Development), 2000 [http://www.nea.fr/html/dbdata/nds_jefreports/jefreport-18/jeff18.pdf]

[48] M. Tribus, *"Rational descriptions, decisions and designs"*, Pergamon Press, 1969.

[49] H. Jeffreys, *"Theory of probability"*, Oxford University Press, 1961.

[50] E. Schrödinger, *"The foundation of the theory of probability – II"*, Proc. R. Irish Acad. **51A** (1947) 141; reprinted in *Collected papers* Vol. 1 (Vienna 1984: Austrian Academy of Science) 479.

[51] Particle Data Group (PDG), C. Caso et al., *"Review of particle properties"*, Phys. Rev. **D50** (1994) 1173.

[52] New Scientist, April 28 1995, pag. 18 *("Gravitational constant is up in the air")*. The data of Table 3.2 are from H. Meyer's DESY seminar, June 28, 1995.

[53] P. Watzlawick, J.H. Weakland and R. Fisch, *"Change: principles of problem formation and problem resolution"*, W.W. Norton, New York, 1974.

[54] R. von Mises, *"Probability, Statistics, and Truth"*, Allen and Unwin, 1957.

[55] D.C. Knill and W. Richards (eds.), *"Perception as Bayesian Inference"*, Cambridge University Press, 1996.

[56] C. Glymour, *"Thinking things through: an introduction to philosophical issues and achievements"*, MIT Press, 1997.

[57] J.O. Berger, *"Statistical decision theory and Bayesian analysis"*, Springer, 1985.

[58] C.P. Robert, *"The Bayesian choice"*, Springer, 1994.

[59] G. D'Agostini, *"Confidence limits: what is the Problem? Is there the solution?"*, Workshop on Confidence Limits, Geneva, Switzerland, January 2000, CERN Report 2000-005 [hep-ex/0002055].

[60] A.L. Read, *"Modified frequentistic analysis of search results (the CL_s method)"*, Workshop on Confidence Limits, Geneva, Switzerland, January 2000, CERN Report 2000-005 [http://ep-div.web.cern.ch/ep-div/Events/CLW/papers.html].

[61] P.L. Galison, *"How experiments end"*, The University of Chicago Press, 1987.

[62] G. D'Agostini, *"Limits on electron compositeness from the Bhabha scattering at PEP and PETRA"*, Proceedings of the XXV Rencontre de Moriond on "Z° Physics", Les Arcs (France), March 4-11, 1990, p. 229 (also DESY-90-093).

[63] A.K. Wróblewski, *"Arbitrariness in the development of physics"*, after-dinner talk at the International Workshop on Deep Inelastic Scattering and Related Subjects, Eilat, Israel, 6–11 February 1994, Ed. A. Levy (World Scientific, 1994), p. 478.

[64] C.E. Shannon, *"A mathematical theory of communication"*, *Bell System Tech. J.* **27** (1948) 379, 623. Reprinted in the *Mathematical Theory of Communication* (C.E. Shannon and W. Weaver), Univ. Illinois Press, 1949.

[65] R.E. Kalman, *"A new approach to linear filtering and prediction problems"*, Trans. ASME J. of Basic Engin. **82** (1960) 35.

[66] P.S. Maybaeck *"Stochastic models, estimation and control"*, Vol. 1, Academic Press, 1979.

[67] G. Welch and G. Bishop *"An introduction to Kalman filter"*, 2002 `http://www.cs.unc. edu/~welch/kalman/`.

[68] C.F. Gauss, *"Theoria motus corporum coelestium in sectionibus conicis solem ambientum"*, Hamburg 1809, n.i 172–179; reprinted in Werke, Vol. 7 (Gota, Göttingen, 1871), pp 225–234.

[69] F. Lad, *"Operational subjective statistical methods - a mathematical, philosophical, and historical introduction"*, J. Wiley & Sons, 1996.

[70] G. Coletti and R. Scozzafava, *"Probabilistic logic in a coherent setting"*, Kluwer Academic Publishers, 2002.

[71] T. Bayes, *"An assay towards solving a problem in the doctrine of chances"*, Phil. Trans. Roy. Soc., **53** (1763) 370

[72] P. Astone and G. Pizzella, *"Upper limits in the case that zero events are observed: An intuitive solution to the background dependence puzzle"*, Workshop on Confidence Limits, Geneva, Switzerland, January 2000, CERN Report 2000-005 [`hep-ex/0002028`].

[73] G.J. Feldman and R.D. Cousins, *"Unified approach to the classical statistical analysis of small signal"*, Phys. Rev. **D57** (1998) 3873 [`physics/9711021`].

[74] J. Orear, *"Enrico Fermi, the man"*, Il Nuovo Saggiatore **17**, no. 5-6 (2001) 30

[75] A. Gelman, J.B. Carlin, H.S. Stern and D.B. Rubin, *"Bayesian data analysis"*, Chapman & Hall, 1995.

[76] D.G.T Denison, C.C. Holmes, B.K. Mallick and A.F.M. Smith, *"Bayesian methods for nonlinear classification and regression"*, Jonh Wiley and Sons, 2002.

[77] G. D'Agostini, *"Inferring $\overline{\rho}$ and $\overline{\eta}$ of the CKM matrix – A simplified, intuitive approach"*, May 2001, `hep-ex/0107067`.

[78] M. Ciuchini et al. *"2000 CKM-Triangle Analysis: A critical review with updated experimental inputs and theoretical parameters"*, JHEP **0107** (2001) 013 [`hep-ph/0012308`].

[79] Particle Data Group (PDG), C. Caso et al., *"Review of particle physics"*, Eur. Phys. J. **C3** (1998) 1 (`http://pdg.lbl.gov/`).

[80] G. D'Agostini, *"On the use of the covariance matrix to fit correlated data"*, Nucl. Instrum. Methods. **A346** (1994) 306.

[81] CELLO Collaboration, H.J. Behrend et al., *"Determination of α_s and $\sin^2\theta_w$ from measurements of total hadronic cross section in e^+e^- annihilation"*, Phys. Lett. **183B** (1987) 400.

[82] G. D'Agostini, *"Determination of α_s and $\sin^2\theta_W$ from R measurements at PEP and PETRA"*, Proceedings of XXII Rencontre de Moriond on "Hadrons, Quarks and Gluons", Les Arcs, France, March 15-25, 1987.

[83] S. Chiba and D.L. Smith, *"Impacts of data transformations on least-square solutions and their significance in data analysis and evaluation"*, J. Nucl. Sc. Tech. **31** (1994) 770.

[84] M. L. Swartz, *"Reevaluation of the hadronic contribution to $\alpha(M_Z^2)$"*,

Phys. Rev. **D53** (1996) 5268 [hep-ph/9509248].

[85] T. Takeuchi, *"The status of the determination of $\alpha(M_Z)$ and $\alpha_s(M_Z)$"*, Prog. Theor. Phys. Suppl. **123** (1996) 247 [hep-ph/9603415].

[86] S. Forte, J.I. Latorre, L. Magnea and A. Piccione, *"Determination of α_s from scaling violations of truncated moments of structure functions"*, Nucl. Phys. **B643** (2002) 477 [hep-ph/0205286].

[87] V. Blobel, *"Unfolding methods in high energy physics experiments"*, Proceedings of the "1984 CERN School of Computing", Aiguablava, Catalonia, Spain, 9–12 September 1984, Published by CERN, July 1985, pp. 88–127.

[88] G. Zech, *"Comparing statistical data to Monte Carlo simulation - parameter fitting and unfolding"*, DESY 95-113, June 1995.

[89] G. D'Agostini, *"A multidimensional unfolding method based on Bayes' theorem"*, Nucl. Instrum. Methods **A362** (1995) 487.

[90] K. Weise, *"Mathematical foundation of an analytical approach to Bayesian Monte Carlo spectrum unfolding"*, Physicalish Technische Bundesanstalt, Braunschweig, BTB-N-24, July 1995.

[91] S.F. Gull and J. Skilling, *"Quantifying Maximum Entropy"*, manual of *MemSys5* package, `http://www.maxent.co.uk/documents_1.htm`.

[92] B. Buck and V.A. Macauly (eds.), *"Maximum Entropy in action"*, Oxford University Press, 1991.

[93] International Workshops on Maximum Entropy and Bayesian Methods (22 editions till 2002), proceedings often published by Kluwer Academic Publishers. See also `http://omega.albany.edu:8008/maxent.html`

[94] K.M. Hanson, *"Introduction to Bayesian image analysis, Medical Imaging: Image Processing* M.H. Loew ed., *Proc. SPIE* **1898** (1993) 716 [*http://public.lanl.gov/kmh/publications/medim93.pdf*].

[95] G. Polya, *"Mathematics and plausible reasoning"*, Volume II: *Patterns of plausible inference*, Princeton University Press, 1968.

[96] A. Franklin, *Experiment, right or wrong"*, Cambridge University Press, 1990.

[97] D.A. Berry, *"Teaching elementary Bayesian statistics with real applications in science"*, Am. Stat. **51** (1997) 241;
J. Albert, *"Teaching Bayes' rule: a data-oriented approach"*, ibid., p. 247.
D.S. Moore, *"Bayes for beginners? Some reasons to hesitate"*, ibid., p. 254. Pages 262–272 contain five discussions plus replies.

[98] K.S. Thorne, *"Black holes and time warps: Einstein's outrageous legacy"*, W.W. Norton & Company, 1994.

[99] M. De Maria and A. Russo, *"The discovery of the positron"*, Rivista di Storia della Scienza, **2** (1985) 237.

[100] See e.g. Y.L. Dokshitzer, *"DIS 96/97. Theory/Developments"*, Proc. 5th International Workshop on Deep Inelastic Scattering and QCD, Chicago, April 1997, J. Repond and D. Krakauer eds. (AIP Conf. Proc. 407) [hep-ph/9706375].

[101] See e.g. G. Altarelli, *"The status of the Standard Model"*, talk at 18th International Symposium on *Lepton–Photon Interactions*, Hamburg, Au-

gust 1997, CERN-TH-97-278, Oct. 1997 [hep-ph/9710434].

[102] R. Feynman, *"The character of the physical law"*, The MIT Press, 1967

[103] B. Efron, *"Why isn't everyone a Bayesian?"*, Am. Stat. **40** (1986) 1, with discussion on pages 6-11.

[104] D.V. Lindley, comment to Ref. [103], Am. Stat. **40** (1986) 6.

[105] A. Zellner, *"Bayesian solution to a problem posed by Efron"*, Am. Stat. **40** (1986) 330.

[106] B. Efron, reply to Ref. [105], Am. Stat. **40** (1986) 331.

[107] J.M. Bernardo, *"Non-informative priors do not exist"*, J. Stat. Plan. and Inf. **65** (1997) 159, including discussions by D.R. Cox, A.P. Dawid, J.K. Ghosh and D. Lindley, pp. 177–189.

[108] E.T. Jaynes, *"Probability theory: the logic of science"*, book in preparation, see http://omega.albany.edu:8008/JaynesBook.html .

[109] G. Zech, *"Objections to the unified approach to the computation of classical confidence limits"*, physics/9809035 (see Ref. [11] for more extensive argumentations).

[110] R.D. Cousins, *"Why isn't every physicist a Bayesian?"*, Am. J. Phys. **63** (1995) 398.

[111] G. Feldman, Panel Discussion at Workshop on Confidence Limits, Geneva, Switzerland, January 2000, CERN Report 2000-005, p. 277. [http://ep-div.web.cern.ch/ep-div/Events/CLW/papers.html].

[112] G. Gabor (gabor@is.dal.ca), private communication, 1999.

[113] A. de Rujula, *"Snapshots of the 1985 high energy physics panorama"*, Proc. of the International Europhysics Conference on High-Energy Physics, Bari (Italy), July 1995, L. Nitti and G. Preparata eds.

[114] G. Salvini, Welcome address to the International Workshop on Deep Inelastic Scattering and related phenomena, Roma (Italy), April 1996; World Scientific, 1997, G. D'Agostini and A. Nigro eds.

[115] J.O. Berger and W.H. Jefferys, *"Sharpening Ockham's razor on a Bayesian strop"*, Am. Scientist **89** (1992) 64 and Journal of the Italian Statistical Society **1** (1992) 17 [http://quasar.as.utexas.edu/Papers.html].

[116] T.J. Loredo and D.Q. Lamb, *"Bayesian analysis of neutrinos observed from supernova SN 1987A"*, Phys.Rev. **D65** (2002) 063002 [atro-ph/0107260].

[117] M.V. John and J.V. Narlikar, *"Comparison of cosmological models using Bayesian theory"*, Phys.Rev. **D65** (2002) 043506 [astro-ph/0111122].

[118] M.P. Hobson, S.L. Bridle and O. Lahav, *"Combining cosmological datasets: hyperparameters and Bayesian evidence"*, 2002, astro-ph/0203259.

[119] C.E. Rasmussen and Z. Ghahramani, *"Occam's Razor"*, Neural Information Processing Systems **13** (2001) [http://www.gatsby.ucl.ac.uk/~zoubin/papers.html], see also http://www.gatsby.ucl.ac.uk/~zoubin/talks/cmu-talk.pdf.

[120] P. Astone, G. D'Agostini and S. D'Antonio, *"Bayesian model comparison applied to the Explorer-Nautilus 2001 coincidence data"*, Class. Quant. Grav. **20** (2003) 769 [gr-qc/0304096].

[121] See, e.g, J. Pearl, *"Probabilistic reasoning in intelligent systems: networks of plausible inference"*, Morgan Kaufmann Publishers, 1988.
F.V. Jensen, *"An introduction to Bayesian networks"*, UCL Press (and Springer Verlag), 1996.
D. Heckerman and M.P. Wellman, *"Bayesian Networks"*, Communications of the ACM (Association for Computing Machinery), Vol. 38, No. 3, March 1995, p. 27.
L. Burnell and E. Horvitz, *"Structure and chance: melding logic and probability for software debugging"*, ibid., p. 31.
R. Fung and B. Del Favero, *"Applying Bayesian networks to Information retrieval"*, ibid., p. 42.
D. Heckerman, J.S. Breese and K. Rommelse, *"Decision-theoretic troubleshooting"*, ibid., p. 49.
R.G. Cowell, A.P. Dawid, S.L. Lauritzen and D.J. Spiegelhalter *"Probabilistic Networks and Expert Systems"*, Springer Verlag, 1999.
`http://www.auai.org/`
`http://bayes.stat.washington.edu/almond/belief.html`.

[122] J.B. Kadane and D.A. Schum, *"A Probabilistic analysis of the Sacco and Vanzetti evidence"*, J. Wiley and Sons, 1996.
P. Garbolino and F. Taroni, *"Evaluation of scientific evidence using Bayesian networks"*, Forensic Science International **125** (2002) 149, and references therein.

[123] F.B Cozman, *"JavaBayes version 0.346 – Bayesian networks in Java"*, January 2001, `http://www-2.cs.cmu.edu/~javabayes/Home/`

[124] `http://www.roma1.infn.it/~dagos/bn/`

[125] D.J. Spiegelhalter, A. Thomas and N.G. Best (et al.), *"Bayesian inference Using Gibbs Sampling"*,
W.R. Gilks, S. Richardson and D.J. Spiegelhalter, *"Markov Chain Monte Carlo Methods in Practice"*, Chapman and Hall, 1996.
`http://www.mrc-bsu.cam.ac.uk/bugs/welcome.shtml`.

[126] `http://www.statslab.cam.ac.uk/~mcmc/`

[127] NA 48 Collaboration, J.R. Batley and al., *"A precise measurement of direct CP violation in the decay of neutral kaons into two pions"*, Phys.Lett **B544** (2002) 97 [`hep-ex/0208009`].

[128] G. D'Agostini, *"Sceptical combination of experimental results: General considerations and application to ϵ'/ϵ"*, CERN-EP/99-139, October 1999, `hep-ex/9910036`, and references therein.

[129] M. Fabbrichesi, *"Estimating $\epsilon\prime/\epsilon$. A user's manual"*, Nucl. Phys. Proc. Suppl. **86** (2000) 322 [`hep-ph/9909224`].

[130] V. Dose and W. von der Linden, *"Outlier tolerant parameter estimation"*, Proc. of the XVIII International Workshop on Maximum Entropy and Bayesian Methods, Garching (Germany), July 1998, V. Dose et al. eds., Kluwer Academic Publishers, Dordrecht, 1999 [`http://www.ipp.mpg.de/OP/Datenanalyse/Publications/`].

[131] W.H. Press, *"Understanding data better with Bayesian and global statistical methods"*, Conference on Some Unsolved Problems in Astrophysics,

Princeton, NJ, 27–29 Apr 1995 [astro-ph/9604126].

[132] C. Pascaud and F. Zomer, *"QCD analysis from the proton structure function F_2 measurement: issues on fitting, statistical and systematic errors"*, LAL 95-05, June 1995 [http://www-h1.desy.de/h1work/fit/h1fit.info.html].

[133] S. Alekhin, *"Extraction of parton distributions and α_s from DIS data within the Bayesian treatment of systematic errors"*, Eur. Phys. J. **C10** (1999) 395 [hep-ph/9611213].

[134] M. Botje, *"A QCD analysis of HERA and fixed target structure function data"*, ZEUS Note 98-062 DESY-99-038, December 1999 [hep-ph/9912439].

[135] R.S. Thorne et al. *"Questions on uncertainties in parton distributions"*, Conference on Advanced Statistical Techniques in Particle Physics, March 2002, Durham, J. Phys. **G28** (2002) 2717 [hep-ph/0205233].

[136] H. Wahl (CERN), private communication, 1999.

[137] G. D'Agostini and M. Raso, *"Uncertainties due to imperfect knowledge of systematic effects: general considerations and approximate formulae"*, CERN-EP/2000-026, February 2000 [hep-ex/0002056].

[138] P. Astone and G. D'Agostini, *"Inferring the intensity of Poisson processes at the limit of the detector sensitivity (with a case study on gravitational wave burst search)"*, CERN-EP/99-126, August 1999 [hep-ex/9909047].

[139] P. Astone at al., *"Search for correlation between GRB's detected by BeppoSAX and gravitational wave detectors EXPLORER and NAUTILUS"*, Phys. Rev. **66** (2002) 102002 [astro-ph/0206431].

[140] T.J. Loredo, *"The promise of Bayesian inference for astrophysics"*, Proc. *Statistical Challenges in Modern Astronomy*, E.D. Feigelson and G.J. Babu eds., Springer-Verlag (1992) 275 [http://astrosun.tn.cornell.edu/staff/loredo/bayes/tjl.html]. This web site contains also other interesting tutorials, papers and links on Bayesian analysis.

[141] ZEUS Collaboration, *"Search for eeqq contact interactions in deep inelastic $e^+p \to e^+X$ scattering at HERA"*, Eur. Phys. J **C14** (2000) 239 [hep-ex/9905039].

[142] CELLO Collaboration, H.J. Behrend et al., *"Search for substructures of leptons and quark with CELLO detector"*, Z. Phys. **C51** (1991) 149.

[143] G. D'Agostini and G. Degrassi, *"Constraining the Higgs boson mass through the combination of direct search and precision measurement results"*, Contribution to the Workshop on "Confidence Limits", CERN, Geneva, 17–18 January 2000 [hep-ph/0001269].

[144] R. Feynman, 1973 Hawaii Summer Institute, cited by D. Perkins at the 1995 EPS Conference, Brussels.

[145] C. Howson and P. Urbach, *"Scientific reasoning - the Bayesian approach"*, Open Court, 1993 (second edition).

[146] J. Earman, *"Bayes or bust? A critical examination of Bayesian confirmation theory"*, The MIT Press, 1992.

[147] R. Jeffrey, *"Probabilistic thinking"*, 1995, http://www.princeton.edu/~bayesway/ProbThink/

[148] M. Kaplan, *"Decision theory as philosophy"*, Cambridge University Press, 1996.

[149] R. Jeffrey, *"Subjective Probability (The Real Thing)"*, 2002 `http://www.princeton.edu/~bayesway/Book*.pdf`. Related essays can be found at `http://www.princeton.edu/~bayesway/`.

[150] J.M. Bernardo, *"Bayesian statistics"*, UNESCO Encyclopedia of Life Support Systems (EOLSS) [`ftp://matheron.uv.es/pub/personal/bernardo/BayesStat.pdf`].

[151] F. Spizzichino, *"Subjective probability models for lifetimes"*, Boca Raton Chapman & Hall/CRC, 2001.

[152] B. de Finetti, *"Filosofia della probabilità"*, il Saggiatore, 1995.

[153] L. Piccinato, *"Metodi per le decisioni statistiche"*, Springer-Italia, 1996.

[154] D. Costantini e P. Monari (eds.), *"Probabilità e giochi d'azzardo"*, Franco Muzzio Editore, 1996.

[155] R. L. Winkler, "*An introduction to Bayesian inference and decision*", Holt, Rinehart and Winston, Inc., 1972.

[156] S. J. Press, "*Bayesian statistics: principles, models, and applications*", John Wiley & Sons, 1989.

[157] G.E.P. Box and G.C. Tiao, *"Bayesian inference in statistical analysis"*, John Wiley and Sons, 1973.

[158] A. O'Hagan, *"Probability: methods and measurements"*, Chapman & Hall, 1988.

[159] P.M. Lee, *"Bayesian statistics - an introduction"*, John Wiley and Sons, 1997.

[160] L.J. Savage et al., *"The foundations of statistical inference: a discussion"*, Methuen, 1962.

[161] A. Zellner, *"Bayesian analysis in econometrics and statistics"*, Eduard Elgar, 1997.

[162] J.M. Bernardo et al., *Valencia Meetings on "Bayesian Statistics"* 1–6, `http://www.uv.es/~bernardo/valenciam.html`.

[163] G.L. Bretthorst *"Bayesian spectrum analysis and parameter estimation"*, Springer Verlag, 1988 [`http://bayes.wustl.edu/glb/book.pdf`].

[164] A. Pole, M. West and P.J. Harrison *"Applied Bayesian Forecasting and Time Series Analysis"*, 1994, Chapman-Hall. More information and related software can be found at `http://www.isds.duke.edu/~mw/books_software_data.html`.

[165] `http://www.bayesian.org/`
`http://www.amstat.org/sections/SBSS/`
`http://bayes.stat.washington.edu/bayes_people.html`
`http://www.ar-tiste.com/blip.html`
`http://www.strauss.lanl.gov/Welcome.html`
`http://fourier.dur.ac.uk:8000/stats/bayeslin/`
`http://astrosun.tn.cornell.edu/staff/loredo/bayes/`.

[166] `http://astrosun.tn.cornell.edu/staff/loredo/bayes/tjl.html`

[167] `http://www.ipp.mpg.de/OP/Datenanalyse/`
`http://public.lanl.gov/kmh/publications/publications.html`

[168] Uncertainty Quantification Working Group, `http://public.lanl.gov/kmh/uncertainty/`

[169] A.F.M. Smith, *"Bayesian numerical analysis"*, Phil. Trans. R. Soc. London **337** (1991) 369.

[170] R.M. Neal, *"Probabilistic inference using Markov Chain Monte Carlo Methods"*, Technical Report CRG-TR-93-1, University of Toronto, 1993, `ftp://ftp.cs.utoront.ca/pub/radford/review.pdf`.

[171] W.R. Gillks, S. Richardson and D.J. Spiegelhalter *"Markov Chain Monte Carlo in practice"*, Chapman and Hall, 1996.

[172] R.E. Kass, B.P. Carlin, A. Gelman and R.M. Neal, *"Markov Chain Monte Carlo in practice: A roundtable discussion"*, Am. Stat. **52** (1998) 93 [`http://www.amstat.org/publications/tas/kass.pdf`].

[173] K.M. Hanson, *"Tutorial on Markov Chain Monte Carlo"*, XX International Workshop on Maximum Entropy and Bayesian Methods in Science and Engineering, Gif-sur-Yvette (France), July 2000, `http://public.lanl.gov/kmh/talks/maxent00b.pdf`

[174] A. Lewis and S. Bridle, *"Cosmological parameters from CMB and other data: a Monte-Carlo approach"*, Phys. Rev. **D66** (2002) 103511 [`astro-ph/0205436`].

[175] `http://cerebro.xu.edu/math/Sources/`.

Index